East African Crops

An introduction to the production of field and
plantation crops in Kenya, Tanzania and Uganda

J D Acland

Published by arrangement with the Food and Agriculture
Organization of the United Nations by Longman Group Limited

Longman

Acknowledgements

Longman Group UK Limited
Longman House, Burnt Mill
Harlow, Essex CM20 2JE, England
and Associated Companies throughout the world.

First published 1971
This impression 1989

Set in Univers Medium

We are grateful to the following for permission to
reproduce photographs:

The East African Railways Corporation for Figs. 45,
47, 56, 68, 76, 77, 93, 107, 110, 111, 112, 120,
125, 146, 150, 160
Kenya Information Service for Figs. 49, 50, 51, 52,
81, 103, 104, 106, 124, 126, 134, 154, 156, 162
The Coffee Research Foundation, Kenya, for Figs.
38, 39, 40, 43, 44, 48, 53, 54
Massey Ferguson Ltd. for Figs. 78, 149
British American Tobacco Ltd. For Fig. 137
R. M. Nattrass for Fig. 92
E. Watts for Fig. 153
G. W. Lock for Fig. 109
World Tobacco for Fig. 138
Paul Popper Ltd. for Fig. 121

Produced by Longman Group (FE) Ltd.
Printed in Hong Kong

ISBN 0-582-60301-3

Contents

Introduction

Whilst teaching crop production in Kenya and Tanzania I found that one of the greatest needs, both of teachers and students, was for a comprehensive textbook on East African crops. This book has been prepared, at the request of the Government of Kenya and with support from the United Nations Development Programme, in order to meet this need.

The scope of the book is indicated by its subtitle: an introduction to the production of field and plantation crops in Kenya, Tanzania and Uganda.

It is introductory in that the undergraduate will need to use it in conjunction with more detailed courses and a study of the bibliography given at the end of the chapters; the college or institute student, studying for a diploma or certificate, will need to use it in conjunction with a great deal of practical instruction. I have written this book primarily for these groups of students but I hope that it will find a wider readership, especially at secondary schools.

Only the production of the crops is considered, although there is an introductory section at the beginning of each chapter indicating the importance of the crop, where it is grown and for what it is used, and a section on relevant plant characteristics. There is no discussion on the principles of crop production in East Africa nor on the economics of the production of each crop. The first should be tackled in a separate book; the second would rapidly become anachronistic.

Only field and plantation crops are included; horticultural crops and fodder crops, each of which requires a separate book, are omitted. The division between horticultural crops on the one hand and field and plantation crops on the other is, of course, indistinct and in some cases it has been hard to decide which crops to include and which to leave out. I have taken the view that a crop ceases to be horticultural when it is grown primarily on a field or plantation scale even though horticultural principles, i.e. care of individual plants rather than of the crop in general, may still be applied to it. Without considering the scale on which they were grown, coffee and tea would surely classify as horticultural crops. Another indistinct division is that between major crops, each of which warrants a separate chapter, and minor crops, which only warrant brief mention in a separate section at the end of the book. I make no apology for my arbitrary definition

of a minor crop as a crop that is important only to a relatively small number of people; nor for the fact that today's minor crop may become tomorrow's major crop and vice versa.

Most authors of books on crops adopt some form of crop classification. I have departed from this practice because although some crops, such as the cereals and the root crops, can easily be classified, others, such as groundnuts and coconuts, fall into more than one group. Pyrethrum, sugar cane and tobacco are especially difficult to classify unless a somewhat unsatisfactory name such as 'cash crops' or 'industrial crops' is used. For these reasons, and for ease of reference, the major crops in this book are listed in alphabetical order. The minor crops are more easily grouped and can be found under such headings as cereals, essential oils, fibres, oilseeds, pulses, root crops, spice crops and tree crops.

The material for this book could not have been collected without the help of very many individuals: Government officials, farmers, staff of commercial concerns, educational institutions, etc. Their number can be judged, although my gratitude for their contribution is in no way lessened, by the fact that there is insufficient space for them to be mentioned by name. I am especially grateful to staff of the Ministries of Agriculture of the three East African countries and to officers of the Food and Agriculture Organization of the United Nations. I have also received invaluable assistance from staff of African Tea Brokers Ltd., Bookers Agricultural Holdings Ltd., British American Tobacco Ltd., C. J. Valentine and Co. Ltd., the Coffee Research Foundation, the Cotton Research Corporation, the East African Agriculture and Forestry Research Organization, the East African Industrial Research Organization, the East African Tanning Extract Co. Ltd., Egerton College, The International Bank for Reconstruction and Development, Kenya Breweries Ltd., Kenya Canners Ltd., the Kenya Sisal Board, Makerere University College, the National Irrigation Board (Kenya), the Pyrethrum Board of Kenya and the Tea Research Institute.

I am very grateful to Mr Michael Sanders, who prepared the line drawings, and to Miss Jean Ball, Mrs Elva Bernardis, Miss Jinnie Chalton, Miss Zephrina Fernandes and Mrs Betty Rossi, who prepared the manuscript.

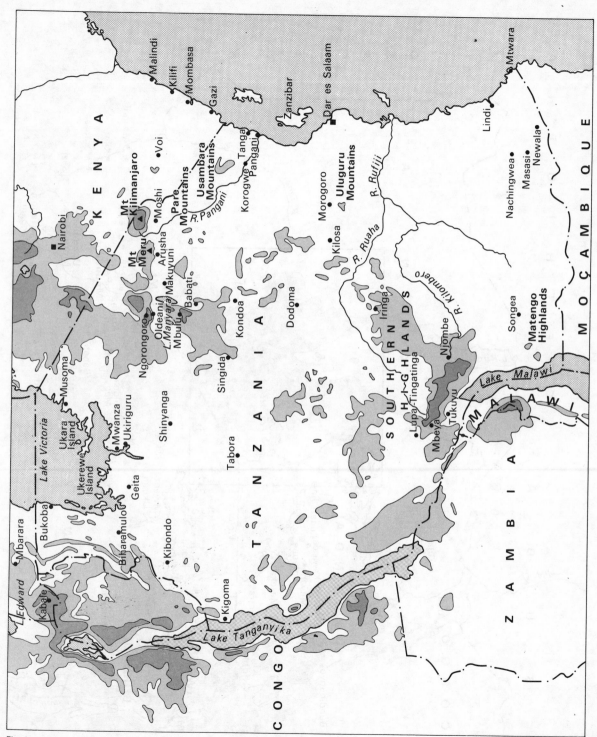

Fig. 1

5

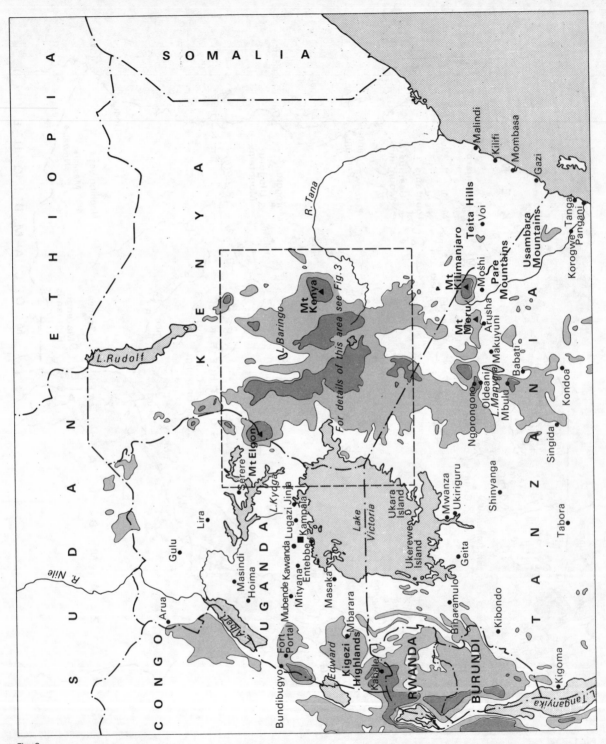

Fig. 2
6

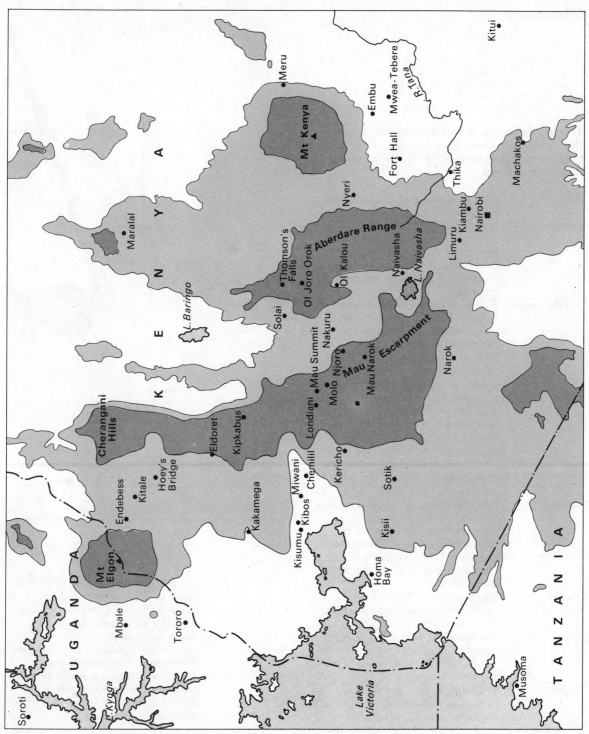

Kitui

Meru

Mt Kenya

Embu

Mwea-Tebere

Fort Hall

R.Tana

Machakos

Thika

K E N Y A

Maralal

Nyeri

Limuru

Kiambu

Nairobi

Aberdare Range

Thomson's Falls

Ol Joro Orok

Naivasha

L. Naivasha

Ol Kalou

L. Baringo

Solai

Nakuru

Mau Summit

Mau Narok

Escarpment

Mau Njoro

Narok

Molo

Londiani

Mau

Cherangani Hills

Eldoret

Kipkabus

Hoey's Bridge

Miwani

Kitale

Chemilil

Kericho

Sotik

Endebess

Kakamega

Kisumu

Kibos

Kisii

U G A N D A

Mt Elgon

Homa Bay

Mbale

L. Kyoga

Tororo

Lake Victoria

Musoma

T A N Z A N I A

Soroti

Fig. 3

Glossary

This glossary contains only the Kiswahili words that appear in the text. They are used in this book because they have no concise, widely accepted English equivalents.

Boma — A cattle *boma* is an enclosure for safeguarding cattle at night.

Debe — A *debe* is a four gallon tin, originally containing kerosene or petrol, but widely used throughout East Africa as a container, for roofing and for a number of other purposes.

Forked jembe — See illustration. Forked *jembes* are especially useful for removing perennial grasses because they pull the rhizomes or stolons out. Normal *jembes* (see below) tend to cut these structures into small pieces, thus aggravating the weed problem. Forked *jembes* are available in different sizes; the one illustrated is fairly large; smaller ones are popular for weeding pyrethrum, whose roots are unusually sensitive to mechanical damage by normal *jembes* or *pangas*.

Jembe — See illustration. A *jembe* is a broad bladed hoe and is the basic implement of cultivation in East Africa. Many different sizes and shapes of blade are made but the kind illustrated is much the most common.

Panga — See illustration. There is no satisfactory English translation for the *panga*. The nearest equivalent is a machette or matchet but this implies a double edged implement as used in Central America or the the West Indies.

Ugali — Dictionaries give 'porridge' or 'bread' as translations for *ugali* but neither of these is satisfactory.

A forked *jembe*.

A *jembe*.

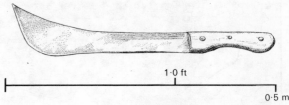

A *panga*.

Ugali is made by boiling flour with water until it becomes a thick paste. It is pliable and is normally eaten by taking pieces between finger and thumb.

Uji — *Uji* is more accurately described in dictionaries; they usually give 'gruel' or 'soup'. It is made in the same way as *ugali* but with a greater quantity of water. It is often drunk through straws.

1
Bananas

Musa spp.

Introduction

Bananas are the staple food in many of the lower altitude, wetter areas of East Africa. They are mostly grown as a subsistence crop although there is much internal trading, often over distances of several hundred miles, between the banana growing areas and the main towns. No banana export trade has developed; among the reasons for this are the distance between the important growing areas and the sea, the distance between East Africa and the major consuming areas, i.e. Europe and the U.S.A., and the fact that most of the export varieties, selected for high rainfall, hot, humid conditions, give poor results in the main banana growing areas of East Africa, which are cooler and drier.

Fig. 4: A pure stand of bananas

Uganda

Bananas are the staple food in all parts of Uganda which do not experience a pronounced dry season, i.e. the area within about 50 miles (c. 80 km) of the shore of Lake Victoria, the Kigezi Highlands, the slopes of Mt. Elgon and the well watered areas of Ankole, Toro and Bunyoro Districts. It has been estimated that there are approximately 1 500 000 acres (c. 600 000 ha) of bananas in Uganda.

Tanzania

Bananas are a staple food in the high rainfall areas of Tanzania, i.e. the wetter parts of Bukoba, Ngara, Kibondo, Kigoma, Arusha, Moshi, Pare, Lushoto, Rungwe and Songea Districts. They are also grown in valley bottoms and along seepage lines in most other areas; in such places they are only a minor constituent of the local diet.

Kenya

Bananas are widely grown in Western, Central Nyanza, South Nyanza and Central Provinces, on the slopes of Mt. Kenya and at the coast. Maize, however, is the staple food in all these areas.

Plant characteristics

The banana plant is shallow rooting; most of the roots are found in the top 6 in (c. 15 cm) of the soil profile. The roots are adventitious, arising from the rhizome which is a thick underground structure showing little lateral growth (see Fig. 5). Leaf scars and buds are found on the rhizome; the buds develop into secondary rhizomes which emerge close to the parent plant giving rise to a cluster, or 'stool', of suckers. Suckers which originate from the lower part of the parent rhizome at first have small, pale, pointed leaves; they are usually called sword suckers. Those which originate near the soil surface, however, immediately have large, green and wide leaves, similar in form to those of mature plants; they are usually called water suckers. The 'stem' of each sucker consists of the fleshy leaf sheaths which encircle one another; the botanical

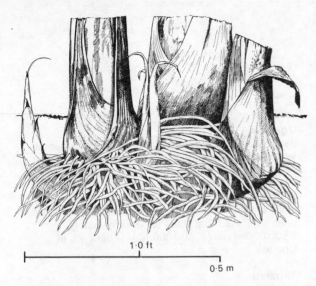

1·0 ft

0·5 m

Fig. 5: The underground structure of a banana plant. Note the sword sucker arising on the left.

name for such a structure is a 'pseudostem'. The true stem, which later bears the inflorescence, begins to grow from the rhizome through the middle of the pseudostem when the sucker has almost reached its full height.*

Mature suckers are from 5–20 ft (c. 1·5–6·0 m) high, depending largely on the variety . Each sucker produces only one inflorescence. Inflorescences vary in many ways, also depending largely on the variety, e.g. in colour, geotropism of stem and fruits, size and shape of fruits, compactness of bunch etc.

The time of emergence of the inflorescence depends mostly on the size of the planting material, the altitude and the variety; it is usually between 9 and 18 months. After emergence the inflorescence usually turns downwards and bracts open successively to reveal row after row of flowers. The first rows of flowers are female and the last rows are male. The male flowers serve no useful function, as pollination and subsequent seed development hardly ever occur. The fruits of the cultivated banana varieties develop without pollination.

*Pseudostems are referred to as stems for the remainder of this chapter.

Ecology

Rainfall, and water requirements
For optimum yields bananas require a constant supply of soil moisture. Where they are grown as a staple food in East Africa this is provided by a reasonably well distributed rainfall. In some of these areas, e.g. Bukoba District in Tanzania and Kakamega District in Kenya, the annual total and the distribution of rainfall compare favourably with those of the major banana exporting countries of the world. In most of the others, however, it is probable that yields are limited by a lower annual rainfall and uneven distribution, e.g. a dry season at the beginning of the year in Uganda and Kenya and in the middle of the year in southern Tanzania. In areas of poor rainfall bananas are often grown in places which provide a constant supply of ground water, e.g. riversides, seepage lines, drains, ditches, etc. In Machakos District in Kenya, where there is a low and unevenly distributed rainfall, bananas are sometimes grown at the bottom of deep pits; these concentrate the available rainfall around the roots and enable bananas to be grown in an area that would otherwise be unsuitable for them.

Altitude and temperature
Bananas require warm conditions and give best yields at altitudes between sea level and 6 000 ft (c. 1 800 m). Above 6 000 ft bananas are unimportant although plants are occasionally seen as high as 8 000 ft (c. 2 400 m). At high altitudes cool conditions cause slower development and in some cases the inflorescences fail to emerge completely from the tops of the stems.

Wind
Bananas can be considerably damaged by wind. If possible they should be planted in sheltered positions. A common recommendation is that they should be planted in blocks rather than strips so that the plants give each other some degree of self-protection. In Uganda a variety called Kisubi, used for making beer, is often grown around the perimeter of a plot of bananas; this variety is usually left unpruned and therefore makes an effective windbreak.

Soil requirements
Soils should be free draining yet capable of retaining a reasonable amount of moisture. The nutrient status of the soil should be high if bananas are to be grown without the addition of fertilisers

or manure. Bananas are grown successfully around the lakeshore in Bukoba District in Tanzania on soils which were originally heavily leached; in the areas where they are grown the structure and the nutrient content of the soil have been improved by repeated applications of mulch and manure.

Bananas grow well in soils whose pH is between 5·0 and 8·0.

Varieties

Where bananas are a staple food, cooking varieties, brewing varieties and varieties for eating raw are recognised. (Cooking varieties are sometimes called 'plantains' in other parts of the world). Where they are of medium importance, however, varieties are grown which may be used either for cooking or for eating raw, e.g. Muraru in Central Province in Kenya and Shisikame in Western Province, also in Kenya. The East African varieties, whose names are too numerous to catalogue in this chapter, have evolved to suit the local environment. In general, introduced varieties from hotter, wetter and more humid parts of the world have given disappointing results. Dwarf varieties, such as Dwarf Cavendish, are often grown for their fresh fruit.

Field Establishment

Land preparation
It is important that all perennial grasses, especially couch grass, should be eradicated. If there is no couch grass a thorough digging with a 'jembe' or ploughing should be sufficient.

Pure stands and intercropping
The practice of planting in pure stands (see Fig. 4) or interplanting with other crops (see Fig. 6) depends on the locality. In Western Province in Kenya most bananas are grown in pure stands. In Uganda they are both grown in pure stands and with other crops, especially Robusta coffee. In Central Province in Kenya pure stands are rare; bananas are almost invariably interplanted with other food crops, e.g. maize, beans, potatoes, cocoyams, sugar-cane, etc. In most of the high rainfall areas of Tanzania bananas are often inter-planted with coffee. Bananas are not the ideal shade for coffee; the feeder roots of both crops are found in the same part of the soil horizon and the banana plant requires a large supply of nutrients. There have been several campaigns in Tanzania aimed at persuading the smallholders either to

plant their coffee and bananas in pure stands or to reduce the population of bananas when the two crops are grown together. These campaigns have been relatively unsuccessful because most farmers attach great importance to bananas: in fact local surveys have shown that they are potentially more profitable than coffee.

Fig. 6: Bananas interplanted with maize, potatoes and pumpkins.

Planting holes
Each hole should be at least 2 ft (0·6 m) deep and 2 ft in diameter and should be filled with topsoil which has been mixed with organic manure and 8 oz (c. 226 g) of single superphosphate. The recommended dimensions are often exceeded, especially in Uganda where holes are often 3 ft (c. 0·9 m) deep and 3 ft wide. In areas with marginal rainfall, e.g. Machakos and Kitui Districts in Kenya, large pits about 5 ft (c. 1·5 m) in diameter and 3 ft deep are recommended (see section on ecology).

Spacing
The recommendations for pure stands are 12–15 ft (c. 3·6–4·5 m) between planting holes for tall varieties on fertile soils, giving about 200–300 plants per acre (c. 480–750 plants per hectare). On less fertile soils or in drier areas the spacing

recommendation in Uganda is 12 ft × 9 ft (c. 3·6 m × 2·7 m), giving about 360 plants per acre (c. 890 plants per hectare). Dwarf varieties should be grown at 8 ft × 8 ft (c. 2·4 m × 2·4 m), giving about 680 plants per acre (c. 1·680 plants per hectare). There has been almost no experimental work in East Africa on banana spacing and on the related subject of pruning.

Planting material

Many different types of planting material are used in different banana growing areas of the world. These range from pieces of rhizome which include

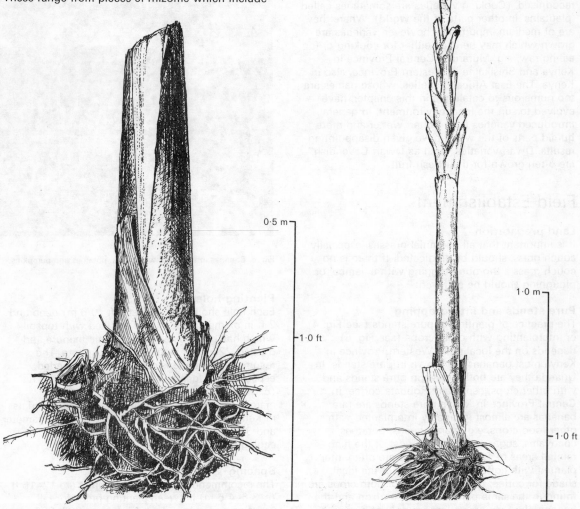

Fig. 7: Banana planting material. This is the type normally used in East Africa.

Fig. 8: Banana planting material. This type of planting material is sometimes used for smaller suckers.

at least one bud, small suckers which have only recently appeared above ground level, sword suckers and large suckers which have reached the broad leaved stage. There is no agreement as to which is the best of these; it is generally accepted, however, that larger suckers produce their bunch more quickly than smaller suckers or pieces of rhizome.

In East Africa large broad leaved suckers are virtually the only type of planting material used. Their size depends on the variety and local practice; 5–7 ft (c. 1·5–2·1 m) high would be a reasonable size for large varieties whilst 2–3 ft (0·6–0·9 m) high would be normal for a dwarf variety. The only limitation on size is that the sucker should not be so well developed that the inflorescence has already started to grow up the middle of the stem. Suckers which originally emerged as water suckers are not usually used; when detached from the parent plant they have a small amount of rhizome and can therefore only produce a limited number of roots in the early stages of growth. Suckers which originally emerged as sword suckers are preferred and can be identified by pushing the stem backwards and forwards; suitable suckers feel well anchored whilst those which originated as water suckers offer less resistance.

In almost all cases either the entire top of the sucker is removed, using a diagonal cut just below the point where the petioles leave the stem (see Fig. 7) or, less commonly, all the leaves are removed at the petioles except the central one which remains rolled up (see Fig. 8). The aim of this is to reduce transpiration. The severity of defoliation usually depends on the size of the suckers; the bigger they are the more leaf is removed. Only in rare instances, e.g. in Kericho District in Kenya, are whole suckers planted without any defoliation.

When selecting suckers for planting it is important that they should be removed with as much rhizome attached as possible. They should be free of weevils. Treatment against both weevils and nematodes is recommended (see section on pests).

Planting

The best time for planting rainfed bananas is the beginning of the rains. Preference is usually given to annual food crops, however. In areas with a continuous supply of ground water any time of the year is satisfactory.

It is recommended that suckers should be planted about one foot (c. 0·3 m) deeper in the soil than their position while they were attached to their parent plant. This gives their initial roots a good chance of growing in moist soil and reduces the risk of wind blowing the young plants over. As the topsoil is returned to the planting holes it should be well compressed around the stem, layer by layer.

Field maintenance

Mulching

It is generally accepted that bananas respond well to mulching. The most common mulching materials are dead or senescent banana leaves, pruned suckers and old stems which have had their bunches removed. Suckers and old stems should be chopped up and allowed to dry in order to discourage banana weevils. These forms of mulch are usually insufficient for optimum yields, especially in view of the fact that suckers and stems are often removed for cattle fodder. The use of other materials is nowadays rare. Other materials were commonly used in Uganda earlier in the century, however, and in Bukoba District in Tanzania grass is still taken from the 'rweya' soils, which are unsuitable for growing bananas, and used as a mulch.

Mulch should not be allowed to come into contact with the banana stems; this would provide moist conditions which would encourage the entry of banana weevils.

Fertilisers and manures

Fertilisers and manures are seldom used on bananas in East Africa. An exception to this rule occurs in the important banana growing areas of Tanzania, especially on the slopes of Mts. Kilimanjaro and Meru, where stall feeding of cattle is common; the manure which accumulates is usually applied to the banana plots.

A steady supply of nitrogen is essential for optimum yields of bananas. 4 oz (c. 113 g) of ammonium sulphate per stool has been shown to be an economic application; it should be scattered in a ring around each stool, between one and two feet (c. 0·3–0·6 m) away from the stems, during the rains. Phosphate is important in eastern Kenya; trials have shown that ½ lb (c. 0·2 kg) of single superphosphate in each planting hole gives economic results. Potassium is considered important in Uganda but at the time of writing there are no definite recommendations.

Weed control

If the land is properly prepared before planting, if the correct spacing is adopted and if mulching is practised, weed control should require little effort. Cultivation should always be shallow in order to minimise damage to the roots. Bananas are very susceptible to most hormone herbicides. Dalapon can safely be used, however, and can be applied to couch grass amongst mature plants.

Pruning

Pruning is one of the most controversial issues of banana husbandry. Recommendations vary slightly in different parts of East Africa; from three to six stems should be left on each stool and these should be chosen so that they are in different stages of development. When removing unwanted suckers they should be cut deeply enough to ensure that they do not grow again.

There have been many experiments done on pruning in the banana exporting countries: the main finding of these is that pruning gives greater bunch size. For export purposes large bunches are essential. In East Africa, however, there have been no long term experiments on the effect of pruning on fruit yield: a criterion of greater importance to East African growers than bunch size. Local observations, however, indicate that pruning not only gives larger bunches but also decreases the time to first harvest and discourages weevil attack.

Pruning practice varies considerably in East Africa. In western Kenya pruning is rare and stools can often be seen with as many as 20 stems. In Uganda and places where bananas are interplanted with other crops, especially coffee, there is usually a limited amount of pruning although it seldom conforms with the recommendations. In Bukoba District in Tanzania, however, bananas are pruned rigorously; there are often only two stems per stool. The reason for this may be the close spacing which is adopted in this district; in the wetter parts the average population is 1 100 stools per acre (c. 2 700 per hectare) which is equivalent to a spacing of approximately 6 ft × 6 ft (c. 1·8 m × 1·8 m).

Staking

Banana stems are liable to break under the weight of a heavy ripening bunch and the fruit may rot on the ground. Forked poles are often used to keep the stems upright.

Removing the male part of the inflorescence

This is sometimes practised in East Africa and it is thought that it shortens the time to maturity.

Experiments in other parts of the world have shown that it has no significant effect on the yield of fruit.

Harvesting

The colour of the fruits when they are ripe varies considerably between varieties. Some, especially cooking varieties, remain green even when ripe. Some turn yellow when they ripen, whilst others turn yellow before ripening.

When bunches are harvested for sale care should be taken to prevent the fruits being bruised. One worker may cut the stem while the other prevents the bunch from falling to the ground. An experienced worker, however, can perform both these operations alone.

Yields

Average yields of bananas are 400–500 bunches per acre per annum (c. 1 000–1 200 bunches per hectare per annum). With good husbandry, 1 000 bunches per acre (c. 2 500 per hectare) have been obtained. Bunches of the large varieties usually weigh about 35 lb (c. 16 kg) although with pruning and good husbandry they can weigh well over 50 lb (c. 23 kg). Average yields expressed as tons of fruit per annum are 6–8 tons per acre (c. 15–20 t/ha) but with good husbandry and irrigation 15–20 tons per acre (c. 38–50 t/ha) have been obtained.

Little is known of the economic life of banana plots or of the factors which cause yields to decline. In the banana exporting countries fields seldom remain under bananas for more than 20 years, but in Uganda there are plots which are over 60 years old. At the opposite extreme, bananas in Embu District in Kenya show a marked drop in yield as early as the second crop.

Pests

Banana weevil. *Cosmopolites sordidus*
This is the most damaging pest of bananas in East Africa; it is found in all areas except to the east of the Rift Valley in Kenya. Movement of planting material is restricted in Kenya to prevent new outbreaks. Damage is worst in neglected plots; on fertile soils and with good husbandry it is seldom serious.

The adult weevil, which lives, feeds and breeds in stems for periods of up to two years, lays eggs against the sides of living stems. Upon hatching the larvae burrow into the stems making tunnels which can easily be seen if the plant is cut up; they weaken the stems, making them liable to wind damage and reducing yields.

Methods of prevention include selection of planting material from uninfected stools, dipping planting material in a dieldrin solution, chopping up old stems longitudinally, covering their bases with soil and encouraging healthy growth by pruning and mulching. The weevil population can be reduced by spreading dieldrin dust around infected stools and laying traps of pieces of stem which have been cut in half longitudinally and treated with dieldrin dust.

Thrips. *Hercinothrips bicinctus*
Thrips discolour or crack the skins of the fruit; they do not harm the flesh. Insecticides, e.g. DDT, BHC and dieldrin, are effective but are of doubtful economic value.

Nematodes. *Radopholus similis*
Nematodes have recently been recorded as doing much damage in several parts of Uganda. Damage may be reduced by dipping the planting material in a nematocide.

Diseases

Panama disease
This disease is caused by the fungus *Fusarium oxysporum* f. sp. *cubense*. It has caused widespread damage in other parts of the world, particularly Central America and the West Indies but in East Africa the pathogen appears to occur in a less virulent form. It has been reported to occur seriously only in Moshi District in Tanzania, at the coast in Kenya and in Ankole District in Uganda. The symptoms are a yellowing of the lower leaves, which later hang downwards, and a purple discolouration of the vascular tissue inside the stems and rhizomes. The only way of preventing damage in infected areas is the use of resistant varieties; most of the cooking varieties and Dwarf Cavendish appear to be resistant. Planting material should not be taken from an infected area to an uninfected one.

Cigar end rot
This disease is caused by the fungus *Verticillium theobromae* and makes the tips of the fruits look like the ash on the end of cigars. It is suspected that damage only occurs in areas which are marginally suitable for banana cultivation, especially in areas with low night temperatures. Cigar end rot can largely be controlled by removing the ends of the inflorescences beyond the developing fruits.

Utilisation

Where bananas form the staple food they are usually cooked by either boiling or steaming. They may be boiled in their skins in which case they are peeled at the table and eaten alone. More commonly, however, they are peeled and then boiled, either whole, in halves (split longitudinally) or in small pieces. In this case they are usually made into a stiff paste (known as 'matoke' in Luganda) which is eaten with a sauce of vegetables, meat or fish. Other less common methods of cooking bananas are peeling and roasting in hot ashes, frying with maize, beans, potatoes or meat and peeling, drying and pounding into flour for 'ugali'. Eating bananas raw is uncommon where they are grown as a staple food but is the most common method of consumption where they are only grown as a minor crop, e.g. in the drier parts of Tanzania and in Kericho and Machakos Districts in Kenya. Where bananas are important yet are not the staple food, e.g. in Western and Central Provinces in Kenya, about half are eaten raw and half are cooked.

Brewing
Bananas are very popular for brewing in Uganda and in the northern banana growing areas of Tanzania. One survey in Masaka and Mengo Districts in Uganda showed an average of half an acre (c. 0·25 ha) of brewing bananas per small-holding. South of Moshi in Tanzania and in Kenya banana beer is rare. In the main method of brewing, as practised in all areas except Arusha and Moshi Districts in Tanzania, bananas are first buried in deep pits or are hung in the roof above the household fire for a period of about five days. They are then peeled and placed in a hollowed out log which resembles a dug-out canoe or in a specially dug ditch which has been lined with banana leaves. The juice is expressed from the fruits by squeezing or, for large quantities, by trampling. Blades of grass (*Imperata africana*) are included as a form of filter; they retain the pulp but allow liquid to flow through. Coarsely ground and fried sorghum or finger millet is added to the liquid which is then allowed to ferment for a few days. In Arusha and Moshi Districts the procedure is different. The peeled fruits are boiled until they are brown and are then allowed to stand in the water for about three days, sometimes with the addition

of more water, sometimes with the addition of ground boiled bark from a tree known as 'mseseve' (*Rauvolfia inebrians*).

Alternative uses for the banana plant

The use of leaves and stems for cattle fodder has been discussed above. Pieces of leaf and stem are used for a number of purposes: for thatching, for wrapping (especially snuff and meat) and for making mats and plates. String can be made from fibres in the stem.

Bibliography

1 **Baker, R. E. D.** and **Simmonds, N. W.** (1951). *Bananas in East Africa, I—The botanical and agricultural status of the crop.* Emp. J. exp. Agric., *19*: 283.

2 **Baker, R. E. D.** and **Simmonds, N. W.** (1952). *Bananas in East Africa, II—Annotated list of varieties.* Emp. J. exp. Agric., *20*:66.

3 **Harris, W. V.** (1947). *The banana borer.* E. Afr. agric. J., *13*:15.

4 **Masefield, G. B.** (1938). *The production of native beer in Uganda.* E. Afr. agric. J., *3*:362.

5 **Masefield, G. B.** (1944). *Some recent observations on the plantain crop in Buganda.* E. Afr. agric. J., *10*:12.

6 **Simmonds, N. W.** (1966). *Bananas.* Longman.

7 **Wallace, G. B.** (1952). *Wilt or Panama disease of bananas.* E. Afr. agric. J., *17*:166.

8 **Whalley, P. E. S.** (1957). *The banana weevil and its control.* E. Afr. agric. J., *23*:110.

2
Barley

Hordeum spp.

Introduction

Barley is a temperate cereal crop which is grown in the Kenya Highlands (about 27 000 acres, i.e. 11 000 hectares, in 1969) and to a much smaller extent on the western slopes of Mt. Kilimanjaro in Tanzania (300–500 acres, i.e. 120–200 hectares, in the same year). The main barley growing areas in Kenya are Molo, Mau Summit, Londiani, Mau Narok and Timau. A little is grown in some lower areas, e.g. Nakuru and Naivasha, but the rainfall is less certain in these places and this prevents good quality malting barley from being produced in many years.

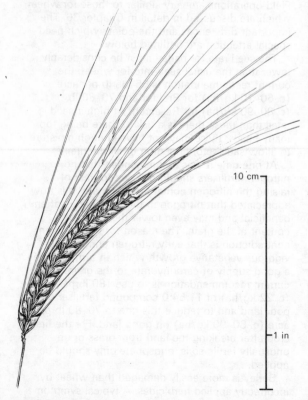

10 cm

1 in

Fig. 9: A head of barley

Malting
Almost all East Africa's barley crop is grown under contract to a Nairobi brewing company which sells bottled beer. Good quality barley is used for making malt which is an essential ingredient of bottled beer.* Sub-standard barley, which usually amounts to about 30% of the East African crop, can only be used as a livestock feed and fetches less than half the price of malting barley. To make malt, barley is first soaked in water and is then spread in a shallow layer in a long box where it germinates; the soaking period is two days and the germinating period is four days. The germinated grains are then dried and the shrivelled roots are removed by screening. Only germination can produce the enzymes which change the starch in the endosperm to sugar. During brewing the action of yeast on the sugar produces alcohol.

Malting quality
This subject is introduced at this stage because it affects almost all considerations in barley growing. At the Maltings, samples are taken to determine quality and the following are measured:

1　The nitrogen content. Barley is normally rejected for malting if it contains over 1·8% nitrogen; ideally it should contain 1·4–1·6%. Barley with a high nitrogen content causes a cloudy suspension of protein in the beer.

2　Coefficient of mealiness. When grains are cut in half they should show a white, mealy endosperm; mealy grains absorb water quickly and therefore germinate quickly and uniformly. If the endosperm appears hard and 'steely' the grains germinate slowly and unevenly. 'Steely' grains tend to have a high nitrogen content. A sample which has 50% or more mealy grains is acceptable; over 80% is ideal.

*Draught beer, which also requires malt, is not produced in East Africa

17

3 Germination. 95% or more is acceptable; 98% or more is ideal.

4 Size. Malting barley must be plump and well filled; this ensures that it is mature and will germinate well. Thin barley which passes through a 2·4 mm (c. $\frac{1}{10}$ in) screen is rejected

5 Moisture content. This must be 14% or lower so that the barley can be stored safely.

In addition, malting barley must not be discoloured (this would cause moulds and poor germination) or be damaged by weevils or by improperly set combine harvesters (this also would cause poor germination). Different varieties must not be mixed because they have different germination characteristics. A finely wrinkled husk is a sign of a well filled mature grain; a coarsely wrinkled skin, however, may be caused by premature ripening and is disliked by the maltsters.

Plant characteristics

In general appearance barley plants resemble bearded wheat. The most obvious differences are that the East African barley varieties have one grain in each spikelet and each head therefore has two rows of grains, one opposite the other. Wheat has three grains in each spikelet and the rows are less obvious. Barley seedlings can be easily distinguished from those of wheat; their leaves are hairless and they have smooth auricles whereas wheat seedlings have small hairs on their leaves and have hairy auricles. In commercial barley the glume and the palea adhere to each seed and cannot be removed by threshing; in wheat they are removed during threshing and constitute the chaff.

Ecology

Rainfall, and water requirements
To produce a plump mealy grain with a low nitrogen content, the crop must grow steadily without check. Periods of drought cause premature ripening, thin and steely grains and a high nitrogen content (due to insufficient accumulation of carbohydrate in the grain). Barley is therefore grown most successfully where there is a reliable rainfall; the annual total is relatively unimportant.

Altitude and temperature
The main barley growing areas are over 7 000 ft (c. 2 100 m). Barley growing is discouraged lower than this because of the unreliability of the rainfall.

Soil requirements
Heavy soils seldom produce good barley because the crop is very intolerant of waterlogging. Light sandy soils should be avoided because they do not usually retain enough moisture for steady growth.

Varieties

Proctor and Research (bred, respectively, in England and Australia) are the two well-established barley varieties in East Africa. Proctor is more widely grown and with its short stiff straw is suited to the high rainfall, high altitude areas. Research is better suited to the lower areas. Several alternative varieties have been introduced, e.g. Pallas and Zephyr; they have yielded well but have been unpopular with the maltsters. Midas and Imbar are new varieties which show promise in East Africa and they may be recommended in the future.

Field operations

Field operations are very similar to those for wheat, which are discussed in detail in Chapter 36. The important differences and the points which need special attention are outlined below.

The seed rate for barley must be considerably lower than the usual seed rate for wheat; the current recommendation is 55–65 lb per acre (c. 60–70 kg/ha) for Proctor and 70–80 lb (c. 80–90 kg/ha) for Research. At higher seed rates than these there is a risk of there being too many heads; these may not have enough moisture to allow satisfactory ripening of all the grains.

At one time it was recommended that no nitrogen fertilisers should be used for fear of raising the nitrogen content of the grain. It is now appreciated that nitrogen in the seed bed is often beneficial and may even lower the nitrogen content of the grain. The reason for this apparent contradiction is that early nitrogen encourages vigorous vegetative growth which in turn provides a good supply of carbohydrate to the grain. The current recommendation is to use 180 lb per acre (c. 22 kg/ha) of 11:54:0 compound fertiliser on poor land and to reduce this rate to 70–80 lb per acre (c. 80–90 kg/ha) on good land. For the first crop after breaking the land from grass or on unusually fertile soils, phosphate only should be applied.

Barley is more easily damaged than wheat by incorrectly applied herbicides. A typical symptom of herbicide damage is incomplete emergence of the ear from the sheath, the awns becoming tied up within the sheath. This causes a curved head and poor grain development. MCPA, being milder in action than 2, 4-D, is therefore the most commonly used herbicide. 2, 4-D and MCPP can be used safely if the sprayer is well calibrated, and if the crop is sprayed at the recommended stage of growth.

Barley must remain in the field until it is fully ripe but farmers are often tempted to harvest it before it has completely filled out. Early harvesting, which would not cause much loss with wheat, usually causes a consignment of barley to be rejected by the maltsters as being thin, steely and high in nitrogen. Proctor and Research mature in $5\frac{1}{2}$–6 months, depending on the altitude. Great care must be taken during combining to ensure that the concave is not set so close to the drum that some of the grains are damaged.

Yields

The average yield is about 1 300 lb per acre (c. 1 500 kg/ha). With good husbandry and in a high rainfall area, 2 500 lb per acre (c. 2 800 kg/ha) should be achieved. The best yields recorded in East Africa are a little over 3 500 lb per acre (c. 3 900 kg/ha).

Pests

Barley is attacked by the same range of pests as wheat. In addition, it is occasionally severely damaged by the barley fly, *Hylemya arambourgi*. The larvae of the barley fly make their way down the centre of the seedlings until they reach, and eat, the growing point. The characteristic symptom of barley fly damage is a dead central leaf on some of the shoots. Serious damage can be avoided by using a dieldrin or aldrin seed dressing.

Diseases

Barley is susceptible to all the diseases which attack wheat. Stem rust is the greatest threat; it not only reduces yields but prevents the flow of moisture, nutrients and carbohydrates to the developing heads, thus causing thin and steely grain. In addition net blotch and spot blotch (*Helminthosporium spp.*) and scald (*Rhyncosporium secalis*) can cause a certain amount of damage. These are fungal leaf diseases which are spread from plant to plant by wind and from year to year by infected debris in the soil or by infected seed.

Serious damage can be prevented by using fungicidal seed dressings and by avoiding continuous growing of barley on the same field.

3
Beans

Phaseolus vulgaris

Introduction

Many names are used for *Phaseolus vulgaris*; these include French beans, kidney beans, haricot beans, snap beans, garden beans, dwarf beans, common beans, field beans, string beans, bush beans and pole beans. In East Africa, however, they are usually referred to simply as beans.

Fig. 10: A pure stand of beans.

Kenya

Beans are by far the most important pulse crop in Kenya. The area is impossible to estimate accurately but it is probably about one million acres (c. 400 000 ha), mostly intersown with maize. Beans are important in all agricultural areas except at the coast.

Considerable quantities of beans are sold in Kenya for canning; they are either exported to Europe or canned locally. They are cooked in tomato sauce and are usually known to the consumer as baked beans. Beans for canning must be small, white and of a certain shape and quality (see section on varieties).

Tanzania

Beans are important in the higher and wetter areas of Tanzania. Seed beans, for export to Europe, have been grown extensively on the plains between Arusha and Makuyuni. The area varied during the 1960s, largely owing to fluctuations in demand; in many years it exceeded 30 000 acres (c. 12 000 ha). Dutch dealers, who distribute to most other European countries, send out small samples of the required varieties; these are bulked up and are returned as twice grown seed. The soils and the rainfall distribution of the Arusha/Makuyuni plains are ideal for seed beans (see section on ecology) and the crops are kept free of the seed borne diseases which plague seed growers in Europe.

Uganda

Beans are important in the wetter bimodal rainfall areas of Uganda, especially Kigezi, Masaka and Mengo Districts. The estimated area of beans grown in Uganda in 1968 was 660 000 acres (c. 270 000 ha).

Plant characteristics

Beans are annual legumes. Nodulation of the roots is very variable; in some areas, e.g. near Kitale in Kenya, large nodules are produced whilst in others, e.g. parts of Uganda, none are produced. Little is

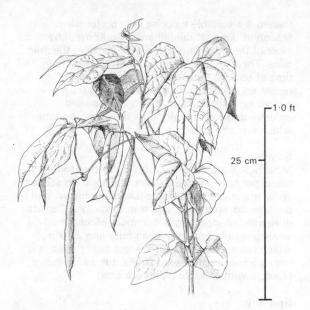

1·0 ft

25 cm

Fig. 11: A bean plant. Note that the first pair of leaves, unlike the others, is not trifoliate.

known about the factors which affect nodulation in East Africa although it is known that the common strains of Rhizobium bacteria belong to the 'cowpea group' and that these are ineffective on *Phaseolus spp.* There is great variation in the growth habit; most beans grown in East Africa are determinate, with bush type growth habit, but there are also indeterminate non-climbing, semi-bush types, and indeterminate climbing types. The predominant colour of the flowers is white, although pink, purple, red and yellow are common. The greatest variation occurs in seed characteristics: many shapes, sizes, colours or combinations of colours can be found. Red, brown, purple, black and white are the common colours and these may occur alone, as stripes or as spots. Beans are nearly 100% self-pollinating under most conditions.

Life cycle

The period from sowing to first flowering of most bush type beans is about five weeks at 4 000 ft above sea level (c. 1 200 m), and flowering continues for about two weeks. It takes about two weeks for a flower to produce a full-length pod and another four or five weeks for the pod to mature and dry out. The life of the crop at medium altitudes is therefore approximately three months. At 6 000 ft (c. 1 800 m) the maturation period is about four months whilst at 8 000 ft (c. 2 400 m)

it is near five. Indeterminate types take longer to mature; at medium altitudes the maturation period is usually four to five months.

Ecology

Rainfall, and water requirements

Beans are not drought-resistant; ideally they need moist soil throughout the growing period. The required length of the wet season depends on the growth habit and on the altitude. In Uganda, for example, bush types are the most suitable in the bimodal rainfall areas where there are two relatively short wet seasons; indeterminate types are more suitable in the northern areas where there is one relatively long wet season. In Kenya bush type beans can be grown in Machakos and Kitui Districts during a two or three month wet season whilst at higher altitudes they need a longer period of rainfall.

Rainfall towards the end of the growing period is undesirable when seed beans or canning beans are being grown. It causes a high incidence of pests and diseases which can discolour the seeds; it also splashes earth onto the pods: when earth-covered pods are threshed the seeds inevitably become stained. Stained and discoloured seeds are unacceptable or fetch a reduced price in the canning and seed bean industries. The plains between Arusha and Makuyuni are one of the few places where rainfed seed beans can be grown successfully. They are sown after the peak rainfall of April but before the abrupt end of the rains which can be forecast with a fair degree of accuracy at the end of May or the beginning of June; they therefore mature in dry weather but have a good supply of moisture owing to the deep, retentive soils and the earlier heavy rainfall.

Altitude and temperature

Beans do not grow well below 2 000 ft (c. 600 m) because high temperatures cause a poor fruit set. They are therefore unimportant along the East African coastal strip. They are best suited to the medium altitude areas, from 3 000 ft up to 7 000 ft (c. 900–2 100 m), although they are often found growing at altitudes as high as 9 000 ft (c. 2 700 m) in parts of Kenya.

Soil requirements

Beans demand free draining soils with a reasonably high nutrient content.

Varieties

Few smallholders in East Africa grow pure line varieties; the best examples of the few pure line varieties are Canadian Wonder, a bush type variety which produces large purple seeds and which is popular in Kenya and Tanzania, and Cuarentino, a indeterminate variety which produces small white seeds and which is grown on sticks or poles in Kigezi District in Uganda. Most smallholders grow natural genetic mixtures, technically termed landraces but usually referred to as 'mixed beans'. In some landraces there is a great diversity of seed type but in others the seed type is fairly uniform, e.g. the plump, pink and purple mottled seeds which are popular in Kenya and which are usually called Rose Coco. Most landraces in East Africa have coloured beans; only in Kigezi District are white seeded varieties preferred.

A breeding programme for food beans started in the early 1960s at Kawanda in Uganda and has since been continued at Makerere University College. The first releases were Banja 2, a bush type variety producing pink and purple mottled seeds and originally showing resistance to anthracnose, and Mutikke 4, an indeterminate semi-climbing variety with large red seeds. A programme of hybridisation is being carried out and higher yielding varieties with greater disease resistance can be expected from this in the future.

The outstanding canning bean, which was selected at Tengeru in Tanzania, is Mexico 142 (in Uganda it is known as No. 212). It has small, white seeds of a suitable shape and has a very small percentage of seeds which remain hard when soaked in water. (It is common for unselected beans to have an undesirably high proportion of seeds which do not soften when soaked and which are therefore unsuitable for baked beans.) Mexico 142 is a bush type variety which is high yielding and rust resistant, matures early and is not prone to shattering or seed splitting.

Field operations

Seed preparation
Beans, having quite large seeds, do not need a fine seedbed. Any seedbed suitable for maize would be equally suitable for beans.

Time of sowing
Short term varieties mature quickly and are therefore a suitable crop for the shorter rainy season in bimodal rainfall areas. In Kenya, however, most of them are intersown with maize in the long rains. There is little information on the optimum time of sowing within the rains; there are no recommendations which have resulted from critical experimentation. Beans for the seed industry and for canning must be sown at a time which will allow them to mature in dry weather.

Sowing
Most beans are sown by dibbling two or four seeds per hole. Where ox-ploughs are used seeds are sometimes dropped into a shallow furrow and are later covered over, whilst in Kakamega District in Kenya broadcasting is common. Mechanised sowing is only possible when pure line varieties, which have uniform seed size, are being used; it is virtually restricted to seed beans, for which maize planters with suitable plates are used.

Spacing
Most beans are intersown with cereals or, in Uganda, cassava or bananas. Only a small proportion are grown in pure stands; notable amongst these are seed beans and canning beans. Pure stands are becoming more popular in Kenya, especially when beans are sown in the short rains.

The correct spacing depends largely on the variety: indeterminate types do not need such a high population as bush types. The usual recommendation for bush types is rows $1\frac{1}{2}$–2 ft apart with plants 6 in apart within the row (c. 0·45–0·6 m × 15 cm), i.e. a plant population of 44 000 to 58 000 per acre (c. 110 000–145 000 per ha). Experiments in Uganda, however, indicate that the optimum spacing is closer: 2 ft rows with the plants spaced 2–3 in apart within the row (c. 0·6 m × 5–7·5 cm) giving a plant population of 87 000–130 000 per acre (c. 215 000–320 000 per ha). In general, smallholders almost always adopt a spacing which is much too wide.

Seed inoculation
Inoculation with *Rhizobium* bacteria has been tried on several occasions in East Africa but has seldom been successful.

There are several methods of inoculation, applicable to all legumes. Pelleting is the safest and most effective method, although it is also the most expensive. Peat inoculant is used and is stuck to the seed using a cellulose solution. Before the sticker dries, the seeds are mixed with lime or phosphate (not superphosphate); this gives each seed a coat of material which protects the bacteria.

Fertilisers

Nodules are sometimes absent and if present are often small; the amount of nitrogen fixed by the root system is therefore usually small. Evidence of this is provided by the fact that leaf yellowing is often seen. Responses to nitrogen fertilisers are common. Responses to phosphate are more common and are usually larger than those to nitrogen. A mixed fertiliser (2 cwt of calcium ammonium nitrate, 2 cwt of single superphosphate and 1 cwt of muriate of potash per acre) has given economic responses at Namulonge in Uganda.

Weed control

Smallholders rely on hand weeding whilst large scale growers aim to cultivate as seldom as possible, killing most of the weeds during seedbed preparation. Beans are susceptible to most herbicides; there are, as yet, no recommendations on chemical weed control.

Harvesting

Smallholders usually harvest beans by uprooting whole plants. These are sometimes dried in the field but are more commonly taken to the homestead where they are dried on bare earth, mats, sacks, tarpaulins or corrugated iron, or on an area which as been smeared with cow dung and allowed to bake hard. The drying period may be anything up to a week, depending on the weather and the amount of drying in the field before harvesting. When they are dry enough they are beaten with sticks, either directly or after putting them into sacks. The haulm and pods are later removed by hand and by winnowing.

Seed bean growers uproot the plants, stook them and leave them to dry in the field for about a week. Threshing is done by heaping the plants on large tarpaulins and driving a tractor around on top of them.

Yields

Actual yields are notoriously low and are usually between 200 and 600 lb of dried seeds per acre (c. 220–670 kg/ha). With improved varieties, good husbandry and good pest and disease control, 1 000 lb per acre (c. 1 100 kg/ha) should be expected. Yields of about 2 000 lb per acre (c. 2 200 kg/ha) have been achieved on a field scale but these are regarded as exceptional.

Pests

Beans are plagued by many pests; these are almost always ignored on subsistence crops but they must be controlled on high value canning or seed crops. Some of the more important pests are discussed below.

Bean fly. *Melanagromyza spp.*
The most damaging and most abundant species is *M. spencerella. M. phaseoli* seems to be more effectively parasitised and is less common. The adult flies lay eggs in the bean leaves or in the hypocotyl; the larvae from these bore downwards and pupate in the stems at ground level. The bases of the stems become thickened and cracked; many plants die and the survivors become stunted and yellow. Damage can be prevented by seed dressing with aldrin or dieldrin. Early planting, crop rotation and the removal of crop residues and volunteer plants are useful cultural precautions. There are varietal differences in susceptibility to bean fly which are probably due to differences in the ability to produce adventitious roots.

American bollworm. *Heliothis armigera*
This pest is discussed in greater detail in Chapter 14. The larvae bore circular holes in the sides of the pods and eat much of the contents. Young larvae are only capable of attacking small pods but they later move to those which are more mature. DDT sprays control American bollworm effectively if they are applied when the larvae are still small.

Spotted borer. *Maruca testulalis*
This is a pest of the drier areas which has caused much damage to seed beans near Arusha in Tanzania. The larvae are olive green with rows of dark spots and are hairy. They eat flowers and pods and can be controlled by DDT sprays.

Bean aphid. *Aphis fabae*
These black, sucking insects cluster around growing points, stems, leaves and flowers. If present in large numbers they can prevent normal growth. Yellowing and distortion of the leaves is common. Sprays of menazon and endosulfan have given good control.

Bean bruchid. *Anthoscelides obtectus*
This is the main storage pest of beans; it also attacks most other pulse crops. Adult beetles often lay their eggs in the field on the developing pods. The larvae bore through the pod walls and into the

seeds; they are so small that their entry holes are almost invisible and as the seeds grow the holes disappear. The larvae feed inside the seeds and each makes a tunnel almost to the surface; only the seed coat is left intact, forming a 'window' at the end of the tunnel. After pupation the adult beetle emerges by pushing out the flap of seed coat, leaving a circular hole. Although some eggs are laid in the field, most are laid by adults emerging in the stores; they lay their eggs loosely among the beans and can cause a very rapid increase in the number of infested seeds. Build-up of bean bruchids can be prevented in stores by mixing the seeds with gamma-BHC dust. It is most important that canning beans are not attacked by bruchids in the field otherwise the larvae appear in the final product. DDT sprays give good bruchid control in the field.

Diseases

Diseases probably account for more crop loss in beans than do pests; some think that diseases are the main cause of yield fluctuation in East Africa. There are probably several virus diseases that have not yet been identified.

Bean rust
This disease is caused by the fungus *Uromyces phaseoli* (Syn. *U. apprendiculatus*). Many small red pustules grow on the undersides of the leaves; a characteristic dark green spot surrounded by a small yellow circle usually appears on the upper side of the leaf above each pustule. Heavily infected leaves become deformed and are shed early. Bean rust is rapidly spread from plant to plant by air-borne spores and probably also by rain splash. Wet weather encourages spore germination and lesion growth. Rust can be avoided by growing resistant varieties, e.g. Mexico 142, but resistance is made complicated by the existence of different strains of the pathogen. Dithiocarbamate sprays have given excellent control but they are only justified when beans are being grown as a canning or seed crop.

Anthracnose
The symptoms of this disease, which is caused by the fungus *Colletotrichum lindemuthianum*, are brown or black lesions; these occur most characteristically on the undersides of the leaves but more obviously on the stems, petioles and pods. Stem and petiole lesions are sunken and longitudinal. Pod lesions are also sunken but are circular, seldom growing larger than $\frac{1}{4}$ in (c. 0·6 cm); each has a well defined black margin around a brown centre. Leaf lesions appear as black marks on the undersides of the veins and are often associated with leaf distortion. Anthracnose is spread in three ways: by infected seed, by infected trash in the soil and by spores splashed from the lesions. Spread by rain splash may begin at a very early stage of the crop's growth because plants grown from infected seed and plants infected via the soil before emergence often have lesions on their cotyledons. Growing resistant varieties, as is planned in Uganda, will probably prove to be the most satisfactory way of reducing crop losses by this disease. Crop rotation, destruction of crop residues and the use of clean seed are important precautions. Dithiocarbamate sprays have given good control.

Halo blight
This bacterial disease, caused by *Pseudomonas phaseolicola*, is common in the cooler and wetter areas. It causes irregular dark spots on the leaves or pods; each spot is usually surrounded by a broad yellow band or 'halo'. When the lesions are well developed they have a water-soaked appearance and their centres sometimes fall out. These symptoms are easily confused with those of common and fuscous bacterial blights, which are caused by *Xanthomonas* and are more important under warmer conditions. These bacterial diseases are spread in a similar way to anthracnose but particularly by wind-driven rain. Their control is therefore similar, except that little selection of resistant varieties has been done in East Africa, and that fungicides are ineffective.

Angular leaf spot
This fungal disease, caused by *Phaeoisariopsis griseola*, causes brown spots on the upper surfaces of the leaves. These lesions are usually irregular in shape because they are bounded by the veins; on the first pair of true leaves, however, they are often round. The underside of each lesion is black and hairy. Lesions sometimes occur on the pods, especially in wet weather; these are usually larger and paler than those of anthracnose. This disease is spread by infected seed and trash, by wind, and locally by rain splash. Using clean seed and crop rotations and destroying crop residues in the field are important precautions. Breeding work in Uganda includes the introduction of varieties which are resistant to angular leaf spot, such as Diacol Nima from Columbia.

Mosaic

Mosaic viruses cause a wide range of symptoms including mottling, brittle leaves, stunting and distortion; in some cases each plant becomes a mass of proliferating stalks. It is not yet known whether these symptoms are caused by the same mosaic virus that is found in Europe and the U.S.A. No control measures are known in East Africa.

Utilisation

The seeds are cooked in a number of ways. The most common is to boil them with maize seeds until both are soft; the mixture is eaten alone or with green vegetables. Alternatively, potatoes, bananas and green vegetables can be added towards the end of the 2–3 hour boiling period; the mixture is then pounded into a paste. Lastly, beans may be boiled alone and then ground into a paste, after removing the seed coats. This paste may be eaten immediately or fried; in both cases it is usually eaten with 'Ugali'. Bean leaves are sometimes used as a pot-herb.

Bibliography

1 **Davies, J. C.** (1959). *A note on the control of bean pests in Uganda.* E. Afr. agric. J., *24*:174.

2 **Davies, J. C.** (1962). *A note on in-sack storage of beans using 0.04% gamma-BHC dust.* E. Afr. agric. for J., *27*:223.

3 **Howland, A. K.** and **Macartney, J. C.** (1966). *East African bean rust studies.* E. Afr. agric. for. J., *32*:208.

4 **Macartney, J. C.** (1966). *The selection of haricot bean varieties suitable for canning.* E. Afr. agric. for. J., *32*:214.

5 **Stephens, D.** (1967). *The effects of ammonium sulphate and other fertiliser and inoculation treatments on beans.* E. Afr. agric. for. J., *32*:411.

4
Bonavist beans

Lablab niger

Introduction

In East Africa there is no commonly accepted name for the pulse crop which is called 'fiwi' in Kiswahili and 'njahi' in the Kikuyu language. The name bonavist beans is widely used, however, in other parts of the world.

Bonavist beans are a traditional crop in Central Province in Kenya, playing an important part in Kikuyu custom. They are also grown in Embu, Meru and Machakos Districts in Kenya and, occasionally, in Nandi District and in the Tugen Hills in Baringo District. Kikuyu workers and settlers have also spread the crop to most parts of the Kenya highlands. In Tanzania bonavist beans are grown in the Southern Highlands, the Pare and Usambara Mountains and on the slopes of Mt. Meru. They are of no importance in Uganda.

Bonavist beans have declined in importance in Kenya with the increasing popularity of beans (*Phaseolus vulgaris*).

Plant characteristics

The bonavist bean plant is a perennial legume although it is sometimes treated as an annual. Varieties differ in many characteristics: the growth habit may be bunch, spreading or climbing; the flowers may be white, purple, pink, blue or yellow; the pods may be short and half-moon shaped or long and thin; the seeds, although usually brown or black, may be cream-coloured or white. The pods have pronounced beaks and each seed has a prominent white hilum. In most varieties the inflorescence is an erect, long stalked raceme which is held well above the foliage.

Ecology

Little is known of the ecology of bonavist beans except that they are very drought resistant. They are mostly grown between 4 000 and 7 000 ft (c. 1 200 and 2 100 m) above sea level. Some varieties are sensitive to the photoperiod; bonavist beans at Mlingano in Tanzania flowered two months after June sowing yet with October sowing flowering was delayed until the onset of shorter days.

Field operations

Almost all bonavist beans are intersown with other food crops, e.g. maize, beans, potatoes, peas, bananas, etc. The climbing varieties often use the stems of taller plants for support. Bonavist beans are usually sown in the long rains. Several seeds are sown in each hole and these remain unthinned. Owing to uneven ripening, the pods are removed individually as they ripen. They are dried in the sun and are later beaten with sticks to remove the seeds.

Pests and diseases

No detailed information is available on the pests and diseases which attack this crop. Larvae are often found attacking the pods and aphid infestations sometimes occur.

Utilisation

Bonavist beans are cooked in any of the ways common for beans, e.g. boiled with maize, ground, fried, or included with the traditional Kikuyu dish 'irio': a mixture which may include maize, beans, bananas, potatoes or green vegetables and which is boiled and made into a paste. The leaves are occasionally used as a pot-herb although cowpea leaves are more popular for this purpose. The green pods are popular amongst the Asian community as a vegetable.

5
Bulrush millet

Pennisetum typhoides

Introduction

Bulrush millet is a cereal crop. It is important in Central Province and around Shinyanga in Tanzania and is the staple food of the Wakara on Ukara Island in Lake Victoria. In Kenya it is important in the lower altitude areas of Kirinyaga, Embu and Meru Districts; it is also grown in parts of Machakos District and in the Kerio Valley. In Uganda it is occasionally grown in the sandy soils in Teso, Lango, Acholi and Karamoja Districts.

Bulrush millet is very drought resistant and gives reasonable yields on infertile sandy soils which would be unsuitable for the other important cereals. Despite these advantages it has declined in importance in East Africa during the last thirty or forty years. In the 1940s it was the most important food crop of the Wasukuma in the densely populated area to the south of Lake Victoria in Tanzania. Now, however, it has been replaced by maize except in the drier southern part. Bulrush millet compares unfavourably with maize in several respects: it has a lower yield potential and only outyields maize if rainfall is very poor; bird damage sometimes causes a complete crop loss; threshing and winnowing are more laborious and people who are accustomed to a diet of maize find it unpalatable.

Plant characteristics

Bulrush millet grows as high as 12 ft (c. 3·5 m) and most varieties tiller freely. The inflorescence, which is a dense spike, is usually erect with no bristles (see Fig. 13). Some varieties, however, have drooping heads or bristles which arise from the base of each spikelet (see Fig. 12); these varieties are less susceptible to bird damage. Bulrush millet is largely cross-pollinated because the stigmas emerge before the anthers. It is quick maturing and can usually be harvested between three and four months after sowing. Unlike sorghum, its initial growth is rapid.

Ecology

Bulrush millet is both drought resistant and, because of its short maturation period, drought evading. These characteristics suit it to areas of low and uncertain rainfall, e.g. Central Province in Tanzania where the average annual rainfall is only 20–25 in (c. 500–625 mm). It needs a warm climate and is therefore seldom grown above 4 000 ft (c. 1 200 m). Although it is usually grown on impoverished soils of a medium or light texture, it has given good yields on adequately drained black cotton soils in Tanzania.

Varieties

At the time of writing no improved varieties are available. A breeding programme started at Serere

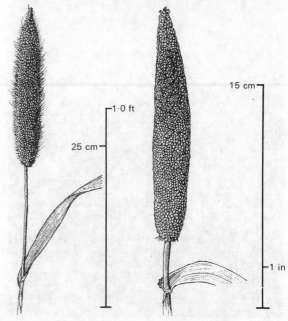

Fig. 12: Bulrush millet; a variety with bristles.

Fig. 13: Bulrush millet; a variety without bristles.

Research Station in Uganda in the late 1960s; high yielding, early or medium maturing, disease resistant varieties have been selected, both with and without bristles, and some of these have been successfully used as the parents of hybrids. Improved varieties will be released during the 1970s.

Field operations

Bulrush millet is grown in pure stands in some areas, e.g. in Kirinyaga, Embu and Meru Districts in Kenya. In others it is more commonly intersown, e.g. with bambarra groundnuts on Ukara Island in Tanzania and with cowpeas, pigeon peas, beans or maize in Machakos District in Kenya.

As it can give reasonable yields on infertile soils it is commonly grown as the last crop in the arable break before the land is rested under a bush fallow. In a bimodal rainfall regime it may be sown in either rains although it is best suited to the shorter of the two, thus allowing slower maturing crops to be sown in the longer rains.

Fig. 14: A pure stand of bulrush millet.

Bulrush millet is usually sown at random and on the flat. The seed is occasionally broadcast. The spacing recommendation from Serere Research Station in Uganda is 24 in × 6 in (c. 60 cm × 15 cm). It is unlikely that spacing is a critical factor because bulrush millet can compensate for wide spacings by increased tillering. Little fertiliser work has been done but Serere recommends 1–2 cwt of ammonium sulphate per acre (c. 125–250 kg/ha) as a split dressing; half should be applied to the seedbed whilst the other half should be top dressed when the crop is about 1 ft (c. 0·3 m) high. Bulrush millet produces tillers so vigorously that most of the weeds are suppressed during the later stages of the crop's growth; weeding should only be necessary before tillering. Bulrush millet is the most resistant of cereal crops to the parasitic weed *Striga hermonthica* (see Chapter 29 for more details on this weed).

Harvesting is done by removing the individual heads with sickles or small hand knives. This is sometimes preceded by breaking the stems; in Meru District in Kenya this may be done by two men who drag a thick pole along two or three rows at once.

Yields

Yields probably average no more than 400 lb of grain per acre (c. 450 kg/ha). With good husbandry yields as high as 1 500 lb per acre (c. 1 700 kg/ha) have been obtained at Serere.

Pests and diseases

Birds, e.g. Quelea, weaver birds and bishop birds, are the greatest pests; they are attracted to the crop when the grains are in the milky stage. Downy mildew, caused by the fungus *Sclerospora graminicola*, rust, caused by the fungus *Puccinia penniseti*, and ergot, caused by the fungus *Claviceps microcephala*, are sometimes found on bulrush millet but they seldom cause much reduction in yield.

Utilisation

Much bulrush millet is made into 'uji'. 'Ugali' is more commonly made in Tanzania and in Machakos District in Kenya. Bulrush millet is sometimes used for brewing; the grain is germinated, dried and ground into a flour and this is boiled in much water and is then allowed to ferment for about a week.

6
Cashew

Anacardium occidentale

Introduction

Cashew is an important cash crop grown along the East African coast. The most important product is the kernels which are used for dessert or confectionery purposes. Also valuable is the cashew nut shell liquid; its main constituents are cardol, cardinol and anacardic acid and these can be used in a number of industrial products including paints, plastics and brake linings.

Tanzania
Cashew is grown in all Tanzania's coastal districts but the bulk of the crop comes from Newala, Lindi, Masasi and Mtwara Districts with a marked concentration on the Mkonde Plateau. For many years cashew nuts have held their position as the fourth most valuable export crop after sisal, cotton and coffee. During the 1960s their annual value ranged from £2 million to £5 million.

Kenya
Cashew nuts are grown in the wetter, southern part of the Kenya coast. Kenya's annual cashew exports have seldom exceeded £500 000 in value.

Smallholders
Almost all East Africa's cashew is grown by smallholders. For most of these it is an attractive crop because it gives a small cash income even when completely neglected. An additional attraction is that, at the time of writing, the market is unlimited.

Plant characteristics

The cashew tree is an evergreen perennial which spreads vigorously (see Fig. 15); old trees commonly have canopies with a diameter of 40 ft (c. 12 m). Root studies in Tanzania have shown that trees can utilise a large volume of soil because their roots grow not only vertically to a considerable depth but also to a radius which is often twice that of the canopy.

The inflorescence is panicle-like and consists of many small pinkish flowers. About 14% of the flowers are hermaphrodite; the remainder are male. The crop is mainly cross-pollinated; insects are considered to be the most important pollinating agents, although the wind may play a part. About 70% of the hermaphrodite flowers fail to produce nuts, and it is common to see only one or two nuts on an inflorescence. The period from flowering to nut fall is 55–70 days. The fleshy cashew apple, below which the nut hangs, is a swollen peduncle (see Fig. 16). It is most commonly left to rot in the field but the juice can be sucked from it or can be fermented to make an alcoholic drink. Most fruits have an astringent taste but some yellow ones are sweet.

Flowering occurs from June to January with a peak from August to October. Harvesting therefore starts in August and lasts until March with a peak from October to December.

Fig. 15: A cashew tree.

Fig. 16: Two cashew nuts from the same inflorescence. The one on the right has a fully developed apple, whilst that on the left has an apple which has not yet developed.

Ecology

Rainfall, and water requirements
Cashew is an unusual crop in that it flowers, sets fruit and maintains a full leaf cover during the driest part of the year. In the most important producing area of East Africa, i.e. southern Tanzania, this dry season is pronounced and may last for as long as six months. The drought resistance of cashew has recently been explained by the great horizontal growth of the roots; this enables it to yield well in areas which receive an average annual rainfall of only 30–35 in (c. 750–900 mm). It can only do this, however, if the spacing is wide enough to prevent overcrowding of the trees.

For good yields of cashew nuts the weather must be dry during the peak flowering period. Rain causes a high incidence of fungal infections on the inflorescences and can do physical damage to the flowers. In addition the main harvesting period should be dry otherwise the nuts become an undesirable brown or black colour or even rot completely.

Altitude and temperature
Cashew is grown between sea level and 2 500 ft (c. 750 m). It grows poorly at higher altitudes, where the temperatures are too low.

Soil requirements
Cashew's main soil requirement is that there should be unimpeded drainage. It is often grown satisfactorily on infertile sands.

Varieties

No breeding work has been done in East Africa and no improved varieties are available. There is great variation between the performance of seedling trees owing to the high proportion of cross-pollination. There is therefore great scope for the selection of high-yielding clones and the adoption of vegetative propagation.

Field establishment

Seed selection
It is recommended that seed should be selected from high-yielding trees; there have been no experiments, however, to evaluate the benefits of this practice. The nuts should be subjected to a flotation test; they should be immersed in a solution of 1½ lb of sugar or 1 lb of salt in 1 gallon of water (150 g sugar or 100 g salt in 1 litre) and those which float should be rejected. These have been shown to give poorer germination, weaker growth and lower yields during the first three years of bearing than those which sink.

Planting
Cashew is usually planted at stake, i.e. the seeds are planted directly into the field. Transplanting seedlings from nurseries into the field is only occasionally practised. When planted at stake three seeds should be planted about 9 in (c. 23 cm) apart in a triangle in each hole; the holes should be about one foot (c. 0·3 m) wide but need be no deeper than 2–3 in (c. 5·0–7·5 cm), as this is the

depth at which the seeds should be placed. In favourable conditions germination occurs in 2–4 weeks. Some form of protection from the sun, e.g. two or three coconut leaves, is beneficial in the early stages of growth. The seedlings should be thinned to leave one (the most vigorous) per hole; this is usually done about a year after planting.

Cashew should be planted as near the beginning of the rains as possible.

Spacing
If spaced as close as 20 ft × 20 ft (c. 6 m × 6 m) cashew trees suffer from severe moisture stress about five years after planting unless they are thinned; they defoliate, the flowers dry up and virtually no crop is borne. An alternative is to plant at 30 ft × 30 ft (c. 9 m × 9 m); at this spacing thinning is necessary after seven years. The current recommendation in Kenya is to plant at 40 ft × 40 ft (c. 1·2 m × 1·2 m) or 50 ft × 50 ft (c. 1·5 m × 1·5 m); this does not give such high yields in the early stages of growth (other crops can be grown between the rows to compensate for this), but it does eliminate the expensive operation of thinning.

Cashew trees are usually grown at random between coconut palms, mango trees, banana plants, cassava, etc. An exception occurs in southern Tanzania where they are planted in a pure stand at an average spacing of 47 ft × 47 ft (c. 14 m × 14 m).

Field maintenance

Field maintenance of cashew requires less time and money than the maintenance of most other perennial crops. Good yields can be obtained without the addition of fertilisers or manures and only a little pruning is necessary to allow enough space beneath the trees for nut collection and weed control. Uncontrolled weed growth conceals the fallen nuts and constitutes a fire hazard; if tall dry grass is burned many trees, especially young ones, can be killed. The vegetation under the trees should either be checked by slashing or should be grazed down by cattle. Cattle should not be allowed in young plantations, however, because they may trample on young plants and break small branches. In the early stages of growth it is desirable to keep a circle of weed-free soil around the young trees.

When cashew is planted at a close spacing the trees must be thinned as soon as their canopies meet. Removing five year old trees is a very labour-consuming operation.

Harvesting

It is recommended that nuts are allowed to drop to the ground before they are collected; this ensures that no unripe nuts are harvested. In fine weather regular collection at intervals of no longer than a week is essential if good quality nuts are to be produced. If they are left on the ground for much longer than a week the seed coats become brown and rotten instead of the desirable grey or greyish blue. In wet weather they rot very quickly and daily collection is recommended. Discoloured nuts are rejected or only fetch about half the price of nuts suitable for export.

Nuts fall to the ground with their apples attached; the two must be separated with a twisting action during collection. A piece of apple flesh invariably adheres to each nut and this must be removed immediately by trimming with a sharp knife. Finally the nuts should be sun-dried for a few days until the kernels can be heard rattling inside when the nuts are shaken.

Yields

Average yields from pure stands in southern Tanzania are about 530 lb of nuts per acre (c. 590 kg/ha) and similar yields have been obtained on the Kenya coast. On well managed plots near Dar es Salaam, however, 900 to 1 000 lb per acre (c. 1 000–1 100 kg/ha) are common, possibly owing to the higher rainfall. Experiments have given yields as high as 1 500 lb per acre (c. 1 700 kg/ha) and if vegetative propagation was used to overcome the genetic variation between seedlings, yields of 2 000 lb per acre (c. 2 200 kg/ha) should be possible. Cashew trees grow quickly and start producing nuts in their third year in the field. Full yields should be obtained between eight and ten years after planting.

Pests

Helopeltis
Two species, *H. anacardii* and *H. schoutedeni,* cause damage to cashew by sucking the leaves, young shoots and inflorescences. The attacked parts show long black lesions and often die back. Regrowth from lateral buds can give a witches' broom effect. An internal browning, often associated with this type of damage, extends for a considerable distance down the shoot. Dieldrin sprays or dusting with BHC, or DDT plus BHC, give good control of *Helopeltis* in Tanzania, but it

is uncertain whether insecticides give economic returns. During 1969 and 1970 *Helopeltis* was not found on the Kenya coast despite careful searching. The symptoms mentioned above were observed, however, and it was found that they were being caused, at least in part, by *Pseudotheraptus wayi*, the important sucking pest of coconuts.

Thrips, a leaf miner, a leaf roller and red spider mites are often troublesome on young trees.

Diseases

A 'die back' is occasionally observed in Tanzania and a fungus or complex of fungi attack the inflorescences in wet weather. Little is known about either of these and no recommendations have been made for their control.

Processing

Cashew processing, i.e. the removal of the kernels from the nuts, represents one of the greatest problems in the cashew industry. The difficulty is presented by a resin, commercially called cashew nut shell liquid (C.N.S.L.), which is found in the mesocarp. C.N.S.L. has two undesirable effects unless precautions are taken: it contaminates the kernels and it severely blisters the skin of any person coming into contact with it.

The traditional method of processing, whether on a small or a large scale, is roasting. This gets rid of most of the C.N.S.L. and makes the shells sufficiently brittle for them to be cracked. If the nuts are roasted in drums the valuable C.N.S.L. is destroyed; if they are roasted in a bath of C.N.S.L. the liquid is extracted and retained. C.N.S.L. baths, kept at a temperature of 356–365°F (180–185°C) are employed in the mechanical processing plants installed at Kilifi in Kenya and at Dar es Salaam in Tanzania. There is also a mechanical plant at Mtwara in Tanzania which works on the principle of cooling the shell in order to fix the C.N.S.L. and to make the shell brittle.

Most of East Africa's cashew nuts are exported to India where they are roasted and then shelled by hand before being shipped to the U.S.A. and Europe. There is a small hand shelling plant at Kilifi in Kenya. During hand shelling the nuts and the workers' hands are kept dusted with wood ash; the ash protects the skin from blistering and absorbs any C.N.S.L. remaining on the shell, thus preventing it from contaminating the kernels.

Bibliography

1 **Auckland, A. K.** (1961). *The influence of seed quality on the early growth of cashew*. Trop. Agriculture, Trin., *38*:57.

2 **Bigger, M.** (1960). Selenothrips rubrocinctus *and the floral biology of cashew in Tanganyika*. E. Afr. agric. J., *25*:229.

3 **Dagg, M.** and **Tapley, R. G.** (1967). *Cashew nut production in Southern Tanzania, V—Water balance of cashew trees in relation to spacing*, E. Afr. agric. for. J., *33*:88.

4 **Northwood, P. J.** (1962). *Cashew production in the Southern Province of Tanganyika*. E. Afr. agric. for. J., *28*:35.

5 **Northwood, P. J.** (1964). *Vegetative propagation of cashew by the air-layering method*. E. Afr. agric. for. J., *30*:35.

6 **Northwood, P. J.** (1966). *Some observations on flowering and fruit-setting in cashew*. Trop. Agriculture, Trin., *43*:35.

7 **Northwood, P. J.** (1967). *The variation in cashew yields in Southern Tanzania*. E. Afr. agric. for. J., *32*:237.

8 **Northwood, P. J.** (1967). *The effect of specific gravity of seed and the growth and yield of cashew*. E. Afr. agric. for. J., *33*:159.

9 **Northwood, P. J.** and **Tsakiris, A.** (1967). *Cashew nut production in Southern Tanzania, III—Early yields from a cashew spacing experiment*. E. Afr. agric. for. J., *33*:81.

10 **Tsakiris, A.** (1967). *Cashew nut production in the Southern Region of Tanzania, II—An economic study of cashew nut production by peasant farmers at Lulindi*. E. Afr. agric. for. J., *32*:445.

11 **Tsakiris, A.** and **Northwood, P. J.** (1967). *Cashew nut production in Southern Tanzania, IV—The root system of the cashew nut tree*. E. Afr. agric. for. J., *33*:83.

12 **Turner, D. J.** (1956). *Some observations on the germination and grading of cashew nuts*. E. Afr. agric. J., *22*:35.

13 **Wheatley, P. E.** (1961). *Insect pests of agriculture in the Coast province of Kenya, II—Cashew*. E. Afr. agric. for. J., *26*:178.

7
Cassava

Mostly *Manihot esculenta,* but interspecific hybrids are also grown (see section on varieties)

Introduction

Cassava, called manioc in some countries, is an important subsistence root crop. It is grown very widely throughout East Africa in areas below 5 000 ft (c. 1 500 m). The only exceptions occur east of the Rift Valley in Kenya, where the rainfall distribution is unfavourable, and in some of the wetter areas of Tanzania where bananas form the staple diet, e.g. Meru, Moshi, Rungwe and Bukoba Districts.

Fig. 17: Cassava in its first year. Note the weed-free conditions.

The main advantages of cassava are its drought resistance, its ability to give good yields on poor soils, its resistance to pests (especially to locusts which were very important earlier in the century) and its ability to remain in the soil as a famine reserve. Other factors which make cassava popular with growers, are that it requires little labour, that there are no labour peaks because the necessary operations can be spread throughout the year and that yields fluctuate less than those of cereal crops. Cassava's limitations are its poor nutritive value (it consists almost entirely of carbohydrate and fibre), its cyanogenic glucoside content (this can lead to hydrocyanic poisoning unless precautions are taken during preparation of the tubers) and its low yielding potential at altitudes above 5 000 ft.

Cassava is exported from Tanzania and Kenya; it is shipped to Europe where it is incorporated into livestock feeds. In Uganda there is a factory at Lira which extracts starch from cassava.

Plant characteristics

Cassava tubers are swollen lateral roots (see Fig. 18); they are usually cylindrical and unbranched, numbering 5–10 per plant and measuring up to 6 in (c. 15 cm) in diameter and up to 3 ft (0·9 m) in length. The stems are woody and much branched; they sometimes grow as high as 15 ft (c. 4·5 m). Flowers and seeds are produced but seeds are not used for field planting.

Ecology

Rainfall, and water requirements
Cassava is very drought resistant and is grown in many parts of East Africa where the rainfall is low and unreliable. In dry periods many leaves are shed, thus reducing transpiration and water requirement. Cassava does not give good yields in areas with a markedly bimodal rainfall, e.g. in the lower altitude areas of Central and Eastern Provinces in Kenya. The reason for this is probably that the rains are too short to allow satisfactory tuber expansion.

Altitude and temperature
Cassava needs a warm climate and is therefore seldom grown above 5 000 ft (c. 1 500 m). It is occasionally found at higher altitudes, e.g. in

Central Province and the Kisii Highlands, both in Kenya, but in such cases it is grown on a small scale, often on the edge of terraces or on wash-stops in gullies or water courses to prevent them from eroding.

Soil requirements

Cassava's nutrient requirements are very small, probably owing to the fact that the bulk of the crop consists of carbohydrate. The main need is that soils should be free draining. Very shallow or stony soils should be avoided because they restrict tuber expansion. Very rich soils have sometimes given poor results; they appear to encourage excessive vegetative growth at the expense of the tubers. A point of interest is that the soil type can affect bitterness. In certain soil conditions a variety may produce tubers which are not particularly bitter whilst in other conditions it may produce extremely bitter tubers.

Varieties

There are many different cassava clones in East Africa; most have a number of local names. The main characteristics of interest to the growers are taste and maturation period. Sweet varieties, in which the cyanogenic glucoside is restricted to the outer layer of the tubers, are more popular and would be grown more widely were it not for one drawback: wild pigs, porcupines and baboons find them appetising and can devastate fields if present in sufficient numbers. Varieties which contain the glucoside throughout their tubers are bitter and

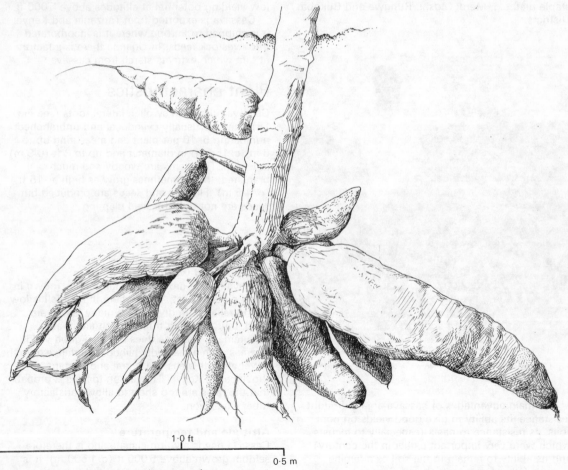

1·0 ft

0·5 m

Fig. 18: The underground structure of a cassava plant.

wild animals ignore them. For this reason bitter varieties are more widely grown, even through their preparation for eating is more labour consuming. Mild hydrocyanic acid poisoning leads to headaches and stomach trouble, and it is not uncommon to hear of cases of severe poisoning which have resulted in death.

As regards the maturation period of different varieties, some, especially sweet ones, can be harvested less than a year after they have been planted; they become fibrous if left in the ground for more than two or three years. Other varieties, especially bitter ones, produce their tubers more slowly; they may not be ready for harvesting until two years after planting and may last as long as six years in the ground without becoming too fibrous.

A cassava breeding programme commenced at Amani in Tanzania in the 1930s. The greatest improvement was achieved by inter-specific crossing rather than by crossing cassava varieties. Cassava was crossed with various *Manihot* species, the most successful of which was Cerea rubber, *Manihot glaziovii*. This cross, with subsequent back-crossing to cassava, produced several high yielding and virus resistant clones, e.g. 46106/27 which is recommended for sandy soils at the coast, and 4763/16, which is recommended for inland red soils. Two introduced cassava varieties which have given good results in many parts of East Africa are Aipin Valenca and Binti Misi. There have been several campaigns in various parts of East Africa to persuade farmers to discard virus susceptible varieties and to plant improved varieties. There is no direct evidence to indicate the success of these campaigns. The general opinion, however, is that either the proportion of improved material now grown is extremely low, or that it is being subjected to such a high level of infection from nearby unimproved stock that it is succumbing to the virus diseases.

Rotations

Cassava is one of the few crops which fit into a definite rotational pattern on smallholdings. Owing to its undemanding nature it is often planted as the last crop in the arable break before the land reverts to a bush fallow. The cassava crop and the fallow almost always overlap because weeding is only carried out in the year of establishment (see Fig. 17); after this, weeds are allowed to grow freely below the cassava plants (see Fig. 19). On Ukerewe Island in Tanzania the common rotation

Fig. 19: Old cassava. Note the heavy weed growth.

is three years of cotton followed by a three year rest of cassava-cum-weeds. (On this heavily populated island maize has been almost abandoned in favour of cassava, the latter being capable of feeding more mouths per acre). A recent study on the effect of cassava fallows on subsequent cotton crops shows that it is not only the weeds which restore the fertility of the soil, for even when the cassava was clean weeded the following crops still benefited considerably. This effect is possibly due to the fact that cassava removes very few nutrients from the soil so they are allowed to accumulate and benefit later crops.

As an alternative to being the last crop in the arable break, cassava may be planted as the first crop after clearing the land, although this is not such a common practice. The reason for this is that the soil is often temporarily infertile when land is cleared if large quantities of organic matter are incorporated instead of being burned off. Cereal crops often give poor yields in these conditions, suffering from a lack of nitrogen, but cassava, being less demanding, gives satisfactory results.

Land preparation

Most of East Africa's cassava is grown on the flat but in the Southern Highlands and the lakeshore areas of Tanzania planting on 5–6 ft (c. 1·5–1·8 m) ridges is usually practised. In these areas ridging has sometimes been shown to increase yields by as much as 100%, but in other parts of East Africa it can cause reductions in yields. An advantage of planting cassava on ridges, whatever the area concerned, is that harvesting is made easier. It is a common practice in all parts of East Africa to plant on the flat but to draw up mounds or ridges during weeding to encourage root development.

Planting material

Cassava is vegetatively propagated by means of stem cuttings. Cuttings should be between 12 and 18 in (c. 30–45 cm) long and should be between 1 and 1½ in (c. 2·5–4·0 cm) thick. Any mature part of the stem can be used but best results are said to be given by the middle portion. The young soft growth is discarded and it is common for about the bottom six inches (c. 15 cm) to be omitted. It is very important that all planting material is taken from virus free plants. Unfortunately many growers use infected plants for planting material because they realise that they will give virtually no yield; few are aware of the consequences of planting infected material.

Planting

A common practice is to plant cassava cuttings at an angle which is usually below 45°. This is usually done in Tanzania and Kenya. At least half of the cutting must be buried in the soil; it must be the right way up as indicated by the buds (see Fig. 20). In Uganda cuttings are usually buried in a horizontal position 3–4 in (c. 7·5–10·0 cm) deep.

Spacing

This is not a critical factor in cassava growing as the plants effectively compensate for wider or closer spacings. In pure stands a spacing of 5 ft × 3 ft (c. 1·5 m × 0·9 m) strikes a reasonable balance between economy of planting material and sufficiently rapid ground cover to suppress weeds. However, the practice of most peasant farmers when planting a pure stand is to plant at random unless ridges have been made. It is common to plant other food crops amongst the cassava during its establishment year, e.g. sweet potatoes, maize and beans. These are later removed to leave a pure stand of cassava. Interplanting with other perennial crops, e.g. bananas, yams, or sugar cane,

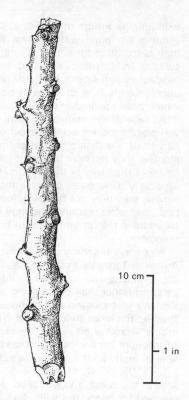

Fig. 20: A cassava cutting. The buds are above the leaf scars in this figure so the cutting would be the right way up for planting.

is common along the East African coastal strip.

Time of planting

This is not an important factor in cassava, provided that the soil is reasonably moist when the cuttings are planted. Experiments, however, have shown that planting at the beginning of the rains gives the best results; late planting gives yields which never equal those from early planting, no matter how long the crop is left in the ground. Planting is often delayed owing to competition with annual food crops which have a more critical response to time of planting.

Fertilisers and manures

These are not recommended for cassava because the expense involved is unlikely to be justified. The responses are small and irregular and if a farmer can afford to buy fertilisers he should apply them to crops which give better cash returns. Nitrogen should be avoided because there is a risk that it may lead to excessive vegetative growth.

Weed control

It is important that weeds should be well controlled during the early stages of cassava growth as the plants grow slowly at first. As mentioned in the section on rotations, weeding is seldom done after the first year although experiments have shown that clean weeding throughout the crop's life gives the highest yields.

Harvesting

Harvesting, which is done either piecemeal or by uprooting whole plants, depends on several factors, one of which is local custom. Sweet varieties are more commonly harvested piecemeal, as and when required, because one tuber is sufficient for a meal. If flour or beer are to be made, however, whole plants are usually uprooted. Young plants are usually harvested piecemeal, whilst old plants are more commonly uprooted to prevent the tubers becoming too fibrous. Harvesting is usually done with various types of 'jembe' or with sticks, but if the plants have been grown on ridges and the soil is reasonably friable and moist it is possible to remove the tubers by pulling the stems.

Yields

Average yields are between three and four tons of fresh tubers per acre (c. 7·5–10·0 t/ha) although reasonable care and attention should give yields of at least ten tons per acre (c. 25 t/ha). The ratio of fresh tubers to peeled and dried chips is about 3:1 (see section on utilisation for description of chips).

Pests

Cassava is troubled very little by pests. White scale sometimes occurs but as it is highly immobile growers can effectively guard against it by selecting clean planting material.

Diseases

Mosaic

This is much the most serious cassava disease in East Africa and is one of the main factors limiting cassava yields. It is caused by a virus which is mainly spread by the use of infected planting material, although it may be transmitted by a white-fly vector (*Bemisia spp.*). Symptoms vary according to the resistance of the variety and the severity of infection. The plant may show only a mild twisting of the leaves, or there may be severe stunting, accompanied by distortion and yellow mottling of the foliage. A hot water treatment is known which will disinfect cuttings, but this method of control is impracticable. The only feasible way of avoiding the disease is ensuring that cuttings come from mosaic-free, and preferably from mosaic-resistant, stock. As mentioned in the section on varieties, the cassava breeding programme at Amani in Tanzania successfully produced a number of mosaic-resistant clones.

Brown streak

This virus disease is less damaging than mosaic. Brown streaks occur on green stems, the marks remaining and appearing as sunken areas on mature stems. Brown horseshoe-shaped marks can be seen by cutting away the leaf scars and black necrotic lesions are found on the roots. This disease is controlled in the same way as mosaic.

Utilisation

There are two main methods of preparing cassava. The first, which can only be used for the sweet varieties in which the cyanogenic glucoside is restricted to the outer layer of the tubers, consists of peeling and boiling. Tubers are usually cut into small pieces after peeling and before boiling, but occasionally, as is common east of the Rift Valley in Kenya, they are only split longitudinally to remove the central fibrous portion. The second method, important in the western part of Kenya and prevalent in all the main cassava growing areas in Tanzania and Uganda, is to grind the cassava into a flour. Tubers are first peeled and cut into chips, which are usually about an inch thick and three or four inches long (c. 2·5 cm × 7·5–10·0 cm). These may be fermented for several days in water or in a covered heap. They are then dried thoroughly in the sun for about a week and are ground, either alone or with cereal grains, to make a fine flour. The flour is boiled with varying quantities of water to make 'ugali' or 'uji'. This method can be used for the bitter varieties which contain the glucoside throughout because the fermentation and later the boiling of minute particles ensures that all the toxic substance is removed. In most areas a small proportion of cassava is eaten raw after peeling, especially by children in the field; only the sweet varieties can be used in this way. Finally, tubers may be peeled and roasted in the ashes of a fire.

Other uses for cassava are as follows. The leaves can be used as a pot-herb; this is extremely popular throughout Tanzania but it is an unusual practice in Kenya, reserved for times of severe food shortage. The stems are occasionally used as firewood. Cassava is used to a certain extent for brewing in various parts of Tanzania, e.g. Ngara and Masasi Districts and much of Western Region.

Bibliography

1 **Anderson, G. W.** (1944). *Notes on cassava preparation in North Kavirondo and Samia.* E. Afr. agric. J., *10*:111.

2 **Chant, S. R.** (1959). *A note on the inactivation of mosaic virus in cassava by heat treatment.* Emp. J. exp. Agric., *27*:55.

3 **Childs, A. H. B.** (1957). *Trials with virus resistant cassavas in Tanga Province, Tanganyika.* E. Afr. agric. J., *23*:135.

4 **Ghosh, B. N.** (1968). *The manufacture of starch from cassava roots in Uganda.* E. Afr. agric. for. J., *34*:78.

5 **Jameson, J. D.** (1964). *Cassava mosaic disease in Uganda.* E. Afr. agric. for. J., *29*:208.

6 **Jennings, D. L.** (1957). *Further studies in breeding cassava for virus resistance.* E. Afr. agric. J., *22*:213.

7 **Jennings, D. L.** (1960). *Observations on virus diseases of cassava in resistant and susceptible varieties, I—Mosaic disease.* Emp. J. exp. Agric., *28*:23.

8 **Jennings, D. L.** (1960). *Observations on virus diseases of cassava in resistant and susceptible varieties, II—Brown streak disease.* Emp. J. exp. Agric., *28*:261.

9 **Nichols, R. F. W.** (1947). *Breeding cassava for virus resistance.* E. Afr. agric. J., *12*:184.

10 **Nichols, R. F. W.** (1950). *The brown streak disease of cassava: distribution, climatic effects and diagnostic symptoms.* E. Afr. Agric. J., *15*:154.

11 **Scaife, A.** (1968). *The effect of a cassava 'fallow' and various manurial treatments on cotton at Ukiriguru, Tanzania.* E. Afr. agric. for. J., *33*:231.

12 **Swaine, G.** (1950). *The biology and control of the cassava scale.* E. Afr. agric. J., *16*:90.

8
Castor

Ricinus communis

Introduction

Castor is an oilseed which grows wild in East Africa and can be found on waste spaces and roadsides in all areas except the driest and the highest.

Tanzania

There is much self-seeded and semi-cultivated castor in the central part of Tanzania. The seeds are sent to Dodoma where an extracting mill was opened in the late 1960s. Annual seed production was about 20 000 tons in most years during the 1960s and during this period the annual value of castor exports sometimes exceeded £1 000 000.

Kenya

Machakos, Kitui, and Kilifi Districts produce almost all of Kenya's marketed castor. There is no oil extracting mill. A mill must handle a minimum of 15 000 tons of seed per annum in order to be economically viable; Kenya's annual production was only about 5 000 tons annually during the 1960s.

Uganda

Castor grows prolifically in all parts of Uganda but at the time of writing there are no marketing facilities.

Plant characteristics

The castor plant has a well developed tap root which normally extends to a great depth. The height of the plant depends on the variety: perennial varieties may grow as high as 20 ft (c. 6 m) whilst most annual varieties grow to a height of only 3–5 ft (c. 0·9–1·5 m). The inflorescence is a panicle with female flowers at the top and male flowers at the bottom (see Fig. 21). As the female flowers open before the male flowers, there is a high degree of cross-pollination. Each female flower produces a thick-walled spiny capsule with three loculi; each loculus contains one seed.

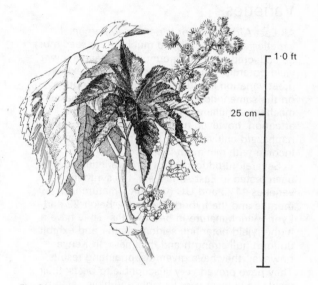

Fig. 21: A castor inflorescence.

Ecology

Rainfall, and water requirements

Perennial castor is very drought resistant, largely owing to its deep root system. As with annual castor, it sets seed poorly if wet weather coincides with flowering. The reason for this is that wet conditions encourage various sucking pests which cause a die-back of the inflorescence. Yields of perennial castor are therefore best in dry areas, e.g. central Tanzania and parts of Machakos and Kitui Districts in Kenya where the average annual rainfall is only in the region of 20–25 in (c. 500–625 mm). Annual castor, with its less extensive roots and its need for reliable rainfall during establishment, is best suited to areas of higher rainfall where it can only yield well if protected from pests by the use of insecticides.

Altitude and temperature

Castor grows well from sea level up to about 7 000 ft (c. 2 100 m).

Soil requirements

Castor is highly intolerant of water-logging. It grows well on most soils provided that they are free-draining.

Varieties

All East Africa's castor is perennial. Perennial varieties shatter readily, so much of the seed must be collected from the ground. They have a lower yield potential than annual varieties and exhibit great variation in hull thickness and strength even on the same inflorescence; this variation makes mechanical hulling difficult. They need little attention, however, and are therefore well suited to haphazard cultivation, providing a small cash income with minimum input.

Several introduced short term varieties have been tested in East Africa, e.g. the synthetic varieties 1123 and UC 53, which mature in 7–10 months and the hybrids Baker 44, Baker 22 and Lynn, which mature in 5–7 months. They have a higher yield potential, seldom shatter and exhibit uniform hull strength and thickness. In Kenya, however, they have given disappointing results. They have proved very susceptible to pests; their yields have been very variable and have seldom been high enough to cover the cost of production.

Field operations

Perennial castor

All East Africa's castor is self-seeded or semi-cultivated. It is sometimes deliberately sown around houses, as a windbreak, as a boundary, along terrace banks, amongst arable crops or even, occasionally, as a pure stand at a spacing of about 6 ft × 6 ft (c. 2 m × 2 m). It usually receives no attention apart from a little protection from weeds in the early stages of growth. It is sometimes ratooned because the stems, if left to grow unchecked, grow to such a height that they are easily blown over; this is especially likely if larvae have bored into the stems. Harvesting is done by collecting seed from the ground and by picking dried capsules from the plants. The capsules are put into sacks and are beaten with sticks to separate the hulls and the seeds.

Annual varieties

The seedbed must be very free of weeds because the young plants are delicate and mechanical cultivation is therefore inadvisable in the early stages of crop growth. The time of sowing must be such that there is unlikely to be any wet weather in the second half of the crop's life. A maize planter with suitable plates can be used and a spacing of 3 ft × 1 ft (c. 0·9 × 0·3 m) has usually been adopted. Trials in Kenya have shown that there are good responses to seedbed applications of 40 lb P_2O_5 per acre (c. 45 kg/ha). Good weed control is important; pre-emergence herbicides have been tried with success on the Uasin Gishu Plateau in Kenya. Because the short term varieties are non-shattering, harvesting is done by collecting the capsules from the plants. The varieties tried in Kenya had such hard hulls that hand hulling was extremely difficult. However, hand operated machines have been developed, and these have proved successful.

Yields

Statistics of yields per acre are not available for perennial castor. They would not be of great relevance because almost all perennial castor plants are scattered rather than grown in blocks. Annual castor, however, has been grown in blocks so yields per acre are relevant. Rainfed trials have shown that yields can be as high as 1 500 lb of seed per acre (c. 1 700 kg/ha). There was great variation, however, and yields were more commonly about 500 lb per acre (c. 550 kg/ha). The biggest problem is that when blocks of castor are established, pests increase rapidly and do far more damage than they do on scattered plants. At current prices and without taking account of the costs of spraying, yields of about 1 000 lb of seed per acre (c. 1 100 kg/ha) are needed to cover the cost of growing annual castor. With current varieties and techniques such yields can only rarely be obtained in East Africa. Annual castor cannot therefore be recommended until there is either a substantial change in the price structure, or a technological breakthrough which increases yields considerably.

Pests

Few crops in East Africa, if any, are damaged by

such a great diversity of pests as is castor. As many as 50 different species of insect have been shown to damage castor in Kenya. Grasshoppers and various larvae may attack the foliage but the most serious pests are sucking insects, e.g. capsid bugs, green stink bugs, *Lygus* bugs and *Helopeltis,* which cause die-back of the inflorescence and growing points. Sucking pests do most damage by puncturing rather than by sucking; the same symptoms can be obtained by pricking the young inflorescences with a pin. Fortnightly sprays of insecticides are required if efficient pest control is to be obtained; they need to be used from flowering to harvesting.

Diseases

Several diseases have been identified on castor in East Africa but they seldom do much damage. They include a leaf spot caused by *Cercospora ricinella,* rust caused by *Melampsora oricini,* and an *Alternaria* leaf spot.

Utilisation

Castor seeds contain 35–55% of a valuable drying oil which can be used in many industrial products, e.g. hydraulic fluids, jet engine lubricants, plastics, synthetic textiles, soap and paint. The industrial properties of the oil are unique because it contains ricinoleic acid which is not known to occur in any other plants. The traditional use of castor oil, as a purgative medicine, absorbs a minute fraction of the total production. The seeds and the residual cake are poisonous; they cannot be fed to livestock unless they are first detoxified. In some countries the cake is used as a fertiliser.

Traditional uses of the oil in East Africa include the curing of hides and skins to make them more flexible (e.g. straps, bedding, clothing, bow strings) and smearing the bodies and hair of women and children, especially prior to dances. In the Tugen Hills and in Elgeyo-Marakwet District in Kenya, castor oil is used on poisoned arrow tips to prevent the poison being inactivated. These traditional uses are declining in all areas and in some are extinct. Finally, castor plants have occasionally been used for shade in smallholder coffee plots west of the Rift Valley in Kenya.

Bibliography

1 **Evans, A.** and **Sreedharan, A.** (1962). *Studies of intercropping, II—Castor-bean with groundnuts or soyabean.* E. Afr. agric. for. J., 28:7.

2 **Peeler, C. H.** (1967). *Castor production in Kenya.* E. Afr. agric. for. J., 33:1.

3 **Weiss, E. A.** (1966). *Dwarf castor: a promising crop for East Africa.* World Crops, 18 (4):43.

4 **Weiss, E. A.** (1967). *Dwarf castor trials in Western Kenya.* E. Afr. agric. for. J., 32:229.

9
Citrus

Citrus spp.

Introduction

The sweet orange (*C. sinensis*) is the most important species of citrus fruit in East Africa. It is followed by limes (*C. aurantifolia*), grapefruit (*C. paradisi*), lemons (*C. limon*), and mandarins (*C. reticulata*). The latter are often called tangerines. A few pummeloes (*C. grandis*) are also grown; the fruits resemble large grapefruit. There are many interspecific hybrids but none of these are commercially important in East Africa.

Citrus is not an important crop in East Africa. The reason for this is not technical, for good quality fruit can be grown; it is rather that the market for citrus is very limited. East Africa cannot compete with the major exporting countries on the world market and the internal market is small. A few farmers in Kenya, however, grow good quality fruit for the local market but at the time of writing none has more than 50 acres (c. 20 ha) of citrus. The main orchards are at the coast and near Machakos, Thika, Naivasha, Elmenteita, Solai and Kitale. In Uganda a large Government citrus nursery was established at Kasolwe in Busoga District in 1965 and this has provided the planting material for a plantation of 160 acres (c. 65 ha) at Kiige, also in Busoga District, and another of 100 acres (c. 40 ha) at Ongom in Lango District. A further plantation of 150 acres (c. 60 ha) is planned at Odina in Teso District. The remainder of East Africa's citrus consists of scattered seedling trees in smallholdings; there are many such trees in the higher rainfall areas, e.g. throughout Uganda (with the exception of Karamoja District) and in the Kenya and Tanzania highlands, but they receive little attention and give low yields of poor quality fruit.

No research has been done on citrus in East Africa other than a few observations on pests, yields and fruit quality. The number of experienced growers is small and their opinions often conflict. It is not generally known, therefore, which techniques give the best results. It would be irrelevant, in this chapter, to catalogue the techniques which have proved successful in the most important citrus growing countries, i.e. the U.S.A., Brazil, Japan, Spain, Italy and Israel. An uncritical account is therefore given of the methods used by the small number of large scale farmers in East Africa who produce good quality fruit.

Plant characteristics

Citrus species and varieties differ in so many characteristics that no attempt is made to list the differences here. The matter is complicated by the fact that variations are common between trees of the same variety; mutations (bud sports) may also arise in different parts of the same tree. A feature of many citrus varieties is that they exhibit polyembryony, i.e. one seed can produce not only a sexual seedling but also several nucellar seedlings which are genetically identical to the mother plant. The latter often grow so vigorously that the sexual seedling is suppressed.

Ecology

Rainfall, and water requirements
A regular supply of soil moisture is essential for high yields of good quality fruit. Rainfed citrus therefore suffers in East Africa because of periodic droughts, notably the December–March dry season. Almost all growers, therefore, have facilities for irrigation. Limes are moderately drought resistant and have a smaller need for irrigation than other species.

With an irregular supply of soil moisture, or with only a little irrigation, flowering is seasonal. For instance, the Washington Navel orange, which is the most important variety in Kenya, produces most of its flowers in the short rains at the end of the year, thus producing a glut of fruit on the local market from May to August. With judicious irrigation, flowering can be encouraged throughout the year, although this may increase pest problems.

Altitude and temperature

Both the species and the varieties within a species differ in their temperature requirements; they are therefore grown at different altitudes and these are mentioned below in the section on varieties.

Temperature has an important effect on the colouring of oranges and grapefruit. Night temperatures below about 57°F (c. 14°C), coupled with a low humidity, produce orange-coloured oranges and yellow grapefruit. At sea level in East Africa, where night temperatures and humidity are high, oranges remain green even when fully ripe although grapefruits colour to a certain extent. At higher altitudes, e.g. 4 500–6 500 ft (c. 1 400–2 000 m) where most of Kenya's citrus is grown, oranges colour to a certain extent but the high humidity usually prevents the development of the full orange colour which is typical of fruit grown in sub-tropical regions. Treating the fruit with ethylene gas destroys the chlorophyll and exposes the orange or yellow coloured carotenoids in the skin; this method of de-greening has been tried successfully in East Africa but is not used at present.

Wind

Citrus fruits are easily damaged by wind; they rub together and develop scars at the point of contact. Flowers, or even small fruits and leaves, may be shed in strong winds. Citrus orchards are therefore situated in sheltered positions, or they are protected by windbreaks.

Soil requirements

The most important requirement for the rough lemon rootstock, which is almost universally used in East Africa, is that the soil must be deep, light or medium textured and well drained. Rough lemon has an unusually high demand for oxygen and suffers when soil aeration is inadequate. Experience in other countries indicates that a pH of 5·5–7·0 is ideal. The nutrient status of the soil is relatively unimportant, as nutrients can be supplied in the form of fertilisers or manure. Saline soils are not tolerated and must be avoided if rough lemon is used as a rootstock.

Varieties

Oranges

Washington Navel is the most commonly grown variety in East Africa. It yields well at altitudes between 3 000 and 6 500 ft (c. 900–2 000 m) but does poorly at the coast. It produces large seedless fruits which ripen between seven and nine months after flowering. Valencia is a low altitude variety which gives best results between sea level and 4 000 ft (c. 1 200 m). The fruits are smaller, on average, than those of Washington Navel but they are produced in greater numbers. The fruits have very few or no seeds, mature later and remain on the tree for several months after ripening. Hamlin is a recently introduced variety which shows promise; it has pale flesh and is almost seedless.

Limes

The Tahiti lime is the most common variety except at the coast. It has a green skin and no seeds and is grown mainly as a garden crop from sea level and 4 000 ft (c. 1 200 m). The fruits are or West Indian lime is the best commercial variety but it is less hardy and should be grown between sea level and 4 000 ft (c. 1 200 m). The fruits have yellow skins.

Grapefruit

Marsh Seedless is the only variety of commercial importance. It grows best from sea level up to 4 000 ft (c. 1 200 m).

Lemons

Lisbon and Villafranca are the two common varieties but Eureka, a new variety which has few seeds, shows promise. Lemons are grown commercially as high as 7 000 ft (c. 2 100 m) and can be found as a garden crop as high as 9 000 ft (c. 2 700 m).

Mandarins

Emperor is the most common variety, growing best in the lower altitude areas. Satsuma, however, is more hardy and can be grown as high as 6 000 ft (c. 1 800 m).

Propagation

Almost all growers who produce good quality citrus in East Africa plant budlings rather than seedlings. The advantages of budlings are as follows:

1 Seedless varieties or varieties which produce few seeds can be grown. These are more popular than varieties with many seeds.

2 Budlings are genetically identical to the tree

from which the bud-wood was taken. They therefore produce trees which vary little in their performance and in the quality of their fruit. Seedlings, on the other hand, often show great differences between plants because of their genetic variability. This variability is reduced or even eliminated in varieties which produce a high percentage of nucellar seedlings.

3 A rootstock can be chosen which has greater vigour, disease resistance and tolerance of certain soil characteristics, (e.g. waterlogging and salinity) than the variety which is being grafted on to it.

4 Budlings are usually less thorny than seedlings.

5 Budlings produce spreading trees whilst seedlings produce trees with predominantly vertical growth. Pruning cannot alter the growth habit of the latter.

6 Budlings come into bearing more quickly than seedlings.

Rootstocks

A rough lemon rootstock is almost universally used in East Africa. The important characteristics of rough lemon are its tolerance to several virus diseases, e.g. *Tristeza* and *Psorosis,* its vigour, its susceptibility to *Phytophthora* fungi, soil salinity and waterlogging and the fact that varieties which are grafted onto it produce inferior quality fruit low in soluble solids. Alternative rootstocks, some of which produce better quality fruit and which are popular in other parts of the world, have received little attention.

Nurseries

Because of the individual attention they need, budlings are raised in nurseries for their first two or two-and-a-half years. Nurseries are divided into seedbeds and seedling plots; transplanting from one to the other allows selection of seedlings with healthy root systems.

Both seedbeds and seedling plots must be sited on well drained, light or medium textured soils and there must be facilities for irrigation. The seeds for the rootstock are taken from vigorous mother trees, are washed immediately they have been removed from their fruits and are sown in the seedbed at a depth of about 1 in (c. 2·5 cm) and at a spacing of 6–9 in × 1 in (c. 15–23 cm × 2·5 cm); they must not be allowed to dry out before sowing. Some

growers apply shade and mulch while others apply mulch only. Rough lemon seedlings are ready for transplanting into the seedling plots 6–8 months after sowing; at this time at least 25% of the seedlings (the least vigorous and those with twisted and bent tap roots) must be discarded. A technique used in Uganda, where the seedlings are left for a relatively long time before transplanting, is trimming the roots to a length of 10 in (c. 25 cm), trimming the top to a length of 16–18 in (c. 40–45 cm) and removing all the leaves, thus reducing transpiration. Elsewhere the seedlings receive less severe treatment; their roots may be pruned a little and about $\frac{1}{3}$ of each leaf may be removed.

The spacing in the seedling plots must allow easy access for budding; it is usually 3–4 ft × 1 ft (c. 0·9–1·2 m × 0·3 m). Seedlings are ready for budding when their stems are pencil-thick at a height of 12 in (c. 30 cm) above the ground; they should reach this stage about six months after planting into the seedling plots. The procedure for budding is described with Figs. 22 to 28.

Fig. 22: A young vertical shoot which has been selected from a healthy, high-yielding mother tree for bud-wood.

Fig. 25: All side shoots and thorns have been removed from the lower part of the rootstock and a T-cut has been made 9–12 in (c. 23–30 cm) above the ground. The vertical part of the cut is 1–2 in (c. 2·5–5·0 cm) long and the bark has been gently separated from the wood. Separation is only possible when the rootstock is obtaining enough soil moisture to be growing actively.

Fig. 23: The shoot in fig. 22 has now had the weak upper part removed. The stem of the upper part has a triangular cross-section, whereas that below it has a circular one. The lowest part, where the buds were not protruding sufficiently, has also been removed. The leaves have been removed by cutting through the petioles as close to the bud as possible with a sharp knife. This is now called a bud-stick. It must be kept moist, preferably between wet sacks, until the buds are removed.

Fig. 26: The bud has been pushed down into the T-cut. The circular projection immediately below the bud is the base of the petiole.

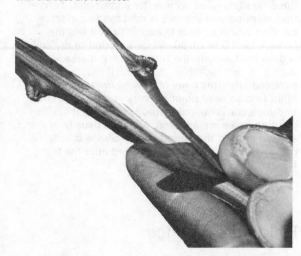

Fig. 24: A bud being removed from the bud stick with a razor-sharp budding knife. The bud-stick is held with its upper end towards the body and the knife is being drawn towards the body, starting below the bud.

Fig. 27: The upper part of the bud-stick, seen projecting above the T-cut in the previous figure, has been removed with a horizontal cut at the same level as the top of the T.

Fig. 28: The bud has been firmly bound with transparent plastic tape to prevent water from entering. The base of the petiole is clearly seen.

About two weeks after budding, the wrapping must be removed in order to inspect the buds. If they are still green they have taken and the stock can be cut through a few inches above the union. (Some growers prefer to cut it only $\frac{2}{3}$ of the way through and to let the top rest on the ground). If they are brown they are dead and the stock must be budded again. A green bud produces a shoot which is known as the scion. When the scion is about 10 in long (c. 25 cm) it is bound loosely to a stake to prevent it breaking in the wind and the stub of the rootstock is pruned off. When it is 3–3½ ft high (c. 0·9–1·1 m) it is pruned to a height of 2½ ft (c. 7·5 m) to encourage the formation of the main 'scaffold' branches. As many as six branches may develop but only three or four of these are selected; the others are pruned off as close to the main stem as possible. Budlings should be ready for transplanting 12–18 months after budding.

Field establishment

Spacing
Spacings in East Africa vary greatly: oranges, limes and grapefruit are grown at spacings which give from 70 to 250 plants per acre (c. 170–620 plants per ha). Lemon trees are usually larger and are grown at spacings which give lower population. In most countries which export citrus the plant populations now favoured are about 80 trees per acre (c. 200 plants per ha).

Planting
Some growers transplant without any defoliation but most cut the branches back to about 8 in (c. 20 cm) from the main stem and remove about $\frac{1}{3}$ of each remaining leaf. In Uganda all leaves are removed after pruning the branches and the stem; all parts above ground are protected from sunscorch by painting them with a whitewash mixture of 12–13 lb (c. 5·5–6·0 kg) of hydrated lime and 1 lb (c. 0·5 kg) of zinc sulphate in 25 gallons (c. 115 l) of water. The tap root is usually trimmed to a length of 12–18 in (c. 30–45 cm) with lateral roots somewhat shorter. With no defoliation, however, all the roots are left. Citrus plants must always be kept moist from the time they are removed from the nursery to the time they are planted out in the field.

Planting holes are 1½–2 ft (c. 0·5–0·6 m) in depth and diameter. In Uganda the holes are three-quarters filled with water immediately before transplanting and the tree is held in the correct position whilst earth is tipped in until it fills the hole. A 12–15 in (c. 30–40 cm) mound of dry soil is then built around the trunk to keep it steady during its first month in the field. (It must be removed after this time, otherwise *Phytophthora* fungi find an easy point of entry at the bud union.) In Kenya conventional methods of filling the planting holes are adopted. Mulch is usually applied around the young plants. Shade is only provided at the coast; it is removed after the first few months.

Field maintenance

Irrigation
Irrigation is essential for high yields of good quality fruit. Furrow irrigation, overhead irrigation and basin irrigation are used.

Deflowering

Many growers remove the flowers or the immature fruits during the first two years after planting in the field. This encourages the development of a strong framework of branches.

Inter-cropping

Annual crops are sometimes grown between the rows of trees during the first few years.

Pruning

Citrus needs very little pruning other than the removal of suckers from young trees, and some growers give it none. Others prevent the centres of the trees from becoming overcrowded with branches and/or prune the tips of the lower branches to prevent them touching the ground.

Fertilisers

All growers recognise that nutrients must be added to the soil if high yields of good quality fruit are to be obtained. They apply these as fertilisers or manures or in the irrigation water. A tentative recommendation on fertilisers in Kenya is $2\frac{1}{2}$ lb (c. 1·1 kg) of both calcium ammonium nitrate and single superphosphate plus $\frac{1}{2}$ lb (c. 0·2 kg) of potassium sulphate per tree; this application should be repeated every six months.

Trace element deficiencies, particularly those of zinc, magnesium and copper, are frequently seen in East Africa. Foliar sprays and fertilisers have been used to correct these deficiencies but some growers have found them ineffective, whilst others have found that they present problems, e.g. sprays of zinc sulphate can increase the incidence of scale insects.

Weed control

A number of methods are used to prevent competition from weeds; these include hand cultivation, disc harrowing, slashing, mulching and the use of contact herbicides.

Harvesting and yields

Fruits must be harvested carefully, otherwise they become bruised and discoloured. After harvesting the fruits should be sorted into grades of different size and quality.

Citrus should reach full bearing about ten years after planting in the field. Average yields of fruit at this stage are about 5 tons per acre (c. 13 t/ha) but well managed plots have produced 10 tons per acre (c. 26 t/ha). (In other parts of the world as many as 25 tons per acre (c. 63 t/ha) are sometimes obtained.) In East Africa the productive life of citrus trees is usually restricted to 20–30 years; poor nutrition and diseases are largely responsible for this.

Pests

Citrus is attacked by a large number of pests, but many of these are found on the fruits and have no effect on the health of the trees. Field hygiene is one of the most important methods of checking citrus pests; infested fruits, which usually colour prematurely and fall to the ground, should be collected at weekly intervals and disposed of in pits, or drums of water covered with a film of oil. If pits are used the fruit should be burned and then covered with a thick layer of soil. Spraying with DDT is discouraged; it can cause a rapid increase of certain pests, especially citrus rust mite, because it reduces the population of their insect predators and parasites.

False codling moth (*Cryptophlebia leucotrea*), **and Mediterranean fruit fly** (*Ceratitis capitata*) Larvae of these pests damage the insides of the fruits. Good field hygiene helps to limit the numbers of both false codling moth and Mediterranean fruit fly. Fenthion, malathion or endosulfan sprays are sometimes used.

Mites

The most important of these are the citrus rust mite (*Phyllocoptruta oleivora*) and the citrus bud mite (*Aceria sheldoni*). The former discolour the surface of the fruit, giving it a rusty appearance, but they do not damage the flesh. The latter are more damaging because they can cause bending and twisting of the twigs and severe disfiguration of the fruit. Chlorobenzilate is the acaricide which is usually used against mites in East Africa.

Scales

The most important of these are the red scale (*Aonidella aurantii*), the green scale (*Coccus viridis*) and, at the coast, the mussel scale (*Lepidosaphes beckii*). Red scales may occur on the fruit, leaves, branches or trunks and may cause die-back of the twigs. Green scales cause less damage; they are sometimes found on leaves and young shoots. Mussel scales occur on leaves, fruits and branches. They can cause premature leaf fall and die-back of the branches. Malathion, diazinon and dimethoate, either with or without white oil, are the most commonly used insecticides.

Aphids. *Toxoptera spp*
These are commonly found on the foliage during dry spells but they usually disappear in wet weather.

Diseases

There are several virus diseases of citrus. Tristeza, which is sometimes called quick decline, leads to stunting and the tree soon dies if it has a susceptible rootstock. Pitting of the wood sometimes occurs, especially in grapefruits and limes, but this symptom is not specific to tristeza; it can be seen, together with the pointed projections of bark which fit into the pits, when a strip of bark is separated from the wood. Other virus diseases include psorosis, greening, exocortis and cachexia. They mostly affect the trunk and the branches and their symptoms include stunting, grooves in the bark, gum deposits beneath the bark, pitting of the wood and bark peeling. Virus diseases are transmitted by sucking insects and by propagation (either by budding or by seed) from infected trees. The only method of control is the choice of resistant or tolerant varieties. Budlings which have rough lemon as a rootstock are tolerant to many of the virus diseases, especially tristeza.

Foot rot, caused by *Phytophthora spp.*, is the most important fungal disease. Both the collars and the roots rot and gum oozes from the trunk and branches. The incidence of foot rot is highest in waterlogged or humid conditions. Heaping soil around the trunk above the union of the rootstock and the scion often encourages infection.

10
Coconuts

Cocos nucifera

Introduction

Coconut palms are very important along the East African coast. Their main commercial products are mature nuts, immature nuts and palm wine.

Mature nuts

Mature nuts yield a thick white flesh; after drying it is called copra. The oil content of copra, on a dry weight basis, is 65–70%. If the quality of the copra is high the oil can be used for edible products, especially cooking oil and margarine. If the quality is low the oil can only be used commercially for making soap. In other parts of the world, notably in the Philippines, Indonesia, India and Ceylon, there is a flourishing copra trade; in East Africa, however, little copra is produced and its quality is poor. The main reasons for this are the poor marketing

Fig. 30: A coconut inflorescence being tapped for palm wine. Note the stick which is holding it down, the string which is keeping it closed and the gourd at the end.

facilities (for instance, there is no premium paid for good quality copra) and the greater returns from other coconut products, i.e. mature nuts sold for their undried flesh, immature nuts and palm wine. When the undried flesh is utilised it is usually grated and then squeezed in a cloth to provide a liquid; this liquid is used in the preparation of a relish to accompany the main carbohydrate food.

Immature nuts

Immature nuts contain only a thin layer of jelly-like flesh; the rest of the central cavity is filled with coconut water which makes a refreshing drink. After drinking the water many people like to scrape out and eat the flesh. Large numbers of immature nuts are sold in towns; people living in the country, however, only use them on festive occasions. In Tanzania their sale has been banned for most of the year because it reduces the amount of copra that can be produced.

Fig. 29: Coconut palms on the East African coast.

49

Palm wine

A large proportion of the coconut palms in East Africa are tapped for palm wine; this is an alcoholic drink which is obtained from the unopened inflorescences. The spathe (the sheath-like covering of the inflorescences) is bound up with string to prevent it from opening, its tip is cut off, a small gourd is tied over the end to collect the sap and the inflorescence is fixed in a horizontal position by means of a stick so that the sap runs into the gourd and not down the spathe (see Fig. 30). Each inflorescence must be visited twice or sometimes three times a day in order to collect the sap and to keep it flowing by making a fresh cut. If managed skilfully an inflorescence can continue to produce sap for several weeks. After collection the sap is stored in large gourds; nothing is added to it and it is ready for drinking after a few days. Palm wine production is more lucrative than nut production and is well organised on the East African coast. Prolonged tapping prevents nut development and weakens the palms.

Tanzania

It has been estimated that there are over 100 000 acres (c. 40 000 ha) of coconut palms in Tanzania. Almost all of these are grown at the coast but there are some, of little commercial importance, on the shores of Lakes Nyasa, Tanganyika and Victoria.

Kenya

In Kenya there are, very approximately, two million palms; if they were all planted in pure stands this would be equivalent to about 40 000 acres (c. 16 000 ha).

Plant characteristics

Roots

The root system of the coconut palm is adventitious and arises from a swelling at the base of the stem called the bole. Most of the bole is below ground level but the top of it is usually visible above ground. Most of the roots grow horizontally and are found in the top 3 ft (c. 0·9 m) of soil; some, particularly those growing from the bottom of the bole, grow downwards and often penetrate to a depth of 6 ft (c. 1·8 m) or more.

Stems and leaves

In East Africa the tall variety of coconut palm usually grows to a height of 50–70 ft (c. 15–20 m).

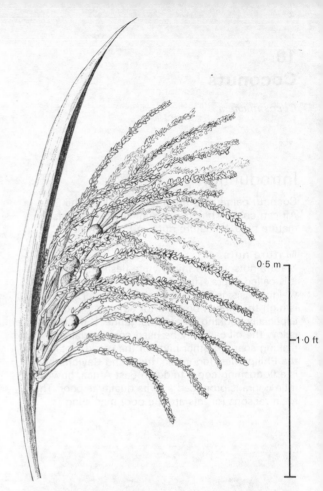

0·5 m

1·0 ft

Fig. 31: A coconut inflorescence. This inflorescence had relatively few female flowers; only five can be seen.

The stem has only one growing point at its tip, above which is a vertical spike of rolled-up developing leaves; the rate at which leaves unroll from this spike is commonly quoted as being one every month but work done in Zanzibar indicates that it ranges from 20 to 33 days. A healthy coconut palm has about 30 unrolled leaves; the time between unrolling and leaf fall for a single leaf is therefore between 1½ and 2½ years.

Inflorescences

Each leaf has an inflorescence in its axil. Immediately before opening it appears as a long, thin, cylindrical structure, about 3 ft (c. 0·9 m) long, whose outer covering is the spathe. The underside of the spathe splits longitudinally to reveal the floral parts about a year after its leaf has unrolled (see Fig. 31). The flowers are borne on a branched structure called the spadix; each spadix has about thirty branches. At the base of some of the branches spherical female flowers are borne; beyond the female flowers are large numbers of smaller, pointed, male flowers. In the tall variety of coconut palm the female flowers become receptive only after the male flowers have shed their pollen; there is therefore almost no self-pollination. The period from pollination to nut maturity is about twelve months. Large numbers of small nuts drop off when they are about three months old; this is largely a physiological response but it can also be caused by sucking pests.

Fruits

The outer skin, i.e. the exocarp, of the fruit in the tall variety is usually green; in the dwarf variety it is usually yellow or orange. In the later stages of maturity the skin dries out and turns brown. The husk of the mature nut, i.e. the mesocarp or coir, is thick and fibrous. The shell, i.e. the endocarp, has three circular 'eyes' at one end; one of these is softer than the others, and beneath it lies the embryo. In an immature nut the central cavity is filled largely with coconut water; at this stage the flesh, i.e. the endosperm, consists only of a thin jelly-like layer (see Fig. 32). As the nut ripens the endosperm becomes thicker and harder; the water is partly used in this process (see Fig. 33).

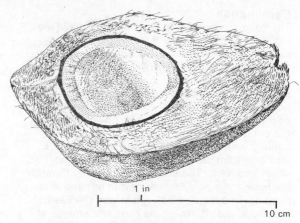

Fig. 33: A mature coconut which has been cut in half; the cavity was approximately two-thirds filled with coconut water. Note that the flesh is fully developed and the shell is black.

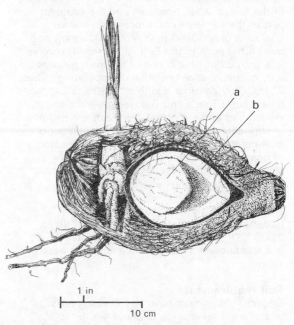

Fig. 34: A germinated coconut which has been cut in half. Note the haustorium (a), and the flesh (b) which is thinner where the haustorium has come into contact with it. A piece of husk has been cut away to reveal the developing shoot and roots; these are connected to the haustorium through a hole in the shell which is not shown in the figure.

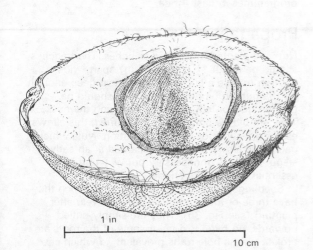

Fig. 32: An immature coconut which has been cut in half; the cavity was entirely filled with coconut water. Note that the flesh is only partly formed and the shell is not yet black.

Germination

The coconut germinates in an unusual way (see Fig. 34). About two months after planting the embryo starts to grow in two directions; the outer end produces the shoot whilst the inner end grows into the central cavity of the nut to form a spherical, spongy body, technically termed a haustorium but commonly referred to as the coconut 'apple'. The haustorium finally occupies the entire cavity, its function being to make contact with the endosperm and to transfer nutrients from the endosperm to the growing shoot and roots. The shoot works its way through the husk and emerges three or four months after planting. The roots, which grow from the base of the shoot, grow through the husk and emerge at the bottom of the nut at about the same time.

Ecology

Rainfall, and water requirements

For optimum production coconut palms need a well distributed annual rainfall of 50 in (c. 1 250 mm) or more. Most of the important coconut producing countries enjoy such conditions. The northern part of the Kenya coast, however, and the southern part of the Tanzania coast receive an annual rainfall of only 30–40 in (c. 750–1 000 mm) and most other parts of the East African coast receive less than 50 in. This is one of the main reasons why coconut yields are often disappointing in East Africa. Only Zanzibar and Pemba Islands and the area around Tanga and Gazi receive a satisfactory rainfall and in these places the yields are higher.

Droughts cause a reduction in the number of nuts per inflorescence, a reduction in the amount of copra per inflorescence and, sometimes, abortion of the inflorescences. These effects are only seen about a year after a drought.

Altitude and temperature

Coconut palms grow well from sea level up to 3 500 ft (c. 1 100 m). Within these limits altitude is a less important factor than a constant water supply.

Soil requirements

Coconut palms must have free-draining soils which are deep enough to allow healthy root development. When they are grown on shallow soils underlain by coral the results are always disappointing. Apart from this important requirement coconuts are undemanding; they grow well on sandy soils which would support few other crops

although in these conditions they only give optimum yields if supplied with fertilisers.

Coconuts tolerate a higher degree of soil salinity than most other crops. They do not prefer such conditions, however, nor do they survive if salinity is excessively high.

Varieties

The only two recognised varieties in East Africa are the tall coconut palm and the dwarf coconut palm (sometimes called Pemba). The dwarf palm is seldom planted except for ornamental purposes; it is unsatisfactory for palm wine production and it produces little copra. It produces excellent immature nuts but they are so easily stolen that it is not even popular for this product.

Selection

Selection of nuts from high-yielding mother trees is recommended. A more important method of selection, and one which is highly recommended, is the selection of the quickest germinating, most vigorous seedlings from the nursery; there is a high correlation between these characteristics and later yields.

Breeding

Coconut breeding is a very complicated and long term task; it is difficult to gain access to the floral parts, and each generation takes a minimum of seven years. There have been no breeding programmes in East Africa.

Propagation

Nuts are planted in shallow trenches. The trenches should be about $1\frac{1}{2}$ ft (c. 0·45 m) apart, wide enough to take a coconut and almost deep enough for the nuts to be buried. The nuts should be planted horizontally, without removing the husk. They should be only two-thirds buried; this allows easy inspection for early germination. Nuts which germinate slowly must be discarded at an early stage. Shade is unnecessary but watering is essential during dry periods.

Seedlings are ready for transplanting when they have three or four leaves, i.e. about a year after planting (see Fig. 35). They are simply pulled upwards; no harm is done if some of the roots are broken unless bole rot is prevalent, in which case root damage should be avoided. There should be a further selection at the three leaf stage when all weak plants are pulled out. Selection should be so

Fig. 35: Coconut seedlings ready for transplanting

rigorous that no more than 50% of the nuts originally planted in the nursery are used for planting in the field.

Field establishment

Replacing unproductive palms
One of the main reasons for poor coconut yields in East Africa is the large number of unproductive, diseased, dying or even dead palms. These should be replaced with selected seedlings. Smallholders are seldom willing to do this; they usually prefer to keep a tree that is producing only five nuts a year rather than spend time and money on removing it and planting a seedling which will be unproductive for the first seven or eight years.*

Transplanting
Planting holes should be about 2 ft (c. 0·6 m) deep and 2 ft in diameter. They should be refilled with topsoil or, ideally, with topsoil mixed with organic manure. When a seedling is placed in the hole its nut should be about 1 ft (c. 0·3 m) below the soil surface and it should be earthed up only to the base of the shoot; the rest of the hole can be filled

*In the Ivory Coast varieties have been selected that come into bearing considerably earlier than usual, about six years after planting.

in gradually as the stem grows. This practice sets the bole well into the ground. Transplanting is best done at the beginning of the rains. If cattle are allowed to graze amongst the coconuts each seedling must be protected to prevent the leaves from being eaten.

Spacing and inter-cropping
The recommended spacings for tall coconut palms are 27 ft × 27 ft or 30 ft × 30 ft (c. 8 m × 8 m or 9 m × 9 m), i.e. plant populations of 60 or 48 plants per acre respectively (c. 148 or 119 palms per ha). The usual practice of smallholders is to adopt a wider spacing than these and to interplant cashew, mangoes, bananas or cassava. Inter-cropping with annual crops for the first two years is a common practice in pure stands; it may help to cover the cost of establishment and it does no harm to the young seedlings if the soil is not cultivated too near to them.

Field maintenance

Fertilisers
Ammonium sulphate is tentatively recommended in Kenya at the rate of two pounds (c. 0·9 kg) a year to each palm. Experiments in Tanzania showed that an application of nitrogen, phosphorous and potassium led to worthwhile increases in the number of nuts produced; trees also started bearing sooner after planting. These results led to a recommendation of 1 lb of calcium ammonium nitrate, 1 lb of double superphosphate and 2 lbs of muriate of potash per palm per year (c. 0·5 kg of C.A.N. and double superphosphate and 0·9 kg of muriate of potash). The effects of fertilisers are only noticed about a year after application. Fertilisers should only be applied if bush growth is well controlled. They are usually spread in a ring around the base of each palm.

Weed control
For the first few years after planting it is only necessary to maintain a weed-free circle around each tree. Under mature trees, however, weed growth and bush regeneration must be checked throughout, otherwise yields are reduced. Clean weeding gives the best results and some estates use disc harrows for this purpose; these implements are no longer recommended because they damage the roots and may thus allow the entry of soil-borne fungi (see section on diseases). Slashing and grazing are less laborious than clean weeding

and are good alternatives. In Tanga Region, in Tanzania, grazing dairy cattle beneath coconut palms has been found to be an efficient method of controlling weeds and undergrowth. Moreover, the yields of both nuts and milk are higher when this method is used than when the two are farmed separately. Bush regeneration, which is very rapid at the coast, is seldom properly controlled by smallholders; many of them only burn the undergrowth each year, thus harming the palms and only giving a temporary check to the bush.

Harvesting

Immature nuts
The coconut water content of nuts is highest when they are immature, i.e. nine or ten months after flowering. If they are to be used as immature nuts they must therefore be picked. It is usually necessary to climb up the trunks in order to cut the base of suitable bunches; if the trees are short, knives on long poles may be handled from the ground. Cutting steps in the trunks should be discouraged as it harms the trees.

Nuts for copra
For the best yields of copra and for the highest oil content, the nuts must be fully mature. They may be allowed to fall to the ground but this has one serious disadvantage: the nuts are very easily stolen. Theft is one of the greatest problems of coconut growers; they therefore prefer to climb the palms and pick the nuts. This is satisfactory if the nuts are almost ready to drop but growers often pick nuts that are not fully mature; they do this to reduce losses by theft from the trees and to obtain a cash income at the earliest possible date. Nuts should be picked at intervals of no more than two months; thus two bunches per tree should be ready at each picking.

Copra drying and quality

In East Africa most copra is sun-dried; this is satisfactory in fine weather but in wet weather good copra can only be produced by kiln drying. Kiln drying is unpopular with growers because of the added expense and because of the absence of any premium for kiln dried copra.

Sun drying
First the husks are removed by impaling the nuts on a sharpened stake whose other end is fixed firmly in the ground. Each nut is then halved by striking it in the middle with a 'panga'; the halves are laid in the sun with the flesh facing upwards. Copra should be dried on a clean surface, e.g. a concrete barbecue, but it is usually dried on bare earth where it picks up much dirt. In two or three days the semi-spherical pieces of flesh should have contracted enough for them to be easily removed from their shells. The aim should be to reduce their moisture content from about 45% when they are split to 6% when they are fully dry. In fine weather this takes about five days after removing them from their shells; in wet weather it takes longer. During this time it is important to turn the pieces occasionally and to cover them at night and during rain.

Smallholders very seldom dry their copra properly; they usually dry it for only about two days and take few precautions to prevent it getting dirty or to protect it from the rain. There are three main reasons for this situation: they want to sell their copra and collect their money as soon as possible; there is no difference in price between good and poor quality copra, and the less they dry their copra the heavier it is and the more money it fetches.

Kiln drying
There are several kiln designs; they all include a pit for the fire, a wire mesh platform over the fire to support the copra and a roof to protect the copra from rain. Coconut shells are the most common fuel because they are in good supply and produce much heat with little smoke. Kiln drying takes about four days.

Copra quality
Well dried copra from mature nuts is brittle, clean and white and smells fresh. It has a moisture content of about 6%, an oil content of 65–70% and a free fatty acid content of below 2·5%. Copra sold by smallholders very seldom reaches these standards. It is usually flabby and elastic because it comes from immature nuts; this leads to poor oil extraction during milling. It is usually dirty from bits of earth and sand picked up during drying; earth and sand can damage machinery in the mills. It is usually discoloured and foul-smelling owing to moulds which thrive in the moist conditions; these moulds cause an increase in the acidity of the oil due to hydrolysis of free fatty acids. If coconut oil is used for soap manufacture a high free fatty acid content is acceptable; if it is to be

used for edible products, however, the oil must be purified to remove the acids and this is an expensive process. Copra sold by smallholders usually has a moisture content and a free fatty acid content of between 10% and 20%. A final feature of much smallholder copra is the high proportion of insect damage; the red-legged ham beetle (*Necrobia rufipes*) readily attacks moist copra causing loss in weight and a large amount of useless 'copra dust'.

Yields

Yields are usually between 15 and 20 nuts per palm per year although 30 is an average figure in the wetter parts of the East African coast. Well managed palms from selected seedlings should yield 40–60 nuts per year but the potential, which has been realised on research stations, is about 100 nuts per year. In Ceylon as many as 140 are sometimes produced. Yields are very variable between palms; some palms yield very poorly even in the best areas whilst other palms yield quite well in generally poor areas.

Two mature nuts should yield one pound of good quality copra.

Pests

Rhinoceros beetle. *Oryctes monoceros*
This is potentially the most damaging coconut pest in East Africa, although it should not prove troublesome if there is a high standard of field hygiene. The adult is a black beetle, about $1\frac{1}{2}$ in (c. 4 cm) long, with a prominent frontal horn which gives the insect its name. The female lays eggs in decaying vegetable matter, showing a strong preference for rotting coconut trunks. The adult is seldom seen as it flies by night; it attacks the terminal buds in the crown of coconut palms, eating unopened leaves and occasionally destroying the growing point. When the unopened leaves are eaten the result is V-shaped notches in the opened leaves; these notches are symmetrically arranged on each side of the leaves. If the growing-point is eaten the palm dies. Coconut palms are particularly prone to damage in the first four to five years after planting.

Chemical protection is impracticable because the crowns are so inaccessible. An important precaution, which has been enforced in Kenya, is to destroy all decaying coconut palms; this involves splitting and drying the trunks to make them burn

as they are very difficult to burn whole. Adult beetles are sometimes killed by poking a piece of wire into their entry holes in the central bud.

Coreid bug. *Pseudotheraptus wayii*
This pest is more important than Rhinoceros beetle where the standard of field hygiene is high. This small bug punctures the very young nuts, causing them to drop off or to develop longitudinal scars. The copra production from damaged mature nuts is greatly reduced. There is no recommended control for this pest. A predator, the maji-moto ant, *Oecophylla longinoda,* is said to be encouraged by planting cashew or citrus trees between the coconut palms. Without interplanting, the maji-moto ant colonises very few palms but even with cashew or citrus nearby it is doubtful whether the bug is effectively controlled.

Diseases

Diseases do little damage to coconut palms in East Africa apart from bole rot, caused by the fungus *Marasmiellus cocophilus,* which has recently been identified in scattered parts of the coast of Kenya and Tanzania. It is soil borne, entering the plant via the roots, and causes a wilting and yellowing or bronzing of the leaves. Precautions against bole rot include transplanting at an early stage without breaking the roots, avoiding transplanting from nurseries in outbreak areas to fields in 'clean' areas, and avoiding cultivation with implements such as disc harrows which can injure the roots.

Alternative uses of the coconut palm

In addition to the flesh of the coconut, its water and the wine that is obtained from the inflorescences, the coconut palm yields a number of useful products. The trunks are used for building houses and boats and for firewood, although they burn rather slowly. Pieces of dried leaf are used to make 'makuti', an important thatching material. Individual leaf blades can be woven into mats or baskets. Leaf mid-ribs are often used for temporary walls or fences. The outer husk of the coconut consists mostly of a coarse brown fibre called coir which can be used for mats, cheap upholstery, mattress stuffing, etc. The shell is used for a large number of domestic products, from cups to fuel and from buttons to combs. The shells are ground

at one estate in Kenya and are exported to Europe
where the powder is included in linoleum.

Bibliography

1 **Anderson, G. D.** (1967). *Increasing coconut
  yields and income on the sandy soils of the
  Tanganyika coast.* E. Afr. agric. for. J., *32*:310.

2 **Childs, A. H. B.** (1958). *The production of
  kiln-dried copra in the Tanga Province of
  Tanganyika.* E. Afr. agric. J., *23*:280.

3 **Childs, A. H. B.** and **Groom, C. G.** (1964).
  Balanced farming with cattle and coconuts.
  E. Afr. agric. for. J., *29*:206.

4 **Sethi, W. R.** (1954). *A practical guide to
  coco-nut planting.* E. Afr. agric. J., *19*:140.

5 **Swynnerton, R. J. M.** (1946). *The improvement of the coco-nut industry on the Tanga
  coast.* E. Afr. agric. J., *12*:111.

6 **Tremlett, R. K.** (1964). *Floral biology of
  coconuts in Zanzibar.* E. Afr. agric. for. J., *30*:74.

7 **Wheatley, P. E.** (1961). *The insect pests of
  agriculture in the Coast Province of Kenya,
  IV—Coconut.* E. Afr. agric. for. J., *27*:33.

11
Cocoyams

Colocasia antiquorum

Introduction

This minor root crop is mistakenly called 'arrowroot' in Kenya. This name should be reserved for the West Indian root crop, *Maranta arundinacea*.

In Kenya cocoyams are widely grown in Central Province and, to a lesser extent, in Embu, Meru, Machakos, Nandi and Kakamega Districts. In Tanzania and Uganda they are occasionally grown in areas with a well distributed rainfall, e.g. the highlands and the north and west shores of Lake Victoria.

Plant characteristics

This crop is perennial but it may be harvested from six months after planting. The corm is cylindrical, growing vertically and seldom branching. A fully mature corm measures about 12 in (c. 30 cm) in length and about 6 in (c. 15 cm) in diameter. Suckers may appear on some varieties if they are left unharvested for long. The growing point is at the top of the corm and a cluster of large leaves rises from it. The leaves grow to three or four feet (c. 0·9–1·2 m) above the ground; each has a long stalk which is attached to the middle of the

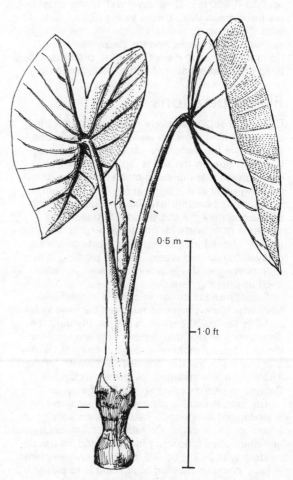

0·5 m

1·0 ft

Fig. 37: An uprooted cocoyam plant. The straight lines indicate the points of cutting for preparing the planting material.

lamina rather than the margin (see Fig. 37). The lamina is shaped like an arrow head and is thickly veined. Flowering hardly ever occurs.

Ecology

Cocoyams need a good supply of water throughout

Fig. 36: Cocoyams.

57

the year and can tolerate continual waterlogging; for these reasons they are usually grown in swamps and along river banks. They can be grown as a rainfed crop in areas with a well distributed, average annual rainfall of over 50 in (c. 1 250 mm). They are mostly grown in medium altitude areas, between 3 000 and 6 000 ft above sea level (c. 900–1 800 m). They grow well at the coast but are uncommon there owing to the lack of suitable land. This crop demands a fertile soil; the rich silt soil along river banks and the deep loams in Central Province in Kenya are ideal, provided they have not been exhausted by previous cropping.

Field operations

Cocoyams are vegetatively propagated from the crowns of the corms, so planting and harvesting are done at the same time. After the whole plant has been dug up the top is cut from the corm, making sure that the cut is made below the growing point and leaving about 2 in (c. 5 cm) of corm on the planting material. The leaves are cut off leaving only the shortened leaf stalks. The crown is now ready for planting (see Fig. 37); it is usually planted immediately but planting may be delayed for up to a week. Delayed planting is said to improve root development. Suckers are rarely used as planting material.

Sometimes the crown is planted in the hole from which the corm was removed but new holes 6–12 in (c. 15–30 cm) deep are usually dug; the dimensions of the hole depend on the size of the crown. Organic manure is sometimes added in the planting holes but this practice varies with locality. There are a few instances of farmers in Central Province in Kenya using superphosphate.

The crop is usually planted as a pure stand on waterlogged sites but on better drained soils intercropping is usual. On rainfed land all cocoyams are intercropped. Spacing in pure stands is usually random with 2–3 ft (c. 0·6–0·9 m) between plants. A fairly common method of spacing is to plant groups of three crowns; each group is separated from the next by 3–3½ ft (c. 0·9–1·0 m).

Maintenance of the crop is easy because the large leaves shade the soil and prevent vigorous weed growth. Harvesting is done piecemeal when the corms are required.

River bank plots may be grown continually or abandoned after a few years, depending on the soil fertility and the amount of manuring. Plots away from river banks are always rested after a period of mixed cropping.

Yields

Yields are difficult to measure because of irregular harvesting. A yield of 6 tons per acre (c. 15 t/ha) would be good.

Pests and diseases

Porcupines and moles are the most serious pests. A few larvae have been reported as attacking the corms but generally the crop is little attacked by insects or diseases.

Utilisation

The traditional method of preparation is to boil whole corms; this takes several hours or even overnight. After cooking they are peeled and eaten alone. A more modern method is to cut the peeled corms into small pieces before boiling or frying; in this case they are always mixed with other foods. Occasionally the corms are roasted in the fire and eaten without peeling.

12
Coffee (Arabica)

Coffea arabica

Introduction

With the exception of oil, coffee is the most valuable commodity in international world trade. For many years Arabica coffee has been Kenya's most important export crop, fluctuating in value from 15% to 35% of the total exports and bringing in as much as £19 000 000 in annual foreign revenue. In Tanzania it has consistently been one of the three most valuable exports but in Uganda, where comparatively little is grown and where Robusta coffee is the most important crop, it is of smaller significance.

Arabica coffee is, potentially, good quality or 'mild' coffee. For its inherent quality to be fully exploited it must be carefully harvested, processed by the wet method and sun dried. Where it is not given this treatment, e.g. around Bukoba in Tanzania and in Brazil, it becomes poor quality or 'hard' coffee and is little better than Robusta coffee (see chapter 13). East Africa produces some of the best quality Arabica in the world; Kenya's coffee and some of Tanzania's is only matched by that from Colombia and Costa Rica in South and Central America. Good quality East African Arabicas enjoy a greater demand and therefore fetch a higher price than most of the world's coffee.

Kenya coffee is solely Arabica. There are approximately 210 000 acres (c. 85 000 ha): 133 000 acres (c. 54 000 ha) of smallholdings and 77 000 (c. 31 000 ha) of estates. The acreage is mainly concentrated along the eastern slopes of the Aberdare range and around the south and east sides of Mt. Kenya. There are considerable smallholder acreages, however, in the Machakos and Taita Hills, in the Kisii Highlands, and in Bungoma and Kakamega Districts. There are also estates around Ol Donyo Sabuk and Machakos, between Nakuru and Subukia in the Rift Valley, between Lumbwa and Songhor, and in Trans Nzoia District, especially on the slopes of Mt. Elgon. The areas west of the Rift Valley have usually given lower yields than the areas in the east, partly owing to the fact that most of the coffee in the former parts is a subsidiary enterprise on mixed farms, but also owing to unfavourable ecological conditions. In Tanzania the main Arabica coffee growing area is in the Northern Region where estates and small holdings are concentrated on the east, south and west sides of Mt. Kilimanjaro and on the south side of Mt. Meru. There are a few estates around Oldeani in the Ngorongoro Highlands. There is a little smallholder production in the Pare, Usambara and Uluguru Mountains, in the hills near Handeni and in parts of North Mara, Kigoma, Kasulu and Kibondo Districts. The Southern Highlands are an important coffee producing area: there are estates and smallholdings near Rungwe, Mbeya and Mbozi and there is a considerable smallholder acreage in the Matengo Highlands. Around Bukoba the coffee is mostly Robusta but there is some Arabica which is processed by the dry method. In Uganda the bulk of the Arabica coffee is grown in Bugisu District on the slopes of Mt. Elgon. There is a little grown in Kigezi, Ankole, Toro and West Nile Districts.*

Coffee has a chequered history in East Africa. It was first introduced from the island of Réunion (formerly called Bourbon) by missionaries in Tanzania; in 1883 seed was brought to Morogoro and two years later to Kilema Mission near Moshi where trees, some of which are still alive, were established. In 1896 seed was sent to St. Austin's Mission in Nairobi via Kibwezi and Taita. This mission, which was responsible for dispensing seed to the early coffee planters, also obtained seed from Bura and from Réunion in about 1900. European settlers, mostly around Nairobi, gradually enlarged the coffee acreage until the start of the first world war in 1914 when expansion was checked. After the war there was a great influx of settlers. This caused a rapid growth in the acreage, making coffee the most valuable of Kenya's exports. Meanwhile, in Tanzania, German planters had established estates on the southern slopes of Mts. Meru and Kilimanjaro but these were abandoned during the war; they were sold only in

---

*The remainder of this chapter deals solely with Arabica coffee; from this point onwards the qualifying word 'Arabica' is seldom used.

1921, by which time they were in a sad state of neglect. A few Wachagga smallholders had started planting coffee on Kilimanjaro before the first world war.

The inter-war years were hard times for coffee growers. The world economic depression during the late 1920s and the early 1930s brought low prices. Pests, especially mealybugs and thrips, and diseases, especially coffee berry disease in western Kenya, caused great damage. Much coffee was uprooted at this time. African growers, however, increased in several areas although their numbers were very restricted in Kenya. Smallholders planted coffee in Meru and Embu Districts and in the Kisii Highlands in Kenya. They were not allowed to plant coffee in Central Province because it was feared that their plots would become a reservoir of pests and diseases which would damage the coffee on estates. In Tanzania the Wachagga smallholders were encouraged to plant coffee; they started the Kilimanjaro Native Cooperative Union whose members, by 1946, numbered nearly 30 000. It was during the interwar years also that the Ugandan smallholders in Bugisu District increased in numbers.

The second world war provided yet another check to expansion but after the war, especially during the 1950s, there was a very great increase in the coffee acreage in East Africa. The expansion, which occurred mainly amongst smallholders, was primarily due to a spectacular rise in prices following a world coffee shortage; in 1946 the average price paid per ton of Kenya coffee was only about £100 but in 1953–54 it reached a seasonal average of £580. An additional factor caused expansion in Kenya; the Government, basing its actions on the Swynnerton Plan of 1954, gave great encouragement to smallholders to plant coffee; land consolidation and registration, the extension of agricultural education and the provision of a large advisory field staff all helped towards this end. In 1953 the Kenya smallholder coffee acreage was only 4 000 (c. 1 600 ha); by the end of 1963 it was 130 000 (c. 53 000 ha). The number of smallholders rose from 5 000 in 1950 to a quarter of a million in 1963.

The 1960s, however, saw yet another change in the trends of expansion and contraction of the East African coffee industry. During the late 1950s prices fell steadily; during the early 1960s the average price paid per ton of Kenya coffee fluctuated around £300. The reason for this fall in prices was world over-supply because coffee had been planted as fast as possible during the 1950s

to take advantage of the high prices. In 1962 the main coffee producing and consuming countries signed the first International Coffee Agreement by which the producing countries agreed to limit their exports to the conventional importing countries to quotas which were to be set by the International Coffee Organisation. To prevent further increases in production Kenya has prohibited new coffee planting and neither Tanzania nor Uganda are encouraging any planting except the re-establishment of dead or unproductive trees.

A further cause of concern during the first half of the 1960s was the increased incidence of coffee berry disease; this spread into very many areas which had previously been free of the disease and caused very serious crop losses. It was only during the late 1960s, with the introduction of improved spraying schedules, that there was any promise of controlling the disease and, with somewhat steadier prices, of stabilising the East African coffee industry.

Plant characteristics

Roots

Root systems of coffee trees vary greatly and are influenced by such factors as spacing, soil characteristics, the presence or absence of mulch and treatment in the nursery and during transplanting. Commonly, however, the following types of roots are found:

1   Several axial roots which originate from the bottom of a short taproot and which grow almost vertically to a depth of 8–10 ft (c. 2·4–3·0 m); sometimes they have been found to extend much deeper than this. The axial roots are fairly thick and branch considerably; their main contributions are water absorption (especially during droughts), and anchorage.

2   A mass of ramifying lateral roots which are the main feeding roots and which form a surface mat or 'plate' seldom deeper than 1 ft (c. 0·3 m).

3   There are always a few roots which can be considered neither as axials nor as surface feeders; they may extend diagonally downwards from the taproot or vertically downwards from the surface feeders.

Branches

There are two kinds of vegetative growth:
1   Vertical growth as formed by the main stems

and suckers. This type of growth only produces further vegetative growth; only very rarely are flowers produced on main stems.

2 Horizontal growth as formed by the crop bearing branches. This type of branch can produce both reproductive growth (i.e. flowers) and vegetative growth. Horizontal branches are usually called laterals. (See Figs. 41 and 42 for both vertical and horizontal growth).

Buds and stems
Each node on vertical growth, i.e. on stems and suckers, give rise to opposite pairs of leaves. If one node has leaves pointing north and south, the next will have leaves pointing approximately east and west. In each axil there are two types of buds. Provided that the stem remains intact the upper bud in each axil will develop, producing a lateral, whilst the lower buds remain dormant. If that lateral is removed or damaged there is no bud in the axil which is capable of replacing it; laterals on main stems, therefore, are not regenerated if removed. Below each lateral is a series of buds, most or all of which are invisible, and which are capable of division and multiplication; one of these buds is stimulated to develop only when the upright stem above it is cut or damaged. The series of buds below each lateral can only produce upright stems; when small these stems are known as suckers; when removed or damaged they can always be replaced by the development of another bud in the same series or by the development of a similar bud from a lower node. Of academic interest, although also of practical interest in some countries, is the fact that bending a stem almost horizontal to the ground and then pegging it into position without cutting or damaging it stimulates the production of suckers from several axils. This is the basis of the 'agobiada' pruning system.

Buds on laterals
On laterals leaves are borne in opposite pairs and on a horizontal plane. In each axil is a series of three to five buds (see Fig. 38). Each bud is capable of developing into an inflorescence or a sublateral (identical in form to a primary lateral which is produced from an upright stem): it may remain dormant but it cannot produce upright growth.

Inflorescences usually consist of about four flowers although occasionally, e.g. at Solai in Kenya, as many as 15 are found. The number of flowers on an inflorescence depends largely on the place of the bud in the series: the bud nearest the lateral branch usually produces 4–5 flowers but those further away usually produce less. At the base of each inflorescence there are minute scale leaves; these protect further buds which may produce branches or flowers at a later date; normally, however, a node only produces flowers once.

The bud nearest to the lateral branch is the one most likely to give rise to a sublateral, although

Fig. 38: Flower buds on a lateral branch of a coffee tree.

this does not usually happen until after flowers have been produced on that node, or unless the tip of the lateral is damaged. Such sub-laterals are normally extended at about 45° from the parent lateral. Laterals were once called primaries (if they grew from a main stem), secondaries (if they grew from a primary), tertiaries (if they grew from a secondary) and so on. This terminology proved rather complicated; the words 'laterals' and 'sub-laterals' are now generally preferred.

Flowering (see Fig. 39) usually occurs in a flush and is stimulated by the onset of the rains after a considerable dry period. In the absence of a dry period there may be small flowerings in almost all months of the year. The white flowers, which are sweetly scented, are short-lived; within a day the petals start turning brown and the flowers fall off soon afterwards. There is only about 10% cross pollination.

Fruit

Each flower produces one fruit, or berry, normally containing two beans. When the berry is ripe the outer skin encloses a slimy mucilage; when the outer skin is removed during pulping some of the mucilage is taken off with the skin whilst some remains on the beans. Each of the two beans is enclosed in a tough membrane called the parchment. Closely adhering to each bean is the very thin testa which is called the silverskin.

Fruit development

The fruit remains in the pinhead stage (see Fig. 40) for 1½–2 months after the calyx and petals have

Fig. 40: Two nodes on a lateral branch of a coffee tree. The fruits are in the pin-head stage.

dropped off. The main period of physical growth is during the next 2–3 months; by the end of this time the parchment has reached its full size although the endosperm (later to be the bean) is still a white jelly. For the next four months the endosperm accumulates dry matter, but only in the last of these four months does the berry begin to change colour from green to red. The final ripening period, from flowering to maturity, is normally eight or nine months although it may be considerably prolonged by a long dry period or by cold weather occurring towards the end of the ripening period. On the other hand, a drought in an earlier stage, before the endosperm has fully developed, can shorten the ripening period.

Cherry

This is a collective term for ripe berries which have been picked but not pulped.

Parchment coffee

This is coffee which has been pulped but which has its parchment still intact. It is sent to mills in this form after fermenting and drying.

Clean coffee or green coffee

This is coffee which has been milled, i.e. the parchment and the silverskin have been removed. Mills are situated at Nairobi and Endebess in

Fig. 39: Coffee flowers.

Kenya, at Moshi and Bukoba in Tanzania and at Mbale in Uganda. 5 lb of cherry should produce about 1 lb of parchment coffee; 6 lb of cherry should produce about 1 lb of clean coffee. If the coffee is very light (see section on quality) considerably more than 6 lb of cherry may be needed to produce 1 lb of clean coffee.

Ecology

Rainfall, and water requirements

Kenya's main coffee growing areas (i.e. east of the Rift Valley) have a lower and more unreliable rainfall than almost any other major coffee producing area in the world. Average annual rainfalls of 35 to 40 inches (c. 900–1 000 mm) are common, especially in the lower parts where most of the estates are situated. In such conditions mulching and good weed control are essential to conserve moisture. West of the Rift Valley, however, and in the coffee growing areas of Tanzania and Uganda, the average rainfall is much nearer the 70 in (c. 1 800 mm) considered ideal in other parts of the world. Such rainfalls lead to better yields provided that the accompanying problems, i.e. increased weed growth, disease incidence and leaching, are dealt with.

The rainfall needs to be fairly well distributed throughout the year with the exception of a $1\frac{1}{2}$–$2\frac{1}{2}$ month dry period. Such a dry spell is generally regarded as beneficial; it hardens the wood and gets the trees into a cycle of flowering and bearing. In continuously wet areas there tend to be many small flowerings which produce inferior yields, which are awkward to pick and process when they ripen, and which may create a coffee berry disease problem because there are almost always susceptible berries on the trees.

East of the Rift Valley in Kenya there is a bimodal rainfall distribution and there are therefore flowerings at the beginning of both the long and the short rains. The short rains crop is usually of rather lower quality because it develops during the January–February drought. It is most important that there should be a good moisture supply until about four months after flowering; if there is a severe moisture shortage during the third and fourth month, i.e. the time when the parchment hardens, the size of the beans may well be limited.

Altitude and temperature

In Kenya the altitude limits set for coffee varied somewhat with district. East of the Rift Valley they were generally between 4 500 ft and 6 500 ft (c. 1 400–2 000 m). West of the Rift Valley they were a little higher; coffee has been successfully established as high as 7 000 ft (c. 2 100 m) on the slopes of Mt. Elgon. In Tanzania the rainfed coffee belt is from 4 000 ft up to 6 000 ft (c. 1 200–1 800 m) but there are some irrigated estates as low as 3 000 ft (c. 900 m). Certain problems are found at high altitudes: there is a tendency towards continual small flowerings, stunted growth and an excessive incidence of the 'Hot and Cold' condition and crinkle leaf. Other problems occur at low altitudes: rainfall tends to be insufficient and erratic, making irrigation essential; some pests and diseases (e.g. berry borer and leaf rust) are more damaging; multiple stem pruning cycles become too short (only two or three years); and the forcing climatic conditions encourage such rapid growth that very careful pruning is necessary and quality is usually lower.

Shade

The early European planters established their coffee under shade trees (see Fig. 000); this practice was copied from Indian estates. Some coffee in East Africa is still shaded but there is a strong trend towards eliminating or reducing shade trees. The most important shade trees in Kenya are *Grevillea robusta* and *Cordia abyssinica* whilst in Tanzanian estates *Albizzia spp.* are more popular. The spacing of such trees depends on their size and varies from 45 ft × 45 ft to 72 ft × 72 ft (c. 14 m × 14 m–22 m × 22 m) All the Tanzanian smallholder coffee, with the exception of that in the Matengo Highlands, and the Ugandan smallholder coffee is grown in the shade of banana plants and scattered forest trees. It has been proved that the banana shade considerably reduces coffee yields and Government officials in Tanzania would prefer to see the two crops grown separately. This subject is dealt with more fully in Chapter 1.

The effects of shade trees are as follows:

1 The potential yielding capacity of the crop is reduced. This is due to two factors: shade reduces the number of flower buds produced, and the output of assimilates is decreased owing to reduced photosynthesis.

2 Reduced photosynthesis reduces weed growth, especially couch grass and *Cyperus spp.*

3 Temperatures are modified. This effect is

noticeable at high altitudes where shade is necessary to prevent the incidence of crinkle leaf. A minor benefit is that reduced day temperatures make more pleasant working conditions; some planters even consider that they get increased work output from labourers under shade.

4 Wind speeds are reduced. This is undoubtedly beneficial but cannot be considered as a justification for shade.

5 Hail damage is reduced, although only to a slight extent.

6 Soil and water conservation are improved as the fall of individual raindrops is broken and drainage is improved by the root action of the trees.

7 Soil temperatures are reduced. The effects of both (6) and (7) can be achieved more efficiently by mulching.

8 The roots of shade trees are said to reach nutrients which have been leached beyond the root range of the coffee trees; these nutrients can then be made available to the crop by means of leaf fall.

9 The microclimate under shade affects some pests and diseases. *Leucoptera meyricki,* an important leaf miner, is discouraged whilst the berry borer is encouraged due to the absence of its natural parasites or predators. As mentioned above, crinkle leaf is virtually eliminated.

10 Shade trees compete for water and nutrients although some, e.g. *Albizzia spp,* are legumes and may contribute nitrogen.

In conclusion, shade is only fully justified at high altitudes where it modifies night temperatures sufficiently to enable coffee trees to grow normally. At low altitudes it may prevent forcing conditions and thus reduce the risk of overbearing; its popularity in such areas, however, lies more in the fact that it reduces the costs of production, especially weeding and applying fertilisers, even though it reduces potential yields at the same time.

Soil requirements
The structure and texture of the soil and the topography must allow free drainage but must also allow reasonable water retention. Medium loams are ideal; heavy loams and clays are unsuitable because they are poorly aerated, whilst sandy soils are equally undesirable because they dry out rapidly.

Deep soils are considered essential as an insurance against drought, and 6 ft (c. 1·8 m) is usually quoted as the minimum depth. In Kenya, however, there is a small acreage of coffee producing fair yields on a soil only 2 ft (c. 0·6 m) deep; in these conditions irrigation, fertilisers, mulch, good weed control and early pruning are all necessary. The presence of a murram hard-pan seldom allows healthy growth, although roots can penetrate a thin or uncompacted layer.

There should be a high nutrient status and the pH should ideally be between 5·3 and 6·0.

Varieties

Bourbon or Bourbon type
The first introductions of seed were from the island of Réunion, then called Bourbon. The early planters in Kenya, however, at first called their coffee 'Mokka' but in the 1930s it was found that East African 'Mokka' bore little resemblance to the variety of the same name in the East Indies; the name was therefore changed to 'French Mission' which is a very mixed type. Many of the best selections made in East Africa, e.g. S.L. 34 and the N series in Tanzania, can be considered of Bourbon type because they were selected from some of the earliest introductions or from French Mission. They only vary in a few unimportant details from the Bourbon varieties of Central and South America.

French Mission
As mentioned above, this was the name given to the early Bourbon type plantings in Kenya. Most of the estate acreage in Kenya still consists of this variety. It gives good quality but the yields are not very high. French Mission is not as susceptible to coffee berry disease as the two important S.L. varieties, largely owing to its intrinsic variability.

S.L. 28
This is a single tree selection made in Nairobi at the Scott Laboratories (S.L.), now the National Agricultural Laboratories, from Tanganyika Drought Resistant. It is quite high yielding and produces excellent quality. Being drought resistant, it is best

suited to the lower areas. Unfortunately it is very susceptible to coffee berry disease and is fairly susceptible to both the common races of leaf rust.

S.L. 34

This is a single tree selection, made at Loresho Estate at Kabete, from French Mission. It is suited to a wide range of ecological conditions, is very high yielding and produces excellent quality. It is one of the most susceptible varieties to coffee berry disease and is also susceptible to the common races of leaf rust. S.L. 28 and S.L. 34 were used for almost all the smallholder plantings in Kenya in the 1950s and early 1960s.

K.7

This variety was selected near Muhoroni in Kenya. It is reputed to have been selected from French Mission but this seems unlikely because it is resistant to one of the races of leaf rust and its bean characteristics and growth habit are more like Kents. Its resistance to leaf rust has made it popular at the lower altitudes where this disease can do much damage. Quality and yielding capacity are both acceptable and there is a fair degree of field resistance to coffee berry disease.

Blue Mountain

This variety was imported from Jamaica where it was grown widely in the Blue Mountains. It has some resistance to coffee berry disease but is only suitable for high altitude areas and is very susceptible to leaf rust. In East Africa it has not produced the good quality coffee for which it made its name in the West Indies and is quite different from the variety of the same name grown in Kivu District in the Congo.

Kents

This variety was originally a single tree selection made by a Dr Kent on his estate in India. It was introduced into East Africa early in the century and from it many good selections were made in Tanzania, e.g. the KP series and the H series. Varieties originating from Kents are generally higher yielding than those from Bourbon and, unlike most other varieties, produce a reasonable crop in the off year during the biennial bearing cycle. There is a high degree of rust resistance (again unlike Bourbon), but the quality is lower than that of Bourbon selections and the keeping properties after processing are poor. Kents was used exclusively for the early plantings in Meru District in Kenya.

Hybrids

A recent breeding programme carried out at the Coffee Research Station at Lyamungu in Tanzania has included crossing some of the existing varieties with Timor and Rumé Sudan which are both very resistant to coffee berry disease. Unfortunately both these varieties produce very low quality coffee but the early results from Lyamungu show that the hybrids are promising and may be released in the future, possibly after further back crossing to improve their quality.

Propagation

Coffee is usually propagated by seed. Vegetative propagation, however, is possible and will be essential when hybrids are released.

Seed

Seed from high yielding trees should be used. It is pulped and fermented in the normal way, the floaters are rejected and it is then dried in the shade until the moisture content is 15–18%. Shade drying ensures that the parchment dries slowly and does not crack. In the past it was recommended that wood ash should be mixed with the beans after pulping to make drying slower still; ash is no longer used because it prevents beans which are not used for seed from being marketed in the normal way. The seed should be sown soon after drying and with the parchment still attached. Viability decreases rapidly and the germination time increases about five months after picking. 1 lb (c. 0·45 kg) of seed contains 1 900–2 300 parchment beans and these should give about 1 600 good seedlings.

Nursery site

The site should be fairly level and sheltered with a good deep soil, with no risk of waterlogging and with a good water supply. Overhanging trees must be avoided or removed and valley bottoms which are likely to be too cold at night must also be avoided.

Bed preparation

Beds should be about 4 ft (c. 1·2 m) wide with paths of a reasonable size between. The beds are usually dug to a depth of about 2 ft (c. 0·6 m) and the topsoil from the paths is piled on top of them. Compost, cattle manure or superphosphates are often incorporated. A fine tilth must be prepared, making sure that all stones, tree roots and couch grass are removed.

Arrangements for shade must be made well before sowing. The most popular method is to place short fence posts along each side of each bed. These should be about 3 ft (c. 0·9 m) high and should have wire trained across the top of them; lengths of split bamboo or split sisal poles, about the same width as the beds, are placed across these wires. The advantages of this type of shade are its easy adjustment and construction. Some people prefer a much higher shade, 6–7 ft (c. 1·8–2·1 m) above the ground, making a roof above the entire nursery; this is less easily adjusted and is more expensive to construct.

Seedbeds

In Kenya seedbeds are recommended; the seedlings only remain in these until they have one pair of true leaves and they are then transplanted into seedling beds. The aim is to save space, because some seeds will fail to produce healthy seedlings, and to create as much uniformity as possible in the seedling beds. In Tanzania seedbeds are not recommended, the argument being that too much transplanting often causes bent or twisted tap roots; direct sowing into seedling beds is therefore recommended.

If seedbeds are used, rows of seeds are sown about 6 in (c. 15 cm) apart and running from one side of the bed to the other. In each row two or three lines of beans are sown; the beans should be end to end within the lines, with their flat sides downwards and about ½ in (c. 1·3 cm) deep. The soil should be covered with a short grass mulch and maximum shade should be provided. Watering is necessary on alternate days for the first two weeks; it can then be reduced to once or twice a week. The seedlings should emerge 6–8 weeks later and are ready for pricking out into the seedling beds when they have one pair of true leaves.

Pre-germination

Pre-germination is an alternative to using seedbeds and thus avoids the necessity of two transplantings. A thin layer of seed is spread between two layers of hessian which are kept in the sun and watered frequently. In two or three weeks, when germination should start, the seeds are inspected daily and those with their radicles emerging are selected and planted into the seedling beds. Alternatively, pre-germination can be done in polythene bags containing moistened vermiculite and coffee seed at a ratio of 3:1.

Seedling beds

Non-germinated seed, pre-germinated seed and seedlings from seedbeds can all be treated in the same way in seedling beds. In Tanzania seedlings are transplanted into the field when they are quite small and the recommended spacing is only 6 in × 6 in. (c. 15 cm × 15 cm). In Kenya the seedlings are left in the nursery until they are larger and the recommended spacing is therefore 8 in × 8 in (c. 20 cm × 20 cm). Full shade is applied for at least the first month. The seedlings are watered every week, making sure that the whole root range is wetted. In Tanzania, where nursery growth tends to be rapid, root pruning has proved successful; a piano wire is pulled along the length of the beds, about 9 in (c. 23 cm) below the soil surface, guided by the lower surfaces of boards which form the sides of the beds; vertical cuts between each plant are made with a 'panga'. This process is repeated each month and encourages lateral root development. The advantages are seen in the field, namely higher survival rates and faster early growth.

The seedlings are hardened off towards the end of the nursery period by gradually reducing shade and watering but the shade should not be entirely removed. If necessary, the seedlings must be protected from cutworms, scales or leaf rust by applying, respectively, aldrin, dimethoate or copper. Aldrin is best dug into the beds when they are being prepared.

In Tanzania growth is rapid and seedlings are considered ready for transplanting when they are about 1 ft (c. 0·3 m) high and have been in the nursery for about a year. Capping seedlings of such a size is not recommended. In Kenya growth is slower and transplanting is recommended at a later stage, i.e. when the seedlings are about 2 ft (c. 0·6 m) high with stems about pencil thickness. About 18 months are needed in the nursery before this stage is reached. Seedlings of this size are often capped at roughly 15 in (c. 0·4 m) (at the top of the brown wood) about a month before transplanting; this is the first stage of either single stem or multiple stem pruning. Capping must be followed by fairly heavy shading to prevent sun scorch on the leaves.

Polythene sleeves

These are a possible alternative to the traditional seedling beds. The sleeves should be 7–8 in (c. 18–20 cm) in diameter and about 1 ft (c. 0·3 m) long. They should be filled with a potting mixture, e.g. a mixture of topsoil, sand and fibre, with added superphosphate and aldrin dust.

Vegetative propagation
Softwood cuttings are the easiest material for vegetative propagation. They must be taken from suckers or main stems in order to ensure upright growth; either single leaf cuttings or tips 4–6 in (c. 10–15 cm) long should be used.

Field establishment

Land preparation
The land must be prepared at least six months before planting and preferably longer. Couch grass and star grass eradication need special attention. Planting holes should be dug at least three months before planting in order to ensure good weathering of the walls of the holes; the topsoil and subsoil should be piled separately. The holes should be about 2 ft (c. 0·6 m) wide and 2 ft deep. About two or three weeks before planting, the topsoil of each hole is mixed with at least one 'debe' of cattle manure or compost and two ounces of double superphosphate; the holes are then refilled using additional topsoil if necessary. Mulch is usually placed around the young trees. Although the system described above is a good insurance, giving good opportunity for early root development, it has not proved to be necessary in all cases; excellent coffee has occasionally been established by subsoiling in a grid pattern and planting a tree wherever the lines crossed, or by planting in a plough row.

Spacing
The traditional coffee spacing is 9 ft × 9 ft (c. 2·7 m × 2·7 m) or, to improve access between the rows, 10 ft × 8 ft (c. 3·0 m × 2·4 m); both of these spacings give plant populations of about 540 plants per acre (c. 1 330 plants per hectare). Most of East Africa's coffee is planted at one of these spacings. However, with a good supply of nutrients and soil moisture, coffee yields best at a closer spacing, e.g. 9 ft × 4½ ft (c. 2·7 m × 1·4 m) giving 1 075 plants per acre (c. 2 660 plants per hectare). This has proved to be true both in areas with good rainfall, e.g. Tanzania and the higher coffee growing areas of Kenya, and under irrigation. Rows much closer than 9 ft apart do not allow easy access by tractors.

Planting
Planting should be done at the beginning of the main rains. The holes are re-opened immediately before planting. The roots must not be bent during planting. If the bare root method is used, the roots must be protected from the sun at all times and the lateral roots must be carefully spread out during planting. Mulch should be applied around the seedlings but it must not touch the stems. Bracken or banana leaf shade applied over each seedling immediately after planting is beneficial.

Pruning

For the following reasons it is essential to prune coffee:

1 *To control cropping*
   Coffee trees, like most fruit trees, have a marked tendency to bear biennially, i.e. to produce a heavy crop one season and a light one the next. To understand this it is important to appreciate that berries are borne on second season wood, i.e. on nodes which were formed in the previous season. When coffee trees carry a very heavy crop, most of the available carbohydrates are channelled to the developing berries; there is therefore a reduced carbohydrate supply for vegetative growth and a considerable reduction in the numbers of nodes for next season's crop. In the following season, when there is only a light crop, the developing berries make small carbohydrate demands and there is consequently a large amount of vegetative growth, leading to a heavy crop later. If there is no crop control, there is a serious risk that very heavy crops may cause overbearing and die-back. When this happens the leaves are unable to supply sufficient carbohydrates to the growing crop so the carbohydrate reserves in the wood are used. The result is defoliated trees and, in severe cases, die-back, i.e. a blackening and death of the laterals from their tips inwards towards the main stems. Sometimes the roots also die back. Trees may take two years to recover from this condition. Overbearing is characterised by berries which, when ripe, are yellow rather than red; this inevitably leads to poor quality. Good pruning limits the amount of bearing wood and thus prevents overbearing.

2 *To facilitate picking*
   If left unchecked coffee trees grow as high as 30 ft (c. 9 m) and can only be harvested with

difficulty. Picking must be made easy in East Africa where at some times of the year the operation must be done every 10–14 days if good quality coffee is to be produced. Coffee is picked primarily by women and children, few of whom can reach higher than 6½ ft (c. 2 m). It is essential to limit the height of coffee trees by pruning in order to avert any danger of damage by pickers climbing into trees or bending stems over too far.

3 *To make spraying more efficient*
This is an important consideration now that coffee berry disease is such a problem. Few hand or tractor operated sprayers achieve a good cover above about 7 ft (c. 2·1 m); it is therefore important to limit the height of the stems. It is also important not to allow trees to become too bushy, otherwise sprays may not reach the inner parts.

4 *To make a less favourable microclimate for certain pests and diseases*
This is especially true of coffee berry disease, leaf rust and Antestia, all of which thrive in the environment of unpruned trees. Pruning off old heads is particularly important because coffee berry disease tends to build up in this part of the tree; this point is discussed more fully in the section on diseases.

Single stem pruning
The aim of single stem pruning is to have one permanent stem, capped at five or six feet (c. 1·5–1·8 m) above ground level, with a permanent framework of laterals (see Fig. 41). The crop is borne on sub-laterals and the main operation in maintaining single stem trees is the annual selection of suitable sub-lateral growth.

Formation
When the young tree is 27 in (c. 69 cm) high it is capped, i.e. its main stem is cut at a height of 21 in (c. 53 cm). One sucker is selected and is capped at 45 in (c. 114 cm) when it reaches 51 in (c. 130 cm). One sucker is again selected and is capped at 66 in (c. 168 cm) when it reaches 72 in (c. 183 cm). In simpler terms, 6 in (c. 15 cm) is taken off when the plant is approximately knee high, then slightly above waist high, then head high. After the final capping all suckers are removed. After the first and second capping, when the chosen sucker has grown 4–5 in (c. 10–13 cm) tall, the primary immediately above it should be

Fig. 41: A single stem coffee tree. All the branches on the nearest side have been removed in order to show the permanent framework of laterals.

Fig. 42: A multiple stem tree. For clarity, the third stem was not included.

carefully removed to prevent impeding upright growth. The object of the repeated cappings is to stimulate a permanent framework of strong primary branches.

Annual pruning

This consists of removing old wood which has borne a crop and selecting new vigorous growth for future cropping. This involves considerable skill and no hard and fast rules can be given, but inward growing, downward growing, broken, dead, spindly, bare and crossing branches should definitely be removed. Thinning should allow an open centre to the tree, and should suitably limit the number of bearing sublaterals. Unless pruned carefully, single stem trees tend to form an 'umbrella', i.e. a dense layer of sub-laterals at the top of the tree which prevent healthy development of lower layers.

A simplified single stem system has been introduced in Tanzania; it only involves opening up the centre of the tree and removing all weak and drooping sub-laterals which have already borne a crop. This is easier and cheaper than the traditional system but it does not control cropping as efficiently.

The main pruning (as with all methods of pruning) is done as soon as possible after the main crop has been picked. The sooner pruning is done the less moisture is removed from the soil. Generally, the only implement needed is a pair of secateurs; only occasionally is a pruning saw needed for large branches. The main pruning is followed up three or four months later, well after the onset of the rains, by a light pruning, usually called handling; the purpose of handling is to thin out the flush of young shoots which will form the following season's bearing wood.

Initially, the single stem method was used for all coffee in East Africa. Multiple stem pruning, however, has become increasingly popular, mainly because of its ease and cheapness. Single stem pruning is only important now in the low altitude areas of Tanzania where growth is so rapid that multiple stem cycles are reduced to two or three years, and in places where suckers are not readily produced, especially under shade.

Multiple stem pruning

The aim of multiple stem pruning is to encourage the growth of two or more main stems which are replaced by selected suckers every 4–6 years (see Fig. 42); the crop is borne mainly on laterals. Each lateral bears two crops and is then removed; crops are therefore borne from higher and higher up the stems in successive years until a new cycle is commenced.

Formation

The recommendations for multiple-stem formation differ in Kenya and Tanzania. In Kenya capping is recommended in the nursery at about 15 in (c. 38 cm), as mentioned above; all laterals below the point of capping are later removed. In Tanzania capping is recommended after planting in the field, at a height of about 21 in (c. 53 cm); the laterals below the point of capping are allowed to remain and to bear only one crop. Whatever the recommendation for the height of capping or for the 'skirt' below the point of capping, both of the two suckers which grow up after capping are allowed to develop. When they are 4–5 in (c. 10–13 cm) long the two lateral branches immediately below them are carefully cut off.

Annual pruning

Pruning each year consists mainly of cutting up, i.e. removing the laterals from the lower parts of the stems. The most common method of measuring how many laterals to cut off is to hold a stick 3–5 ft (c. 0·9–1·5 m) long against the stem being pruned, with the top of the stick against the tip of the stem; all primaries below the bottom of the stick are removed. This leaves a 3–5 ft bearing head on each stem. The length of the head depends mainly on the growth rate; if growth is slow, i.e. at high altitudes, the head should be long and if growth is fast it should be short. The current recommendation for smallholders in Kenya is to remove all laterals which have borne two crops.

At the same time as cutting up, some of the following may be done—removing laterals which cross in the middle of the tree between the main stems; removing unwanted suckers from the main stem near ground level; spiral removal of alternate laterals within the bearing heads if there is overcrowding of these branches.

Handling

A secondary pruning or handling must be done during the main rains. The main aim of handling is to remove unwanted suckers and to limit the number of sub-laterals.

Changing the cycle

In areas where growth is slow, i.e. at high altitudes, the stems can continue to produce crops for 7 or 8 years without getting too tall; 4 to

6 year cycles, however, are the general rule. As far ahead as 18 months before cutting down the old stems, sucker removal should cease and three suitable suckers should be selected; if possible they should be borne on the original stem or 'leg' and should be spaced well apart from each other. Three is the most common number of stems in the second and subsequent cycles although this is not a fixed rule. To encourage the growth of healthy suckers the inward-pointing laterals should be cut up higher than the outward-pointing laterals towards the end of a cycle; this encourages the stems to bend outwards, owing to the greater weight of laterals on one side, and allows the developing suckers plenty of light. It is sometimes rather difficult to get suckers to come away, especially in the first cycle after planting, and this aspect has often been ignored by smallholders. If suckers are not forthcoming it may be necessary to cut down one of the stems.

Implements
Secateurs are used for cutting up and handling and thin-bladed pruning saws are used for removing old stems; diagonal cuts, sloping away from the centre of the tree, should be made as near the base of the stem as possible, taking care not to damage the new stems. Paint or wound dressings are applied by some growers to the larger cuts but this is only necessary in high rainfall areas to prevent rotting.

Conversion from single stem to multiple stem
In Kenya it is recommended that below 5 500 ft (c. 1 700 m) the first 7 years after planting should be single stem and then the trees should be converted to multiple stem. Suckers are encouraged by removing all the primaries on one side of the tree. This is a common practice on estates.

Single stem v. multiple stem
The following are the important advantages and disadvantages of the two main systems of pruning:

1   Single stem pruning is a slow job, needing skilled operators, and is therefore expensive. The main annual pruning of multiple stem coffee, however, is a quick job and little skill is required; about 100 trees can be pruned by one man in a day; the operation is therefore much cheaper. This factor has been the most important in persuading growers to abandon single stem pruning.

2   Multiple stem coffee is higher yielding in the first years of bearing but the advantage is less noticeable in later years. Experiments at Lyamungu and Ruiru showed that the two systems eventually gave very similar yields.

3   Single stem coffee is the only feasible system at the lower altitudes of Tanzania where multiple stem cycles last for only two or three years.

4   Single stem coffee, being lower, is more easily picked and sprayed.

Variations
There are many variations and two of the important ones are briefly described below.

Capped multiple stem
This system has recently become very popular on estates; it involves capping multiple stem trees at a convenient height, usually 5–6 ft (c. 1·5–1·8 m). The main advantage is that spraying is more efficient and that old heads, which tend to harbour more coffee berry disease than other parts of the tree, are avoided. Laterals become semi-permanent rather than temporary so single stem principles are necessary in selecting suitable sub-laterals for bearing the bulk of the crop. When the trees become unproductive they are stumped back to allow a new cycle; this is done about every seven years.

Side pruning
This system, which is still in the experimental stage, can be applied either to single stem or capped multiple stem coffee. It involves cutting back the laterals on one side of the tree leaving only two or three nodes on each; sub-laterals are then selected to provide the bearing branches. Sides are pruned alternately; each side is pruned every 2–4 years, depending on the growth rate. The main attractions of this system are its simplicity, efficient spraying, and the possibility of partially mechanising pruning; work at Lyamungu, using a tractor mounted vertical cutter bar, has shown promise. The main disadvantage of side pruning is that it can be dangerous in areas where there is a risk of *Fusarium* bark disease.

Weed control

Good weed control is essential. With its shallow feeder roots coffee is highly intolerant of competition. Competition for water is probably the most important factor, especially in the lower coffee growing areas of Kenya where moisture is often limiting. Competition for water is especially severe in the dry season but it is also important in the wet season, when rainfall should be going into reserve rather than being transpired by weeds. Weeds use a great deal of nitrogen, which is the most important element in coffee nutrition. The leaves of poorly weeded coffee are often an unhealthy yellow colour. A further disadvantage of weeds in coffee is that they can create a moist micro-climate near the base of the trees and this may encourage leaf rust.

Annual broad leaved weeds are common but the most damaging weeds are couch grass (*Digitaria scalarum*), star grass (*Cynodon dactylon*) and sedges (*Cyperus spp.*), all of which are difficult to eradicate and which compete strongly with coffee. Excessive competition from such weeds can cause severe debility and even die-back.

Fig. 43: A rotary cultivator in operation.

Cultivation

Weed control is by far the most important reason for cultivating between the rows of coffee. Incidental benefits can be improved water infiltration or soil aeration.

Continuous cultivation, i.e. maintaining a weed-free soil, at first gives better yields than slashing, cover cropping or only cultivating at certain times of the year. However, this advantage usually decreases with time because a continually bare soil prevents good water infiltration and encourages soil wash. Continuous cultivation is therefore unacceptable, although cultivation of some sort is almost always necessary in conjunction with other methods of weed control.

A 'forked jembe' is the most effective hand implement; it leaves a rough cloddy surface which resists erosion but its use is slow and expensive. In Central Province in Kenya much weeding is done with 'pangas'; these are used not for slashing but for shallow cultivation. The work is fast and cheap and is often done several times a year by smallholders to back up one or two 'forked jembe' operations. Disc harrows are sometimes used but cause deterioration of soil structure if used too much and unless the rows are on the contour they may encourage erosion by creating small channels for the water to run down. The Coffee Research Station at Ruiru has tested and modified a rotavator (see Fig. 43) which produces a rough cloddy surface if the soil moisture is suitable. It works well east of the Rift Valley but in the Rift Valley and west of the Rift the soils have proved

too abrasive, causing rapid wear of the tines. Most estates which purchased rotavators have now turned over to chemical weed control. Subsoiling is sometimes done by growers who think it may improve the water infiltration and aeration of their soils; experiments east of the Rift Valley have shown that subsoiling, although not harmful, is of no benefit because the subsoils are already well cracked.

Slashing

Slashing can be done by hand or by rotary slashers. The advantages of this type of weed control are its speed (it is often used when there is a shortage of labour during picking) and its suitability in steep areas during the rains when other methods of weed control might encourage erosion. The disadvantages of slashing are the facts that it checks weed competition less effectively than most other methods and that it encourages the development of a sward of grass if used too much.

Mulching

Weed control is an important function of mulching (discussed in more detail in the next section). The soil must be clean before the mulch is applied and there must be no couch grass present because this can easily grow through the mulch. Weeds tend to grow up through the mulch about six or seven months after it is applied and these may be hand pulled if not in large numbers. Alternatively, the mulch may be raked aside, the soil cultivated and the mulch replaced. Weeds, and the operations needed to remove them, tend to break up mulch and it seldom lasts as long as a year. If herbicides are used, however, it lasts longer.

Chemical control

Long term experiments have shown that chemical control is more expensive than conventional methods of controlling weeds during the first two years of use; in the following years, however, it becomes cheaper as the weed population is reduced and as it becomes necessary only to apply spot application.

Paraquat has become popular on estates and, recently, on smallholdings in Nyeri District, Kenya. It can, however, cause damage to suckers so it should be very carefully applied when suckers are needed for a new cycle. *Fusarium* bark disease can easily enter the wood via damaged suckers. There are a few weeds, e.g. *Commelina benghalensis*, *Cyperus spp.* and some grasses, which are not fully controlled by paraquat and which tend to multiply when all the other weeds are killed. Where paraquat is used it is usually necessary to follow a programme of minimum cultivation to get rid of the resistant weeds when necessary.

Other herbicides which are sometimes used are residuals, e.g. diuron, simazine and atrazine, and dalapon for couch grass control. Many growers find it difficult to control couch grass using dalapon; it is mostly removed by hand.

Cover crops

No cover crops have been found in East Africa which do not compete severely with coffee. This method of weed control is not used or recommended.

Shade

As mentioned in a previous section, shade suppresses weed growth to a certain extent, especially the growth of couch grass and *Cyperus spp.*

Mulching

Mulching (see Fig. 44) is very beneficial in coffee, especially in the drier areas, where it can produce almost double the yields of unmulched coffee. Responses are much lower under irrigation or in high rainfall areas or where there is dense shade; in high rainfall areas mulch may even cause yield reductions. Mulching often improves bean size.

The main beneficial effects of mulching are generally considered to be improved soil and water conservation. Further benefits are weed control and

Fig 44: Coffee which has been mulched in alternate inter-rows. The contrast is most clearly seen at the far end of the field.

the addition of considerable amounts of organic matter and nutrients. This last advantage is especially important where soils are poor.

The drawbacks of mulching are as follows:

1 The land needed to produce the mulch. One acre of well managed Napier grass should produce enough mulch for one acre of coffee. If vlei grasses are used, e.g. *Hyparrhenia spp.*, *Cymbopogon spp.*, etc. up to five acres of grassland are needed to mulch one acre of coffee. Competition for land is no problem where there are considerable acreages of soils which are unsuitable for planting coffee, e.g. the vlei soils in the lower coffee growing areas east of the Rift Valley in Kenya; in such areas mulch is usually obtained from the areas with poor soils. Where there are no poor or marginal soils, however, many coffee growers have preferred to plant as much coffee as possible rather than to use the land for mulch production.

2 The labour and expense involved in producing and applying mulch.

Very few smallholders mulch their coffee adequately; many consider it more profitable to feed grass to cattle than to apply it as a mulch; their decision may not be unwise if they are in an area with reasonable rainfall and if livestock products are a more attractive financial proposition than coffee.

Methods of application

The most common methods are applying mulch to each or alternate tree rows or to alternate inter-rows. The first method is usually recommended for the first two years after planting when the roots have not reached far into the inter-rows. Where there is plenty of material mulch may be applied to every tree row or to every inter-row; this is especially beneficial on steep slopes or on infertile soils or where the rainfall is marginal. It is important to get rid of all weeds before mulching and to apply a reasonable depth of mulch; an initial depth of 6 in (c. 15 cm) should give a settled depth of 3–4 in (c. 7·5–10·0 cm) which is quite enough.

Time of application

Where the rainfall is rather low and erratic, e.g. in the lower coffee growing areas east of the Rift Valley in Kenya, mulch gives much higher yields if it is applied before the rains start; this encourages maximum water infiltration. Where the rains are very heavy, e.g. on the slopes of Mt. Kilimanjaro where about 40 in (c. 1 000 mm) usually fall in April and May, mulch makes the soil too wet if applied before the rains so it should be applied as the rains end.

Mulching materials

Napier grass is popular because it produces more bulk than any other grass: up to 70 tons of green matter per acre (c. 175 t/ha) from two cuts per year. It has several disadvantages, however. Its production cannot be mechanised because it can only be planted and harvested by hand; this makes it suitable only for small scale growers. It is a very heavy consumer of potassium (as much as 5% K in the dry matter) and if applied repeatedly may cause induced magnesium deficiency which must be corrected by magnesium foliar sprays and/or magnesium fertilisers. Napier grass decomposes rather quickly, i.e. in six or seven months, unless it is completely mature when cut, and it tends to take root in the coffee unless it is very thoroughly dried. Other grasses, whether growing wild on vleis, e.g. *Hyparrhenia spp.*, or whether specially sown, e.g. Rhodes grass, are lower yielding than Napier grass but can be harvested mechanically and can be baled for convenient transport and application. They usually last about 9 or 10 months, although they last longer when herbicides are used. Banana trash is an ideal mulch because it is very persistent, often lasting more than a year. It is used where coffee is interplanted with bananas, but there is seldom enough material for efficient mulching and growers often concentrate the trash around the banana stools rather than around the coffee trees. Other materials, applied when available, include sisal waste, coffee pulp, wheat straw, sawdust and wood shavings. These materials are seldom available in large enough quantities to make much of a contribution. Sisal waste contains large quantities of calcium and can make the soil too alkaline, causing iron deficiency.

Fertilizers and manures

Nitrogen

Nitrogen is by far the most important element in coffee nutrition and good yields are never maintained without regular applications. Positive responses to nitrogen fertilisers are almost always found; only under heavy shade do the responses become so small that applications are unprofitable. The recommended nitrogen fertiliser programmes should raise the yields of clean coffee by $1\frac{1}{2}$ to $2\frac{1}{2}$ cwt per acre (c. 190–310 kg/ha).

The type of nitrogen fertiliser to be used depends on the soil reaction. The indiscriminate use of sulphate of ammonia has, in the past, led to considerable reductions in the soil pH; in some cases the pH was reduced as far as 4·0. If the soil pH is below 5·3 calcium ammonium nitrate is recommended for two years in three with ammonium sulphate nitrate for the third year in order to add sulphur. If the soil pH is between 5·3 and 6·5 (as it is in most coffee soils) calcium ammonium nitrate and ammonium sulphate nitrate are recommended in alternate years. If the soil pH is over 6·5 sulphate of ammonia should be used. Foliar sprays of urea are fairly common on estates and give good results. The use of compound fertilisers is being adopted on an increasing number of estates although there is no experimental evidence to support their use.

The amount of nitrogen to be applied depends on the anticipated size of the crop. A small crop of 6–7 cwt of clean coffee per acre (c. 750–880 kg/ha) needs only about 80 lb of nitrogen per acre per year (c. 90 kg/ha); a 10 cwt crop (c. 1 300 kg/ha) needs about 100 lb of nitrogen per acre per year (c. 110 kg/ha), whilst a heavy crop may need 120–140 lb of nitrogen per acre per year (c. 135–155 kg/ha).

The best results are thought to be given by splitting the nitrogen into three dressings, although there is no experimental evidence to support this. The first should be at the beginning of the long rains; the second should be about a month later and the third should be during the short rains. In areas with a unimodal rainfall distribution the three dressings should be applied at the beginning, in the middle and towards the end of the wet period.

Nitrogen fertilisers must always be applied to bare soil. They must never be placed on top of mulch. In mature coffee they are usually spread in a band along the inter-rows. In young coffee they can be spread in a ring around, but not too near, each trunk.

Phosphate

Early fertiliser experiments showed that there were no responses in mature coffee either to super-phosphates or to basic slag. The recommended use of phosphate fertilisers was therefore limited to their inclusion in planting holes.

Potassium

Responses to potassium fertilisers are rare although they have occasionally been obtained in the absence of Napier grass mulching in places where the potassium content of the soil is unusually low, e.g. in some areas west of the Rift Valley in Kenya and on one or two estates east of the Rift. Positive responses may become more common in the future as more and more potassium is removed from the soil. Some estates are already applying potassium in compound fertilisers as an insurance against such a situation. There is a risk that excess potassium in the soil may cause a reduction in quality; high potassium contents are often associated with brown beans (see section on quality).

Magnesium

Induced magnesium deficiency has sometimes been diagnosed after repeated heavy applications of potassium-rich materials, e.g. Napier grass and manure. In such cases, foliar applications of Epsom salts quickly restore good health whilst soil applications of magnesium fertilisers prevent the recurrence of magnesium deficiency.

Manure

Cattle manure has, in the past, been used in large quantities on coffee. Its use has declined, largely owing to its increased demand for use on other crops, but there are still some estates and some smallholders east of the Rift Valley who obtain manure from the Masai areas. Such manure is usually of very poor quality, often containing as much as 75% soil and as little as 15% organic matter; this compares unfavourably with good quality farmyard manure, made with suitable litter and protected from the sun and rain, which contains 50–60% organic matter on a wet weight basis and which contains little or no soil. Manure of high quality is produced by the Wachagga farmers of Mt. Kilimanjaro, who apply it to their mixture of bananas and coffee.

Most experiments have shown that applications of organic manures to established coffee give no positive responses; exceptions occur only in badly eroded areas where most of the topsoil has been washed away or in the rare instances that the soil is low in potassium (see previous section). Regular applications of manure are now discouraged because of the risk of raising the potassium content of the soil to such an extent that coffee quality is reduced.

Irrigation

Some of the lower estates in Kenya and all the lower estates in Tanzania irrigate their coffee using overhead irrigation during dry periods.

Harvesting

Coffee harvesting, usually called picking, is done entirely by hand (see Fig. 45).

Only uniformly ripe cherry can produce good quality coffee; the entire surface of each cherry should be red although a very small amount of green at the base is acceptable. Over-ripe berries (dark coloured) may be difficult to pulp. Under-ripe berries (green) may not have enough mucilage for efficient pulping. Berries of mixed ripeness cause uneven fermentation. Berries which are yellow in colour, possibly owing to drought, produce poor quality 'light' coffee and may lead to coated beans and pales in the roast (see section on quality). Berries which have been attacked by diseases may cause pulping problems or may produce a poor quality end product. It is essential, therefore, that high standards of picking are maintained. Picking must be done every 10–14 days during the harvesting season, sometimes once a week, to prevent the berries becoming over-ripe; thorough sorting of desirable and undesirable berries must be carried out before delivery to the factory: this may be done either during or after picking or, preferably, on both occasions.

Women and children do almost all the coffee picking in East Africa. In order to reach the higher berries they are sometimes equipped with a piece of string with a hook at one end and a short wooden rod at the other; they attach the hook to the upper part of a tall stem and then stand on the wooden rod; this bends the stem over and enables

Processing

Bukoba and North Mara Districts and the Uluguru Hills, in Tanzania, are the only areas in East Africa where Arabica coffee is dry processed. There are two methods of dry processing. Firstly, the berries can be left to dry out in the field; they are collected from the trees and from the ground and are called 'buni'. Secondly, the berries can be picked when they are ripe and can be dried in the sun on raised trays, protecting them from rain and dew. The product is called 'sun-dried cherry', is of superior quality to 'buni' and fetches a higher price. Both are sent to the mills where their outer skin, parchment and silverskin are all removed in one operation. Dry processing produces poor quality or 'hard' coffee.

Almost all of East Africa's Arabica coffee is processed by the wet method, i.e. the cherry is fed, with water, into a pulping machine which separates the beans from the skins; the parchment coffee is then left for two to four days in fermenting tanks where the sticky mucilage is broken down by naturally occurring micro-organisms and enzymes; the beans are then washed to get rid of the degraded mucilage and then dried in the sun. If wet processing is done inefficiently, good quality cherry produces an inferior end product; the reverse, however, is untrue, for however well a factory is run, it cannot produce top quality coffee from poor quality cherry. In experiments the mucilage has been removed chemically and mechanically and has produced as good quality coffee as the wet method. Both were more expensive, however, so wet processing remains the only practical way to produce good quality coffee.

Each coffee estate and, in Kenya, each co-operative society has its own factory. In Tanzania and Uganda, however, most smallholders process their own coffee, using small hand pulpers; basins, buckets or boxes are used for fermenting. Strenuous efforts have been made, both in Tanzania and Uganda, to introduce central factories in order to get greater quality in the final product. These factories, however, have usually been unpopular; the distances from farm to factory have often been greater than the corresponding distances in Kenya; factory management has often been poor and the prices paid for factory coffee have seldom been much greater than for home-processed coffee. When considering transport problems it should be remembered that dried parchment coffee weighs only one-sixth of the cherry from which it was produced.

Fig. 45: Picking arabica coffee.

them to pick with both hands. The standard measure for pickers on task work is the 'debe'; this was once a heaped square four gallon kerosene tin but in most instances it has been replaced by a standard cylindrical 20 litre (c. 11·3 pint) tin which holds approximately 30 lb (c. 14 kg) of cherry. Hired pickers are usually paid by the 'debe', based on a task of three 'debes' a day. In the flush season, however, experienced pickers often pick as many as ten 'debes' a day, whilst during 'fly picking', i.e. picking either very early or very late in the season from trees which have very few berries, pickers may have a hard job to fill half a 'debe'; in the latter instances pickers are usually paid on a daily rate rather than a task rate.

Cherry must be delivered to the factory for processing on the same day that it is picked. Failure to do this, as was common in the mid-1960s in Kenya when there were not enough factories for the cooperative societies, inevitably leads to poor quality coffee.

Outlined below are the different operations in a coffee factory.

Inspection

Good inspection is essential to ensure that over-ripe, under-ripe, yellow or diseased berries are not included with the good quality ripe cherry (see section on harvesting). Poor quality cherry must be processed in separate batches. Generally speaking, the cooperative societies in Kenya have maintained very high standards of inspection and this has been largely responsible for the high quality coffee which they often produce. Estates are usually unable to insist on quite such high standards of picking; if they did they would often find that they lost their pickers to less demanding estates. They are also reluctant to discard part of their own coffee, whereas the organisers of a cooperative are quite willing to penalise a grower, by discarding his coffee, if it is below the quality of that delivered by other members.

Cherry hopper

A cherry hopper is a large tank; its purpose is to supply a steady flow of cherry to the pulper. There are two main types of hopper: dry feed and siphon feed. Dry feed hoppers are much the more common; at the bottom of each is a rotary paddle, driven from the pulper shaft, which ensures a constant supply of cherry. Siphon feed tanks are full of water; the best, heavy cherry sinks to the bottom and is fed to the pulper, as the name implies, by a siphon.

Pulper

The function of the pulper is to squeeze each berry; this ruptures the skin and expels the beans. The parchment coffee is fed into the sieve or the pregrader and the skins are led off to a skin pit.

There are two types of pulper: drum and disc. Hand operated drum pulpers (see Fig. 46) are used by smallholders in Tanzania and Uganda. Engine driven drum pulpers are sometimes used as repassers (see below). Drum pulpers are easy to adjust but they are not as robust as disc pulpers, nor do they have such a wide range of adjustments; for these reasons disc pulpers are much more common in East Africa. Disc pulpers (see Fig. 47) have 1–4 discs; each disc can pulp about 2 000 lb (c. 900 kg) of cherry in an hour. A flow of water through the pulper is essential to prevent blockage. Disc pulpers need careful adjustment to ensure that they do not nip the beans, nor pass an excessive amount of unpulped cherry to waste or through to the sieve or the pre-grader.

Fig. 46: Hand pulping. The woman is sorting the beans from the pulp.

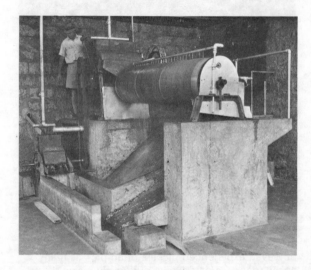

Fig. 47: A disc pulper, immediately in front of the man, and a rotary sieve. Note the repasser at the bottom left.

Sieve

The function of the sieve is to separate the parchment coffee, i.e. the pulped beans, from any unpulped cherry which has passed through the pulper; the parchment coffee falls through the holes of the sieve, aided by a spray of water, whilst the unpulped cherry and any skins pass along the surface of the sieve and are fed into the repasser. There are two types of sieve: the most common is the rotary sieve (see Fig. 47) which is a revolving cylinder; the other type is the shaking sieve which is flat. Both rotary and shaking sieves are mounted at an angle to allow a steady flow of material along them.

Repasser

The repasser (see Fig. 47) is a small pulper, either a single disc or a drum pulper, which receives unpulped cherry from the sieve or from the pregrader (see section on pregrader below). It is always adjusted to give a closer spacing than the main pulper so that it effectively pulps the smaller berries which have passed through.

Pre-washing channels

The name of these channels is an abbreviation of 'pre-fermentation washing channels'; it differentiates them from the final washing channels which are used after fermentation. The functions of pre-washing channels are to wash the parchment coffee, removing bits of skin and any other floating objects, and to grade it according to density. The coffee is fed into one of the long pre-washing channels together with a heavy flow of water; a weir at the far end allows floating objects to pass over the top but retains all the beans which have sufficient density to sink. Floating beans are known as 'lights'; their endosperm is invariably poorly developed and their quality is low; they must be processed separately. In the pre-washing channels the parchment coffee is well stirred with wooden paddles; this is done against the current and encourages the separation of the densest beans which sink quickly and the less dense beans which sink more slowly; the densest beans, called 'firsts', move down the channel slowly whilst the less dense, called 'seconds', flow rapidly down the channel with the current. There are usually two pre-washing channels so that one can be filled whilst grading is being done in the other. It is important that pre-washing is done quickly, otherwise much of the sugar in the mucilage is removed and this may lead to 'onion flavour' developing during fermentation (see section on fermentation below).

'Aagaard' pre-grader

The 'Aagaard' pre-grader (see Fig. 48) performs the functions of both sieves and pre-washing channels, so neither of these are needed in factories which are equipped with these machines.

The pre-grader is a tank which is filled with water and which has a submerged shaking sieve. Parchment coffee enters the pre-grader directly from the pulper and is passed through the tank by a continuous flow of water. Firsts are led off without water, seconds and lights are led off separately with water, and unpulped cherry is sent to the repasser. The main advantages of the 'Aagaard' pre-grader are that it saves labour and water; it also occupies less space, gives a more uniform grading and involves no risk of 'onion flavour' because the beans are in contact with water for no longer than a minute. The only problem of the 'Aagaard' pre-grader is that the short period of submersion does not encourage the removal of browning substances from the beans; it is essential, therefore, that soak tanks are used to avoid poor quality brown coloured coffee. Browning substances and soak tanks are discussed in more detail in a following section.

Fig. 48: A series of three 'Aagaard' pre-graders. One of the pulpers is seen at the bottom right with the electric engine which drives it above. The sieve of the nearest pre-grader is clearly visible; if the machines were operating it would be submerged.

Factory cleanliness

It is most important that at the end of each day's pulping the hopper, the pulper, the sieve, the pre-washing channels, the repasser and the pregrader should all be thoroughly cleaned. If any cherry or beans remain they will start fermenting and will cause an unpleasant flavour in the next batch.

Fermentation

The function of fermentation is the removal of the

Fig. 49: Stirring coffee in the fermenting tanks.

sticky mucilage from the parchment; if the mucilage were not removed there would be handling problems, drying would be impeded and the beans would attract moulds and insects.

Firsts, seconds and lights are led from the pre-washing channels or from the pregrader to separate fermentation tanks. In the conventional, one-stage, dry fermentation the water is drained from the bottom of the tank and the parchment coffee is left for two to three days; it is washed and turned over with wooden spades each morning (see Fig. 49). Tanks are seldom filled completely; this would make it difficult to turn the coffee over. To prevent uneven fermentation the beans must be protected from direct sunlight (strips of hessian are usually laid across the tops of the tanks) and drainage must be efficient. The usual test for completion of fermentation is done by washing some coffee in a basin of water; if it still feels slimy, fermentation is incomplete; if it feels gritty, fermentation is complete.

Fermentation is predominantly the result of micro-organisms attacking the protopectin which is the main constituent of the mucilage. Enzymes, which are already present in the mucilage, give some assistance. The micro-organisms also attack the soluble sugars in the mucilage. The speed of fermentation is dependant upon the suitability of the environment to the desired organisms. If, for example, the water which is used for pulping, pre-washing, pregrading and transporting the beans is drained from the bottom of the fermenting tanks and re-circulated, it contains a higher level of soluble sugars than if fresh water were used and is also warmer; it therefore favours a rapid build-up

of micro-organisms and a rapid degradation of mucilage. At higher altitudes, where the temperature is lower, fermentation may take longer. It can be accelerated either by using re-circulated water or by adding pectic enzymes, in powder form, to the fermentation tanks.

If there is a high sugar content, lactic and acetic acids are formed rapidly and the pH soon falls below 4·3. These organic acids have no harmful effect on the flavour of the coffee. If, however, the sugar content is low, as may happen if pre-washing is prolonged, organic acids form slowly, the pH may not fall below 4·3 and in these conditions propionic acid may develop. If the conditions lead to propionic acid production, they also give rise to 'onion flavour' which severely lowers the quality of the coffee.

Coffee beans contain a browning substance, probably a polyphenol, which can cause poor quality brown beans in the raw. If pre-washing channels are used, most of the browning substance is washed out. If an 'Aagaard' pregrader is used, however, most of the browning substances remain in the beans because contact with water is restricted; therefore there is a considerable risk of brown colours developing during fermentation, even if the beans are washed thoroughly every morning. The best way of avoiding this is to use two-stage fermentation.

The first stage of two-stage fermentation involves the removal of the mucilage in well drained fermentation tanks as described above. After successful testing for complete mucilage removal the coffee is washed to remove the liquor of degraded mucilage, and is then placed in soak tanks where it remains under water for 16 to 20 hours. Sodium metabisulphite may be added to the water to help prevent browning, but this substance is used in only a few factories. Soaking ensures that the browning substances are removed from the bean and therefore gives the best quality coffee. At the time of writing about half of the estates and about half of the cooperatives in Kenya are using two-stage fermentation. Two-stage fermentation is essential when an 'Aagaard' pregrader or re-circulated water is used because the browning substances will have had a smaller chance of diffusing out of the beans.

Final washing and grading
After fermentation, or sometimes in the middle of two-stage fermentation between fermenting and soaking, the parchment coffee is washed in a long channel to remove all the broken-down mucilage

Fig. 50: Final washing. Note the fermenting tanks on the right.

with handles at each end are usually used (see Fig. 51). The coffee must be frequently stirred by hand to prevent uneven drying and parchment cracking; the latter can cause some discolouration of the beans and is especially likely if soak tanks have been used. At the same time as stirring, the beans are hand sorted; broken, diseased and discoloured beans are discarded together with any undesirable objects such as bits of skin or unpulped cherry. After skin drying the moisture content of the beans should have fallen from over 50% to about 45%.

After skin drying the parchment coffee is taken to the drying tables (see Fig. 52). Most drying tables consist of sisal or hessian strips laid on chicken wire; the chicken wire is stretched across a wooden framework. For even drying it is most important that the surface of the table should be flat so that a uniform depth of coffee can be spread over it; unfortunately many tables are

(see Fig. 50). The procedure is very similar to pre-washing; the coffee is pushed against the current of water by means of wooden paddles; a further grading of the densest and the less dense beans is done simultaneously.

Drying

Without some sun drying, top quality coffee cannot be produced. 48 hours is considered the minimum period of sunlight; without this, brown colours appear in the raw. Poor drying can cause poor quality coffee, so drying is a most critical stage in coffee processing.

Immediately after final washing and grading the parchment coffee must be skin dried; this involves the removal of water from the surface of the parchment and from between the parchment and the bean. The coffee is laid on coffee tray wire (hessian or sisal paper would rot) in layers no more than 1 in (c. 2·5 cm) deep. Portable trays

Fig. 51: Parchment coffee being fed onto skin-drying trays from the final washing channel.

Fig. 52: Turning coffee over on the drying tables.

allowed to fall into disrepair, or are constructed with insufficient support for the chicken wire; the result is an undulating surface which conceals deep pockets of coffee.

Final drying usually takes 5–7 days, but in poor weather it may take much longer. During this time the beans pass from the soft white stage (30–44% moisture), through the black stage (12–29% moisture), to the hard grey-green or ideally grey-blue stage of 10–11% moisture. Over-dried coffee is slightly yellow. For efficient drying the layer of coffee must be no more than an inch thick; it must be stirred and hand-picked regularly; it must be protected from bright sun in the middle of the day to prevent uneven drying and parchment cracking; it must be covered at night and when it rains. Coffee must never be completely enclosed by a waterproof layer; only the top should be covered.

Conditioning
After drying in the sun it is important to let the small moisture differences in individual beans even up. This can be done either by heaping the parchment coffee on a wooden floor for two weeks and turning it every day with wooden spades, or by using conditioning bins. Conditioning bins have perforated floors; some have fans which blow air up

through the coffee. They can also be used for storing coffee at a constant moisture if the weather is bad during drying, or for drying coffee from the hard black stage (12–15% moisture) to the fully dry stage. They must never be used for long periods of drying because this would deprive the beans of the sunshine they need to give them a good colour. After conditioning the parchment coffee is put in clean, clearly marked bags and stacked.

Yields

The average yields of clean coffee are about 5 cwt of clean coffee per acre (c. 630 kg/ha); good management (including good disease control) should produce regular yields of about 10 cwt per acre (c. 1 250 kg/ha) whilst a very few estates have reached regular yields of 1 ton per acre (c. 2·5 t/ha).

Multiple stem coffee should come into bearing about two years after planting, and into full bearing about four years after planting.

Quality

Coffee quality is judged at three stages: after the parchment and most of the silverskin have been removed, after roasting and after grinding and mixing with hot water; these stages are called, respectively, the raw, the roast and the liquor.

The raw
Size is important only because large beans often fetch higher prices owing to the whim of the consumer. It has been conclusively shown that size has no effect on the liquor; some coffees, e.g. those from Solai in Kenya, produce excellent coffee despite their small size. Colour, however, has a pronounced effect on both the roast and the liquor; bluish or greyish-blue beans produce the best roasts and liquors; greyish-green or greenish beans are acceptable but any with even a faint trace of brown produce a dull roast and a poor liquor. Brown beans may be the result of poor processing or may be associated with an imbalance of potassium and magnesium in the soil.

One or more of the following comments may accompany the report on the quality of a consignment of coffee:

1 *Coated*
Coated beans pass through the mills with much of their silverskin still attached; they are produced by yellow cherry from overbearing trees or those that have suffered from drought.

2 *Foxy*
This indicates a reddish colour which may be caused by over-ripe cherry or by an undue delay between picking and pulping.

3 *Light*
Light beans have incompletely developed endosperm; they are often the result of poor growing conditions but any batch of cherry contains a certain proportion of light beans; they must be separated from the good quality coffee during grading.

4 *Ragged*
Ragged beans are poorly shaped, often being 'boat-shaped', i.e. thin and with somewhat pointed ends; they are caused by poor growing conditions.

5 *Green water damaged*
Re-wetting during storage, but sometimes during drying, discolours the beans; they are then usually classified as green water damaged.

6 *Stinkers*
A stinker is a bean with a taint so bad that it can spoil the liquor of a good sample of coffee. Stinkers are associated with the death of the embryo. The most likely cause of this, as has been recently revealed, is damage by fruit flies. It can also be caused by overheating during processing.

7 *Amber beans*
These are caused by iron deficiency induced by a high soil pH.

8 *Triple or double centre-cuts*
The centre-cut is the crease on the flat side of each bean; when there is sudden rain after a long drought at the time the cherry is ripening, the beans may develop abnormally with more than one centre-cut; such beans are said to have triple or double centre-cuts; they have more silverskin than usual inside the bean.

9 *Elephant beans*
Elephant beans are formed when two ovules develop in one ovary, causing two beans to grow together where there should be one; during milling they fall apart causing an uneven sample; their cause is genetic.

10 *Peaberry*
When one bean in a berry fails to develop the other one usually develops spherically; if separated from the rest of the consignment peaberry can produce very good quality coffee; its main cause is genetic, but it may be due to faulty pollination.

11 *Triangular beans*
These are formed when, owing to another genetic abnormality, there are three beans present in one berry.

The roast
Roast beans should be shiny; a dull appearance is often caused by a brown raw and invariably results in a poor liquor. The centre-cut should be white and not brown, as sometimes happens after bad fermentation or drying. Soft beans sometimes occur in the roast; they are often the result of under-drying. Pales are yellow roast beans; they are often produced by yellow cherry from trees under severe physiological stress. Stinkers usually appear as yellow beans in the roast.

The liquor
The liquor should have a certain amount of acidity and should be full bodied. There are many taints or undesirable flavours which may be mentioned in a report on quality; they include 'onion flavour', caused by excessive pre-washing or delayed skin drying, 'musty flavour', caused by poor storage, and 'bricky flavour', which is caused by the use of the insecticide BHC. 'Bricky flavour' is hardly ever reported nowadays because the use of BHC on coffee has been virtually eliminated, but it may be produced in railway trucks which have carried this chemical.

Pests

Predators and parasites, mostly wasp-like insects of the super-family *Ichneumonidea* or flies of the family *Tachinidae,* effectively check many coffee pests. Overall applications of persistent contact insecticides, especially DDT and dieldrin, have often upset the balance of these predators or parasites and the pests which they attack. This is mainly due to the residual effect of these insecticides, but is also due to the fact that they are more efficient at killing predators and parasites than they are at killing leaf miners, mealybug, scales and mites. Overall applications of DDT or dieldrin have often resulted in considerable increases in the numbers of these pests.

Parathion, a contact insecticide and stomach poison, is not persistent and is therefore useful in coffee because it has no long term effect on predators or parasites; it is widely used on estates but is much too toxic to mammals for use on smallholdings. Dicrotophos and methomyl are systemic insecticides and therefore have little effect on predators or parasites; they have recently been recommended on coffee estates but cannot be recommended for smallholders owing to their high mammalian toxicity. Fenitrothion and fenthion are insecticides which are trans-laminar, i.e. they penetrate leaves and the epidermis of the berries but are not translocated in the vascular system; they have a low mammalian toxicity and are not persistent on the surface of the tree; they are therefore useful insecticides for smallholders. Malathion and diazinon are now used only to a limited extent in coffee; the former is effective against only a small proportion of coffee pests while the latter has proved to be more expensive than most other insecticides. BHC must never be used on coffee because it causes a 'bricky flavour' in the liquor.

Leaf miner. *Leucoptera meyricki* and *Leucoptera caffeina*

Leaf miner is the most important coffee pest in Kenya, notably east of the Rift Valley. In Tanzania it is a seasonal pest which seldom causes as much damage as it does in Kenya; the reasons for this are that in Tanzania callus tissue develops very quickly in the leaves, trapping the larvae, and the heavy rainfall often drowns some of the larvae in their mines.

Leucoptera meyricki is the most important species, predominating in unshaded conditions. *Leucoptera caffeina* predominates only in shaded coffee. Serpentine leaf mines are caused by the larvae of a fly and are of no economic importance.

Small, white, nocturnal moths, about $\frac{1}{6}$ in (c. 0·4 cm) long, lay eggs on the upper surface of fully hardened leaves. Upon hatching the larvae bore into the leaf and feed on the pallisade tissues; they form communal mines and these are seen as brown blotches on the upper surface of the leaves. The reduction of photosynthetic tissue caused directly by the mines is less important than the leaf fall which even a small mine can induce. When they pupate the larvae emerge, descend on a silken thread and form small, white, H-shaped cocoons on the undersides of the leaves or on the ground in a shaded position.

Leaf miners are normally efficienctly checked by parasites and, to a lesser extent, by predators. If the balance between pests and parasites gets out of hand, i.e. when 20 or more moths fly out of a tree when it is shaken, chemical control is necessary. The best time to spray is when as few of the pests as possible are in the pupal stage, protected by their cocoons; this can be arranged by spraying a week after the peak moth population or by spraying when most of the cocoons are dry and empty when squeezed. A second spray may be necessary two or three weeks later to back up the first. Parathion, dicrotophos, fenitrothion and fenthion are all recommended.

Antestia. *Antestiopsis spp.*

This is the most important coffee pest in the wetter areas, i.e. in Tanzania and west of the Rift Valley in Kenya. Antestid bugs, which should not be confused with ladybirds, are shield shaped and are speckled dark brown, orange and white (see Fig. 53); they go through five nymphal stages, each of which resembles the adult stage. They are sucking pests and by this method of feeding introduce a *Nematospora* fungus. They may feed on flower buds (causing blackening and abortion), on the young green berries (causing them to drop), on older berries (causing 'zebra stripes' on the parchment and soft white beans) or, if no fruit or flowers are present, on the growing points (causing short internodes and fan branching). In western Kenya the Antestia population only builds up significantly during May and June; by this time the berries have passed the young green stage, so berry drop is rare; bean discolouration, however, is common.

Fig. 53: Antestia.

Open pruning discourages Antestia. Parathion, dicrotophos, methomyl, fenitrothion and fenthion are all effective insecticides. A method of testing for Antestia has been developed; it involves spreading a sheet around the base of a tree, spraying it with pyrethrum, waiting 15 minutes and then shaking the tree and counting the number of dead or stunned bugs which fall on to the sheet; when there is an average of more than one bug per tree spraying is recommended.

Berry borer. *Hypothenemus* (*Stephanoderes*) *hampei*

This is a low altitude pest that was formerly restricted to Robusta coffee and to the lower Arabica areas of Tanzania. Larvae feed in the beans, as many as twenty per bean, boring tunnels and imparting a blue colour to the beans. Female beetles survive the season when no crop is present by living in 'buni' which has been left on the trees or on the ground. During the 1960s, berry borers reappeared in Kenya, largely owing to the amount of coffee affected by coffee berry disease which had been left in the field to become 'buni'.

Regular picking and field hygiene are almost always enough to control berry borer. Heavy shade, however, may cause outbreaks of this pest by discouraging its natural enemies; in this case overall dieldrin sprays are effective although they involve the risk of a build up of scales, leaf miners, mites and mealybugs.

Berry moth. *Prophantis smaragdina*

This is a seasonal pest which seldom causes serious damage. Larvae feed on the berries, hollowing them out and spinning silken webs around clusters of berries. If present in large numbers they may need to be controlled by using parathion or fenitrothion sprays; these must be applied at an early stage before the larvae are protected by their webs.

Mealy bug. *Plannococcus kenyae*

Mealybugs form a white mealy mass around flower clusters, fruits and growing tips. They posed a great threat to Kenya's coffee industry east of the Rift Valley in the 1920s and 1930s. In 1938 predators which had been collected from Uganda were released in Kenya and one of these, a small wasp *Anagyrus* near *kivuensis*, has controlled mealybug effectively ever since. In cold weather or after overall spraying of persistent insecticides or after a very dry, dusty period, mealybugs may multiply due to lack of predators; they should then be controlled by preventing their attendant ants climbing the trees; this is done by painting a dieldrin band around each trunk just before the main rains, i.e. the time that the adult beetles lay their eggs on the bark; methylene blue is usually added to the dieldrin as a marker. Dieldrin banding is effective for two to five years.

Giant looper, (*Ascotis selenaria reciprocaria*) and Green looper, (*Epigynopteryx stictigramma*)

These are large leaf-eating caterpillars which walk with a characteristic looping motion; they feed mainly on suckers and new flushes of leaves. Loopers are often found on estates which have used repeated parathion sprays because the larvae are markedly resistant to this insecticide. The giant looper can be found at most altitudes but the green looper occurs only on high shaded estates. Methomyl is the only insecticide recommended for loopers as an overall spray. An alternative method of control is to spread a ring of DDT powder around the trunk of each tree and then to spray with pyrethrum; the pyrethrum causes the larvae to fall to the ground and they then pick up a lethal dose of DDT when they return to the trees.

Scales

Green scales, *Coccus alpinus,* are quite common on shoot tips and main leaf veins. Dieldrin banding controls their attendant ants, thereby allowing parasites and predators to control the scales. Without ants the scales also drown in their own honey-dew. Star scales, *Asterolecanium coffeae,* feed on rough mature bark and often occur where dust from a nearby road has discouraged their natural enemies. Star scales can be controlled by painting the affected bark with tar oil, or they can be discouraged by checking the dust; barriers of vegetation or oil on the road are recommended for the latter. White waxy scales, *Ceroplastes brevicauda,* and brown (or helmet) scales, *Saissetia coffeae,* are less frequently seen than green and star scales.

Thrips

Several species of thrips attack coffee but *Diarthrothrips coffeae* is the most common. Thrips used to be a major pest, doing much damage to leaves when trees suffered from moisture stress in hot dry periods. Outbreaks become much less widespread with the use of mulch which prevents severe moisture stress; thrips can still be a considerable problem, however, in unmulched coffee, and have caused considerable damage in

Kenya during the late 1960s. Dicrotophos controls thrips effectively; parathion, fenitrothion and fenthion may also be used.

Diseases

Coffee berry disease (CBD)
This fungal disease is caused by *Colletotrichum coffeanum*. CBD was first identified in Kenya in 1922 near Hoey's Bridge and in the next decade it spread throughout western Kenya down to Sotik, causing most of the coffee in that area to be abandoned. There was a small outbreak in Nyeri in 1939 and it became serious in upper Kiambu from 1951, but until the mid-1950s it was considered to be only a disease of the coffee growing areas west of the Rift Valley and of the high altitude areas east of the Rift Valley. However, in the mid-1950s and especially in the 1960s CBD spread rapidly until all but the lowest coffee growing areas in Kenya were severely affected. In 1966 it was reported for the first time on the slopes of Mt. Kilimanjaro in Tanzania and it soon spread to Mt. Meru and the Pare Mountains. At the time of writing it has not appeared in the Southern Highlands of Tanzania but there is little doubt that it will. CBD has caused more concern than any other plant disease in East Africa; during the mid-1960s many people thought that it might cause the collapse of East Africa's coffee industry. During the 1960s, however, an intensive research programme led to the discovery of effective, if expensive, control.

Wherever coffee is grown in the world *C. coffeanum* is found on the tree as a harmless saprophyte. In East Africa, the Congo, Cameroon, Angola and Guatemala, however, a parasitic strain of *C. coffeanum* has developed and it is this strain which causes CBD. The parasitic strain causes much more damage in East Africa and Cameroon than it does in the other parts of the world; the reason for this is not yet known.

Symptoms
1 Flowers. Dark brown blotches or streaks appear on the flowers. The whole flower soon becomes invaded and dies. Flowers are very short lived so they usually escape infection; damage to flowers is therefore relatively unimportant.

2 Green berries. Small dark sunken patches appear; in wet conditions and if the berry is expanding (i.e. between approximately the 8th and 18th week after flowering) the sunken

Fig. 54: CBD lesions on green berries.

patches spread rapidly and the berry soon becomes blackened and mummified; beans fail to develop inside such berries. Pink spores can often be seen on blackened berries during wet weather. When berries have completed their expansion phase they are moderately resistant to CBD until they ripen; the dry or cool weather of January/February or July/August/September usually coincides with green berry maturity and further discourages the disease. In dry or cool conditions, which are unfavourable for the spread of the disease, CBD lesions on mature green berries do not develop beyond the brown scab stage; the fruiting bodies (*acervuli*) of *C. coffeanum* can easily be seen on these lesions as small raised black spots.

3 Ripe berries. Again, dark sunken patches appear and the fruiting bodies can usually be seen on them. In wet weather the entire surface of the berry is soon blackened; the beans are unharmed but pulping is very difficult, if not impossible. Some growers have successfully hulled coffee which was attacked by CBD when the berries were ripe; such coffee is saleable, though not of the highest quality because it has not been processed by the wet method. When CBD attacks ripe berries it is sometimes called brown blight; it was once thought to be caused by a different pathogen.

4 Leaves. Brown patches occur on the margins of the leaves; the patches are often characterised by concentric rings. Leaf lesions are also called

brown blight but they are uncommon in East Africa.

Life of the fungus

C. coffeanum lives mainly in the first eight internodes of maturing bark on lateral branches; this enables it to survive the period when there is no crop on the trees; it does no apparent damage in this area. Fruiting bodies appear on the surface of the young bark and in wet weather these produce pink spores. At least 24 hours of wet conditions are needed before fruiting bodies (either on branches or on berries) produce spores. Initial infection of berries is from the branches, but later in the season the berries themselves provide a much more potent reservoir for infecting other berries; it has been shown that two berries can produce more spores than the entire branch surface of one tree. Spores are spread by rain splash; this method is only important within the tree; the movement of the spores is mostly in a downward direction. CBD is spread from tree to tree, from farm to farm, from estate to estate or even from one coffee growing area to another by pickers' hands, birds' feet or by infected seedlings.

Spores can live for several days while they are still attached to the fruiting bodies, but once they have been removed they live for less than 24 hours. The infected surface must remain wet for about 5 hours. During this period the temperature must be between 59° and 82°F (c. 15·0–27·7°C), otherwise the spores fail to germinate. Infection can occur during the day or the night because night temperatures seldom fall below 59°F during the rainy season except in the higher altitude coffee areas. The lesser importance of the disease in low altitude areas is considered to be due more to the smaller amount of pathogen in the bark than to lack of suitable infection periods.

Multiple stem pruning and its effect on CBD

Multiple stem pruning, which was widely introduced in the 1950s, played a large part in helping the increase of CBD. (Some considered that a change in the rainfall regime was entirely responsible but few people now agree with this theory). Multiple stem pruning has two major disadvantages as far as CBD is concerned. The first is that it often produces two crops a year; if the early crop becomes infected it releases spores which infect the main crop and the main crop can later infect the next early crop; there is thus a continuity of high potential infection. With single stem pruning there is virtually no early crop so there are no berries on the trees for three or four months each year; this means that initial infection of the main rains crop must come from fruiting bodies on the branches and, as mentioned above, these produce fewer spores than do berries. The second disadvantage of multiple stem pruning is the presence of tall old stems; the heads of these stems almost always harbour more CBD than other parts of the trees and the spores from these heads are in an ideal position for being washed down on to the crop below. Several reasons have been suggested to explain the increased incidence of CBD in old heads: it may be due to the dense nature of the foliage in old heads (this would create a moist microclimate suitable for CBD development); it may be due to the fact that old heads collect more dew than other parts of the trees; it may be due to the fact that old heads catch the early morning sun more than other parts of the trees, thus enabling them to reach a temperature favourable for spore germination more quickly; it may be due to poor spray penetration (both the height and the density of the heads may play a part here); it may be due to fungicide being washed out of old heads more readily (lower parts of the trees, in contrast, receive fungicide from the heads); lastly it may be due to poor nutrition of old heads (overbearing stress is more noticeable in this part of the tree and this fact lends support to the nutrition theory). The relative importance of these possible reasons has not yet been explained but many estate owners have already taken the precaution of eliminating old heads by adopting a capped multiple stem pruning system (see section on pruning).

Spray timing

After some early controversy and after some highly unsuccessful recommendations it is now generally agreed that CBD can be effectively controlled by the timely application of fungicides during wet weather. The exact recommendations have yet to be finalised but it is unlikely that they will ever be rigid; fungicide applications will only be necessary when the weather is sufficiently wet and when berries are in a susceptible stage. The current recommendation in Kenya, where almost all the CBD work has been done, is to spray at flowering at the beginning of the long rains and then to protect the crop until it is past the berry expansion phase, or until the weather becomes unfavourable for CBD. Using captafol, the most effective fungicide, this means that four or five applications are necessary during the long rains. Applications should commence again with the onset of the short rains (or in mid-October if they have not started by then); only two or three applications should be needed in these rains.

Fungicides

Captafol has proved to control CBD very effectively. It has two important advantages when compared with the other important fungicide used for CBD, namely copper: its effect is longer lasting and applications are only necessary once every four weeks (c.f. every three weeks if copper is used). In addition it has a tonic effect which cannot be fully explained; berries expand more rapidly than usual and the general health of the tree often improves after captafol sprays. Copper controls CBD slightly less effectively than captafol; it must be applied more frequently and has less tonic effect, although it does improve leaf retention. 'Benlate' shows considerable promise and has recently been given a general recommendation. 'Benlate' also has a tonic effect.

Leaf rust

This disease is caused by the fungus *Hemileia vastatrix*. Before CBD spread beyond the high altitude coffee growing areas, leaf rust was considered the most important coffee disease in East Africa. It has been seen less frequently since the widespread adoption of intensive spraying programmes against CBD but it is still a potential hazard, especially in warm, wet areas.

Symptoms

Yellow spots first appear on the leaves; these can be seen from above and from below. They enlarge rapidly, forming circular lesions which have orange pustules on the lower surface. Leaf rust causes a direct and an indirect reduction in photosynthetic tissue; the indirect reduction is caused by leaf fall which can be caused by the presence of only one lesion on a leaf; leaves may be retained on the tree for as long as a year in the absence of rust but they may last for only four months if rust is severe.

Dispersal

Unlike all other rust diseases, leaf rust is spread by rain splash and not by air. Outbreaks of the disease occur only after rainstorms of sufficient intensity to wet the lower surfaces of the leaves. Such storms splash spores on to the lower surfaces; the spores then germinate, provided that it is sufficiently warm, damp and dark, and enter the stomata. Lesions appear four or five weeks after infection. It has been shown that rainstorms of

0·3 in (c. 7·5 mm) or more are needed to cause an outbreak. The severity of an outbreak depends not only upon the amount of rain one month before but also on the amount of inoculum present when heavy rain started; the last fact greatly affects the anti-rust spraying programme.

Long range dispersal of leaf rust may be by infected seedlings or by a wasp predator of a minute midge which inhabits pustules; the wasp becomes covered in spores as it oviposits.

Control

Copper is almost 100% effective in controlling the disease; captafol is much less effective. In areas where both CBD and leaf rust occur captafol should not, therefore, be used alone. The important times to spray against leaf rust are just before the rains start and, in areas where leaf rust is serious, about three weeks after the first rainstorm of over 0·3 inches. It should be stressed that copper has also been used as a tonic spray, to improve the health of the tree and to reduce leaf fall, in areas where neither rust nor CBD occurred, e.g. the higher altitude coffee areas of Tanzania and, before the increase of CBD, of Kenya. Cultural measures to discourage the disease include open pruning and good weeding; these allow free air circulation and rapid drying.

Elgon die-back

This disease occurs mostly in the coffee growing areas west of the Rift Valley and at high altitudes. The blackened leaves remain on the lateral branches, in contrast to the shed leaves of branches affected by other forms of die-back. The disease is thought to be caused by the same fungus that causes CBD, *Colletotrichum coffeanum*. The fungus normally penetrates no further than the bark, but in trees which have been affected by Elgon die-back it is likely that it penetrates into the wood. Regular copper sprays, shade and resistant varieties (Kapretwa Series 'A' and Geisha) give some reduction of the disease although they do not control it fully.

Fusarium bark diseases

There are three different types of attack: scaly bark, bark disease and collar rot. *Fusarium* bark diseases can cause much damage in warm conditions and much coffee has been abandoned owing to serious attacks around Tukuyu in the Southern Highlands of Tanzania and Taita in Kenya. It is thought to occur only when husbandry standards are very poor or where trees have been damaged by herbicides. Affected trees must be uprooted and burned.

Crinkle leaf
This disorder is often confused with, and sometimes improperly called, 'Hot and Cold' (see below). The symptoms are distorted margins of leaves; the distortions are often symmetrical when opposite pairs of leaves are considered. Leaves are considerably reduced in size. Shade or captan sprays reduce the incidence of crinkle leaf, but the latter are expensive and should be confined to suckers. The cause of the disease is unknown.

Hot and Cold
The symptoms of this condition are small leathery leaves with yellow margins. 'Hot and Cold' is a poor name for this disease, because although it occurs at high altitudes and after cold spells, it is also common whenever plants are under stress, e.g. after transplanting. Shade prevents the occurrence of 'Hot and Cold'.

Armillaria root rot
This fungal disease is caused by *Armillaria mellea.* It sometimes occurs in coffee at the higher altitudes but is more important in tea. It is discussed in more detail in chapter 33.

Bibliography

General
1 *A Handbook on Arabica Coffee in Tanganyika,* Editor **J. B. D. Robinson.** Tanganyika Coffee Board, 1964.

2 *Coffee Growers' Handbook,* Editor **C. J. Ombwara,** Equitorial Publishers, Nairobi, 1968.

Physiology
3 **Wormer, T. M.** (1964). *Normal and abnormal development of coffee berries.* Kenya Coffee, 29:91.

Shade
4 **Robertson, J. K.** (1954). Acacia spp. *as shade trees for coffee.* E. Afr., agric. J., 19:272.

Varieties
5 **Blore, T. W. D.** (1965). *Arabica coffee selections and genetic improvement in Kenya.* Kenya Coffee, 30:491.

6 **Jones, P. A.** (1956). *Notes on the varieties of* Coffea arabica *in Kenya.* Kenya Coffee, 21:305.

Propagation
7 **Fernie, L. M.** (1950). *The preparation of nursery beds and the sowing of coffee seed.* E. Afr. agric. J., 16:56.

8 **Fernie, L. M.** (1961). *The possibilities of a new nursery technique for coffee.* Kenya Coffee, 26:379.

Pruning
9 **Bird, C. E. F.** (1966). *Systematic pruning by blocks in multiple-stem coffee.* Kenya Coffee, 31:167.

10 **Bucher, H.** (1966). *The Burka pruning system.* Kenya Coffee, 31:175.

11 **Fernie, L. M.** (1966). *Coffee pruning.* Kenya Coffee, 31:153.

12 **Solly, N. R.** (1966). *Pruning multiple-stem coffee in areas east of the Rift Valley in Kenya between the altitudes of 5 400 and 6 000 ft.* Kenya Coffee, 31:163.

13 **Von Roretz, F. E.** (1966). *Pruning on Selian Estate, Arusha.* Kenya Coffee, 31:497.

Weed control
14 **Jones, P. A.** and **Wallis, J. A. N.** (1963). *A tillage study in Kenya coffee, II—The long-term effects of tillage practices upon yield and growth of coffee.* Emp. J. exp. Agric., 31:243.

15 **Mitchell, H. W.** (1967). *The possibility of weed control with minimum cultivation.* Kenya Coffee, 32:232.

16 **Mitchell, H. W.** (1969). *Herbicides for weed control in coffee.* Kenya Coffee, 34:130.

17 **Pereira, H. C.** and **Jones, P. A.** (1954). *A tillage study in Kenya coffee, I—The effects of tillage practices on coffee yields.* Emp. J. exp. Agric., 22:231.

18 **Pereira, H. C.** and **Jones, P. A.** (1954). *A tillage study in Kenya coffee, II—The effects of tillage practices on the structure of the soil.* Emp. J. exp. Agric., 22:323.

19 **Pereira, H. C.** *et. al.* (1964). *A tillage study in Kenya coffee, IV—The physical effects of contrasting tillage treatments over thirty consecutive cultivation seasons.* Emp. J. exp. Agric., 32:31.

20 **Wallis, J. A. N.** (1961). *The place of herbicides in the management of Kenya Coffee, I—General weed control.* Kenya Coffee, *26*:77.

21 **Wallis, J. A. N.** (1961). *The place of herbicides in the management in Kenya Coffee, II-Control of perennial grasses.* Kenya Coffee, *26*:159.

22 **Wallis, J. A. N.** and **Blore, T. W. D.** (1964). *Weeds in coffee. Kenya Coffee.* 29:299.

Mulching
23 **Mitchell, H. W.** (1968). *Grasses for mulching coffee.* Kenya Coffee. *33*:327.

24 **Robinson, J. B. D.** and **Chenery, E. M.** (1965). *Magnesium deficiency in coffee with special reference to mulching.* Emp. J. exp. Agric., *26*:259.

25 **Robinson, J. B. D.** and **Hosegood, P. H.** (1965). *Effects of organic mulch on fertility of a latosolic coffee soil in Kenya.* Expl. Agric., *1*:67.

26 **Robinson, J. B. D.** and **Wallis, J. A. N.** (1960). *Recommendations for the application of organic mulches in coffee.* Kenya Coffee. *25*:14.

Fertilisers and manures
27 **Jones, P. A.** *et. al.* (1960). *Fertilisers, manures and mulch in Kenya coffee growing.* Emp. J. exp. Agric., *28*:335.

28 **Pereira, H. C.** and **Jones, P. A.** (1950). *Maintenance of fertility in dry coffee soils.* E. Afr. agric. J., *15*:174.

29 **Pereira, H. C.** and **Jones, P. A.** (1954). *Field responses by Kenya coffee to fertilisers, manures and mulches.* Emp. J. exp. Agric., *22*:23.

30 **Robinson, J. B. D.** (1956). *The influence of fertilisers and manure on the pH reaction of a coffee soil.* E. Afr. agric. J., *22*:76.

31 **Robinson, J. B. D.** (1967). *Yield and response to fungicide and fertiliser of peasant grown Arabica coffee on Mt. Kilimanjaro, Tanzania, I—Review and description of experimental sites and methods.* E. Afr. agric. for. J., *32*:426.

32 **Robinson, J. B. D.** (1967). *Yield and response to fungicide and fertiliser of peasant grown Arabica coffee on Mt. Kilimanjaro, Tanzania, II—Effects of cultural conditions on yield.* E. Afr. agric. for. J., *32*:433.

33 **Robinson, J. B. D.** (1967). *Yield and response to fungicide and fertiliser of peasant grown Arabica coffee on Mt. Kilimanjaro, Tanzania, III—Effects of treatments and season on yield'* E. Afr. agric. for. J., *33*:123.

See also references 37 and 41.

Irrigation
34 **Blore, T. W. D.** (1966). *Further studies of water use by irrigated and unirrigated coffee in Kenya.* J. Agric. Sci., *67*:145.

35 **Pereira, H. C.** (1957). *Field measurements of water use for irrigation control in Kenya coffee.* J. Agric. Sci., *49*:459.

36 **Wallis, J. A. N.** (1963). *Water use by irrigated coffee in Kenya.* J. Agric. Sci., *60*:381.

Processing and quality
37 **Blore, T. W. D.** (1965). *Some agronomic practices affecting the quality of Kenya coffee.* Kenya coffee, *30*:553.

38 **Bowden, J.** *et al.* (1964). *Effect of lindane on the liquor of Arabica coffee as evaluated by different liquorers.* E. Afr. agric. for. J., *30*:40.

39 **Brownbridge, J. M. C.** and **Wootton, A. E.** (1965). *Recent developments in coffee processing.* Kenya Coffee, *30*:187.

40 **Ghosh, B. N.** (1966). *A review of the mechanisation problems of coffee processing in in East Africa.* Kenya Coffee, *31*:245.

41 **Jones, P. A.** (1964). *Research into problems of coffee quality in Kenya.* Kenya Coffee, *29*:489.

42 **Northmore, J. M.** (1965). *The use of sodium bisulphite in coffee fermentation.* Kenya Coffee, *30*:285.

43 **Northmore, J. M.** (1965). *Some factors affecting the quality of Kenya coffee.* Kenya Coffee, *30*:373.

44 **Northmore, J. M.** (1966). *Sodium bi-sulphite in coffee fermentation, II.* Kenya Coffee, *31*:217.

45 **Northmore, J. M.** (1967). *Pales in the roast.* Kenya Coffee, *32*:67.

46 **Northmore, J. M.** (1968). *Stinkers. A preliminary investigation.* Kenya Coffee, *33*:131.

47 **Northmore, J. M.** (1969). *Stinkers, II—Over-fermented beans and 'stinkers' as defectives of Arabica coffee.* Kenya Coffee, *34*:302.

48 **Robinson, J. B. D.** (1960). *Amber beans.* Kenya Coffee, *25*:91.

49 **Wallis, J. A. N.** (1968). *Coffee processing. Factory management.* Kenya Coffee, *33*:97.

50 **Wallis, J. A. N.** (1968). *Coffee processing.* Kenya Coffee, *33*:167.

51 **Wootton, A. E.** (1961). *Fermentation and onion flavour.* Kenya Coffee, *26*:126.

52 **Wootton, A. E.** (1963). *The fermentation of coffee, I.* Kenya Coffee, *28*:239.

53 **Wootton, A. E.** (1963). *The fermentation of coffee, II.* Kenya Coffee, *28*:277.

54 **Wootton, A. E.** (1963). *The fermentation of coffee, III.* Kenya Coffee, *28*:317.

55 **Wootton, A. E.** *et al.* (1968). *The sun drying of Arabica coffee.* Kenya Coffee, *33*:261.

56 **Wootton, A. E.** and **Brownbridge, J. M. C.** (1966). *The use of pectic enzyme preparations in coffee processing.* Kenya Coffee, *31*:253.

57 **Wormer, T.M. and Njuguna, S. G.** (1966). *Bean size and shape as quality factors in Kenya coffee.* Kenya Coffee, *31*:397.

Storage

58 **McCloy, J. F.** and **Wallis, J. A. N.** (1966). *Arabica coffee storage, III—The design of coffee stores.* Kenya Coffee, *31*:301.

59 **Wallis, J. A. N.** (1966). *Arabica coffee storage. A review of the problems in Kenya.* Kenya Coffee, *31*:257.

Pests

60 **Crowe, T. J.** (1962). *The star scale.* Kenya Coffee, *27*:9.

61 **Crowe, T. J.** (1962). *The white waxy scale.* Kenya Coffee, *27*:93.

62 **Crowe, T. J.** (1964). *Coffee leaf miners in Kenya, I—Species and life histories.* Kenya Coffee, *29*:173.

63 **Crowe, T. J.** (1964). *Coffee leaf miners in Kenya, II—Causes of outbreaks.* Kenya Coffee, *29*:223.

64 **Crowe, T. J.** (1964). *Coffee leaf miners in Kenya, III—Control measures.* Kenya Coffee, *29*:261.

65 **Crowe, T. J.** and **Leeuwangh, J.** (1965). *The green looper.* Kenya Coffee, *30*:433.

66 **Evans, D. E.** (1965). *The coffee berry borer in Kenya.* Kenya Coffee, *30*:335.

67 **Evans, D. E.** (1967). *Thrips control.* Kenya Coffee, *32*:51.

68 **Foster, R.** (1955). *Laboratory observations on the effects of insecticides on the white coffee borer beetle.* E. Afr. agric. J., *21*:6.

69 **McNutt, D. N.** (1967). *The white coffee borer: its identification, control and occurrence in Uganda.* E. Afr. agric. for. J., *32*:469.

70 **Tapley, R. G.** (1955). *Insecticidal laquers and white coffee-borer control.* E. Afr. agric. J., *20*:145.

71 **Wheatley, P. E.** (1963). *The giant coffee looper.* E. Afr. agric. for. J., *29*:143.

Diseases

72 **Bock, K. R.** (1956). *Investigations on coffee berry disease; laboratory studies.* E. Afr. agric. J., *22*:97.

73 **Firman, I. D.** (1963). *Elgon die-back.* Kenya Coffee, *28*:9.

74 **Firman, I. D.** (1964). *Fusarium diseases of coffee in Kenya.* Kenya Coffee, *29*:353.

75 **Firman, I. D.** and **Wormer, T. M.** (1963). *Recommendations for the control of crinkle-leaf.* Kenya Coffee, *28*:463.

76 **Griffiths, E.** (1968). *Control of coffee berry disease.* Kenya Coffee, *33*:393.

77 **Nutman, F. J.** and **Roberts, F. M.** (1969). *Coffee berry disease: epidemiology in relation to control.* Expl. Agric., *5*:271.

78 **Wallace, G. B.** and **Wallace, M. M.** (1955). *The bark diseases of coffee.* E. Afr. agric. J., *21*:25.

79 **Wallis, J. A. N.** and **Firman, I. D.** (1962). *Spraying Arabica coffee for the control of leaf rust.* E. Afr. agric. for. J., *28*:89.

80 **Wormer, T. M.** and **Firman, I. D.** (1961). *Crinkle-leaf and hot and cold symptoms of coffee in Kenya.* Kenya Coffee, *26*:13.

81 **Wormer, T.M. and Firman, I. D.** (1965). *Control of crinkle-leaf of coffee in Kenya.* Expl. Agric. *1*:1.

13
Coffee (Robusta)

Coffea canephora

Introduction

Robusta coffee is grown in low altitude, wet, humid areas which are unsuitable for Arabica coffee production. It is of lower intrinsic drinking quality than Arabica and most is processed by the dry method; this produces 'hard' coffee which has a bitter flavour. So far instant coffee has been manufactured predominantly from Robusta. Robusta is not bitter if it is processed by the wet method, even though it cannot produce the characteristics of 'mild' coffee, i.e. wet processed Arabica. There is a limited market for wet processed Robusta coffee for blending with Arabica.

Uganda

Robusta coffee has a long history in Uganda. The Baganda have prized it since well before the arrival of Europeans and wild trees can be found in some of the forests. In 1950 only about 20 000 tons were produced from about 160 000 acres (c. 65 000 ha). Between 1950 and 1954 there was a fourfold increase in the price offered to the grower and both the acreage and production rose dramatically. In 1969 production reached a peak of about 220 000 tons from an estimated 600 000 acres (c. 240 000 ha). In most years during the 1960s Robusta coffee contributed almost half of the value of Uganda's exports.

Robusta coffee is almost entirely a smallholder crop in Uganda; the estate acreage in 1969 was only about 10 000 acres (c. 4 000 ha). Most of the crop is grown in the 'fertile crescent': a strip of land within 40–45 miles (c. 65–70 km) of the shores of Lake Victoria between Jinja and the Tanzania border. A little is grown in Ankole, Toro and Bunyoro Districts. At the time of writing, planting Robusta coffee is no longer encouraged owing to the fear of overproduction.

Tanzania

Coffee is the most important cash crop in Bukoba District where it is grown in the high rainfall area within 30–40 miles (c. 50–65 km) of the lake shore. About two-thirds of this coffee is Robusta.

Kenya

Robusta coffee was planted by smallholders in Busia District, and the lower parts of Bungoma District during the 1950s. Because of the threat of over-production, planting has been prohibited since the International Coffee Agreement of 1962. The present production is small and is sold either to the Kenya Planters' Cooperative Union or, illegally, across the border in Uganda.

Plant characteristics

Robusta coffee differs from Arabica in the following ways. It is a more vigorous tree. It has larger, coarser leaves and the surface of these is slightly corrugated instead of being flat. When adequately supplied with water and nutrients, more flowers and berries develop on each node (see Fig. 55). Sublaterals very seldom grow from the primary lateral branches. After bearing a crop for about three years, a lateral dies back and is later shed. (In Arabica coffee the laterals tend to be retained even when they are unhealthy). Robusta coffee is self-incompatible, i.e. the pollen from one tree or clone is almost incapable of pollinating flowers from the same tree or clone; there is therefore seldom more than 3–4% of self-pollination.

Ecology

Rainfall, and water requirements

Robusta coffee yields best where there is a constant supply of soil moisture throughout the year, especially during the period when the crop is maturing. There should, however, be a short dry period to encourage a heavy and uniform flowering at the beginning of the rains. In Uganda and Tanzania, Robusta coffee is grown in areas which usually receive 35–80 in (c. 900–2 000 mm) of rain annually with a dry period in January and February.

Altitude and temperature

Almost all East Africa's Robusta coffee is grown

Fig. 55: A heavily bearing lateral of robusta coffee.

Varieties

There are no recognised varieties of Robusta coffee in commercial use in East Africa because, with 100% cross pollination and a great diversity of genetic material, there is very great variation from plant to plant when seed is used for propagation. Seed from high yielding mother trees has been distributed in the past but this can have been of little benefit, since the male parents were unknown and the superiority of the chosen trees could have been caused as much by the immediate environment of the plant as by genetic worth. During the late 1950s and early 1960s 'polycross seed' was issued in Uganda; this was produced in a block where pollination between different high yielding clones was allowed. With a highly heterozygous crop like Robusta, 'polycross seed' can only produce a highly variable population but even so it was thought that the performance of that population would be better than that of a population resulting from unselected seed. Genetic uniformity can only be achieved in Robusta coffee by growing clones; several high yielding clones have been identified although there has, as yet, been no call for the distribution of this material.

Propagation and establishment

In East Africa Robusta coffee is virtually always propagated by seed; the principles and practices mentioned in the previous chapter apply equally to this crop. In other parts of the world, however, clones are often raised by using vegetative propagation.

Vegetative propagation
Robusta coffee can easily be propagated by vegetative means. Softwood single node cuttings are usually used, although these can be split to make single leaf cuttings if planting material is scarce. It is of interest to note that before the colonial era the Baganda sometimes used long hardwood cuttings.

Spacing
Where pure stands are established, the recommendation is to space the plants at 10 ft × 10 ft (c. 3·3× 3·3 m), i.e. a plant population of about 435 plants per acre (c. 1 075 plants per hectare). In practice, however, a random spacing leading to a considerably lower population than this is usually adopted. Where coffee is interplanted with bananas a wider spacing is obviously needed.

between the altitudes of 3 500 ft and 4 500 ft (c. 1 100–1 400 m). Below 3 500 ft there is seldom adequate or adequately distributed rain, although an exception occurs in the Semliki Valley on the western side of the Ruwenzori Mts. where Robusta is grown at an altitude of only 2 000 ft (c. 600 m). In West Africa, notably in the Ivory Coast, Robusta thrives at sea level. Above 4 500 ft Robusta is seldom grown because it is too cold; at such altitudes Arabica coffee becomes a more rewarding crop.

Shade
Most of East Africa's Robusta coffee is grown amongst, and in the shade of, bananas. Some of the estates use shade trees, mostly *Albizzia spp.*

Soil requirements
Robusta coffee is a vigorous crop which grows well in a wide variety of soils, provided they are well drained. In parts of Uganda it thrives where the soil is underlain by rock at a depth of 2 ft (c. 0·6 m). The loams of the 'fertile crescent' in Uganda supply enough nutrients to allow reasonable crops to be obtained without the addition of fertilisers or manure.

Field maintenance

Pruning

Multiple stem pruning is the only possible method of pruning Robusta coffee. The single stem method cannot be used because it relies on a permanent framework of lateral branches; as noted previously, the lateral branches of Robusta coffee are not permanent, but are shed after a few years. Because of this phenomenon it is sometimes described as being 'self pruning'. The only work involved in pruning is cutting back old stems when their crop is borne so high that it is out of reach of the pickers.

Pruning is recommended because it facilitates picking. Nevertheless, many growers leave their trees unpruned, preferring to climb into the trees to harvest their crop. The other three reasons for pruning do not apply to Robusta coffee. Unlike Arabica, it does not suffer from overbearing, so pruning to control cropping is unnecessary. Unlike Arabica, it does not have to be sprayed at regular intervals, so the trees do not need to be kept short to allow this operation. Finally, there are no important pests or diseases which thrive in the moist microclimate of a dense bush, so pruning to allow a free flow of air amongst the bushes is unnecessary.

Fertilisers

Experiments have shown that nitrogen is the only element which gives worthwhile responses. Recommended rates are 110–160 lb of N per acre per annum (c. 120–180 kg/ha) and these can increase yields by 3–5 cwt of clean coffee per acre (c. 380–630 kg/ha), depending on the general standard of management. Virtually no growers, however, apply fertilisers.

Weed control

Poor weed control is one of the most important reasons for the low average yields of Robusta coffee in East Africa. Couch grass (*Digitaria scalarum*), which thrives in the warm wet areas where Robusta is grown, is by far the most damaging weed.

Experiments on different methods of soil treatment to check weed growth give similar results to those done on Arabica coffee. Mulching, when compared to clean weeding, can give a 20% increase in yield; it is officially recommended but very rarely practised. Cultivating with disc harrows, growing cover crops and repeated slashing all give considerable reductions in yields. Herbicides, e.g. paraquat and dalapon, are effective but are virtually never used.

Fig. 56: Harvesting robusta coffee.

Harvesting and processing

Most Robusta coffee growers in East Africa sell sun-dried cherry, i.e. 'kiboko'. This is sent to central factories for hulling, i.e. removing in one operation the black, dried pulp and the parchment from the beans. To produce good quality 'kiboko' farmers should pick only the ripe berries and they should then dry them in shallow layers on mats or on raised tables or trays. Unfortunately, little attention is given to picking and drying; branches are often stripped of all their fruit regardless of whether it is over-ripe or under-ripe and drying is often done on bare ground where impurities can easily be included with the crop. Growers have no incentive to produce good quality 'kiboko' because they only receive one fixed price for it. Only if it is hulled can the quality be assessed, and this is now being encouraged at the primary buying level in Uganda. A small proportion of Uganda's Robusta crop (about 30 000 tons in 1969) is wet processed by the method described in chapter 12. This demands a higher standard of picking because over-ripe and under-ripe cherry cannot be pulped.

Yields

The low average yields in East Africa, which are quoted as being about 4 cwt of clean coffee per acre (c. 500 kg/ha) are a reflection of the poor general standard of husbandry. Average yields as high as 17 cwt per acre (c. 1 900 kg/ha) are obtained on well managed plots.

Pests

Robusta coffee suffers little from pests. The only routine measure necessary is the collection of fallen berries and over-ripe berries on the tree to prevent prevent the build-up of berry borer. Unfortunately, the precaution is often ignored. Berry borer is discussed in more detail in chapter 12.

Diseases

Robusta coffee suffers little from the two main diseases of Arabica, coffee berry disease and leaf rust. CBD does not occur and leaf rust is only of minor importance.

Red blister disease
This disease, which first appeared in Uganda in the mid-1950s, is caused by the fungus *Cercospora coffeicola*. It is of minor importance and the damage done by it decreased during the 1960s. In the event of severe attacks it is easily checked by spraying with a copper fungicide.

Bibliography

1   **Butters, B.** and **Clegg, D. E. H.** (1963). *Improved methods of weed control in robusta coffee in Uganda*. E. Afr. agric. for. J., *29*:67.

2   **Ingram, W. R.** (1965). *An evaluation of several insecticides against berry borer and fruit fly in Uganda robusta coffee*. E. Afr. agric. for. J., *30*:259.

3   **Stephens, D.** (1967). *Experiments with nitrogen and magnesium fertilisers on coffee in Uganda*. Expl. Agric., *3*:191.

4   **Stephens, D.** (1967). *A note on the correlations between coffee yields and soil analyses in Uganda*. E. Afr. agric. for J., *32*:456.

5   **Stephens, D.** (1968). *Fertiliser trials on small coffee farms in Uganda*. J. Hort. Sci., *43*:75.

14
Cotton

Gossypium hirsutum

Introduction

The most valuable product of the cotton plant is lint, i.e. the hairs which grow from the seed coats. Lint can be spun into a yarn and, throughout the world, is the most important plant product used for textiles. The oil which constitutes about 20% of the seed is a valuable by-product; it is extracted by heating and crushing the seeds after the lint has been removed. Cotton seed oil is used in the manufacture of cooking oil, margarine, soap, etc. Cotton seed cake, which is the residue after milling, has a high protein content and is a valuable stock feed. Linters are a minor by-product; they are the short hairs on the seed coat which are sometimes called 'fuzz'. Linters can be used in carpet manufacture or for upholstery but they are seldom utilised in East Africa.

Species
There are several species in the genus *Gossypium*, e.g. *G. arboreum*, a perennial with short hairs which is grown by some peasant farmers in India. There are only two species, however, which are important: *G. hirsutum* and *G. barbadense*. *G. hirsutum*, commonly called upland cotton, is the cotton of commerce in East Africa. *G. barbadense* is less important and contributes only about one-third of the world's cotton. It produces lint of high quality, notably Sea Island cotton from the West Indies which has a staple length of over 2 in (c. 5 cm). *G. barbadense* is grown commercially in the West Indies, Peru, Brazil, Arizona and New Mexico in the U.S.A., Egypt and the Sudan. Perennial varieties of *G. barbadense* grow wild in East Africa, notably in some of the drier parts of Kenya. They are of no commercial importance although their lint is sometimes collected for local use. The physical differences between *G. hirsutum* and *G. barbadense* are discussed in the next section.

Uganda
Most of Uganda's cotton is produced in the Northern and Eastern Regions with the greatest concentrations in Teso, Busoga, Lango and Bukedi Districts. The country's cotton acreage was estimated as being 1 600 000 acres (c. 650 000 ha) in 1968. Uganda's cotton production reached 344 000 bales* as long ago as 1938 and remained relatively stable from then until the end of the 1960s. In 1968–69 it was 421 000 bales.

Tanzania
All but a small fraction of Tanzania's cotton is produced in an area to the south and east of Lake Victoria including Mwanza Region and parts of Mara, Shinyanga, Tabora and West Lake Regions; this is called the Western Cotton Growing Area. A small acreage is grown in the central areas and at the coast. During the 1960s an estimated 1 250 000 to 1 500 000 acreas (c. 500 000–600 000 ha) of cotton were grown in Tanzania each year. The country's production rose from 133 000 bales in 1956 to 390 000 bales in 1968–69. This increase was mainly due to a great expansion of the cotton acreage.

Kenya
Kenya produces much less cotton than the other two East Africa countries. During the 1960s annual production was about 25 000 bales from approximately 200 000 acres (c. 80 000 ha). The traditional cotton growing areas are the coastal strip and the shores of Lake Victoria (including parts of Busia District) but in these areas the crop is regarded by the growers as being of secondary importance so extension efforts have been hindered. During the late 1960s cotton was the subject of an intensive extension effort in some of the lower altitude areas of Central and Eastern Provinces, especially Kirinyaga, Machakos and Kitui Districts. There has been a great expansion of the cotton acreage in this area but the pest problem is unusually severe; the level of future production will therefore depend mostly upon the efficiency of insecticide applications.

*In East Africa bales of lint weigh approximately 400 lb (c. 180 kg) but in most other parts of the world they weigh approximately 480 lb (c. 220 kg).

Fig. 57: A 3½ month old cotton plant. These are two monopodial branches at the base of the main stem; note that there are no fruiting points at their nodes. The rest of the branches are sympodial; note the fruiting points at the nodes of the lower three.

Smallholders
East Africa's cotton is entirely grown by smallholders.

Plant characteristics

Roots
Most roots are found in the top 12 in (c. 30 cm) in non-alluvial soils. (In alluvial soils they penetrate deeper). The tap roots, however, often penetrate to a depth of 7 ft (c. 2 m) provided that there is no murram layer or hardpan. Under very favourable conditions roots have been found as deep as 15 ft (c. 4·5 m).

Stems and branches
The main stem of the cotton plant, which becomes woody in the later stages of the crop's life, has spirally arranged leaves. Cotton exhibits dimorphic branching: the lower nodes of the main stem produce vegetative branches, usually called monopodia, whilst the upper nodes produce reproductive branches, usually called sympodia. A monopodium can only produce sympodia from its nodes (see Fig. 57). Each sympodium terminates after a short distance in a fruiting point subtended by a leaf. From the axil of this leaf a further branch develops, terminating again in a fruiting point, and so on. This sequence gives the sympodia a zig-zag appearance. The first sympodia usually appears between the 4th and the 8th node on the main stem, depending on the variety, the temperature and the spacing, (close spacing reduces the number of monopodia).

Leaves
The leaves have long petioles, usually with small stipules, and are palmately lobed with three or five lobes. All the currently grown East African varieties have small hairs on their lower surfaces; this character has been bred into the local varieties because it confers jassid resistance. Varieties with hairless leaves are usually heavily attacked by jassids in East Africa.

Squares
This is the name given to unopened flower buds when all that can be seen are the three bracteoles of the epicalyx closely pressed together (see Fig. 58). If bollworms attack at this stage, the squares 'flare', i.e. their bracteoles open out to reveal the small unopened corolla (see Fig. 64).

Flowers
Flowers (see Figs. 60 and 61) are borne singly on short stalks; each is surrounded by three deeply divided bracteoles. There are five large petals; these are pale yellow on the day they open and pink the next day. On the third day the petals usually crumple up; they fall to the ground shortly afterwards. Pollen is shed from the anthers just after the flowers open, thus allowing a small degree of cross pollination. The proportion of outcrossing is normally 5–10%, although higher figures have occasionally been recorded. The sequence of flowering is from the bottom of the plant to the top, and from the centre to the outside; the earliest flowers are therefore nearest to the main stem on the lowest reproductive branches.

Bolls, i.e. fruits
Many bolls are shed shortly after flowering and before they have had time to grow much; this is especially true of bolls which formed late. Boll shedding is a physiological response to strain. The fruits are leathery structures (see Fig. 62) which split down the middle of each carpel (or locule), revealing the seed cotton (see Fig. 63). There are normally four locules in each boll, sometimes three or five, and each contains about ten seeds surrounded by the fibres which grow from their seed coats.

Life cycle
Cotton is a weak perennial. Growth continues

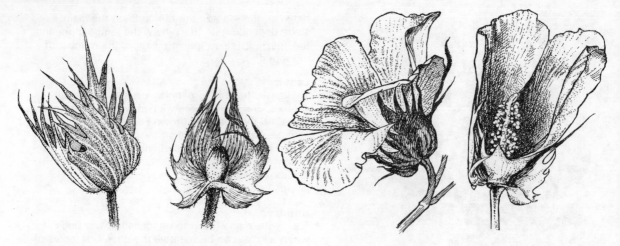

Fig. 58: A cotton square.

Fig. 59: A cotton square with one bracteole removed.

Fig. 60: A cotton flower.

Fig. 61: A cotton flower with one side removed.

indefinitely, with regular cycles of flowering, provided there is sufficient soil moisture. In practice, however, cotton is treated as an annual and is uprooted in less than a year; this is done partly to allow other crops to be grown, and partly because perennial growth would be impossible in some rainfall regimes. In addition, perennial cotton would create severe pest problems.

Flowering usually commences about nine weeks after sowing with squares appearing three or four weeks earlier. In cool conditions, however, flowering may be delayed until twelve weeks after sowing. Flowering continues for a considerable period but the bulk of the yield (the main crop) derives from the first three or four weeks of flowering. (An exception occurs in Central and Eastern Provinces in Kenya, whose special circumstances are discussed in more detail in the section on time of sowing.) The period from flowering to boll splitting is eight to ten weeks in East Africa. During the first part of this period the seed hairs grow to their full length; in the second half spiral deposits of cellulose are laid down on the inside of each hair until, at maturity, there is only a small cavity in the middle. When a boll splits, the hairs dry out and collapse, their cross section becoming flattened rather than round; it is at this stage that bends and twists appear in each hair.

Gossypium barbadense
The most obvious differences between *G. hirsutum*

and *G. barbadense* are as follows. *G. barbadense* has leaves which are darker green and more deeply lobed; they are usually hairless and shiny. Its petals are deep yellow with a purple spot at their base. Its bolls are heavily pitted instead of being smooth walled and its seeds have no fuzz with the exception of a turf at one end.

Ecology

Rainfall, and water requirements
Cotton tolerates droughts to a greater extent than most other annual crops. In Central and Eastern Provinces of Kenya, for example, it is sown in October and usually experiences a severe drought of two months' duration during January and February. This is an exceptional case, however, for in all other parts of East Africa cotton should be sown at a time when an adequate and uninterrupted rainfall is anticipated during the following five months.

The water requirements of cotton vary with different spacings, temperatures, soil conditions, etc., but a general picture of the demands of the crop is given in Fig. 65. During the first two months little soil moisture is required because the leaf area is small. The supply must not be too erratic during the early part of this period, however, otherwise germination may be poor or there may even be a complete crop failure. At flowering the water demand begins to rise, reaching a peak about $1\frac{1}{2}$ months later, but thereafter declining as the leaves become senescent and are shed. Droughts during

Fig. 62: An unopened cotton boll.

Fig. 63: An open cotton boll.

Fig. 64: A 'flared' cotton square.

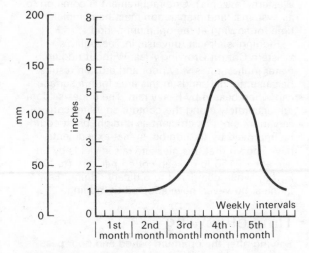

Fig. 65: Approximate water requirements of cotton during three-week periods at weekly intervals.

the three month period of high water demand can cause premature shedding of leaves, flowers, buds and bolls and the development of short, immature lint. The East African cotton varieties can compensate for early crop loss (caused either by moisture stress or by insect attack) by putting on extra growth at the top of the plant; this compensatory growth can only occur, of course, if there is adequate soil moisture.

Very high rainfall is undesirable for cotton; it does most harm by causing discolouration of the lint after the bolls have opened but it can also contribute to decreased yields by causing water-logging, excessive leaching and a high incidence of fungal and bacterial boll rots. 18 in (c. 450 mm) during December, January, February and March is the optimum at Ukiriguru; rainfalls much above or below this figure cause yield reductions.

Altitude and temperature
Cotton needs a warm climate and is therefore seldom grown above 4 500 ft (c. 1 400 m). When night temperatures are low, i.e. much below 60°F (15·5°C), short staple, poor quality cotton is produced, possibly owing to reduced formation of cellulose. There is considerable variation in temperature between different cotton growing areas in East Africa; at Namulonge in Uganda, where the night temperatures are often near the lower limit, cool conditions cause a slow rate of node production, slow vegetative growth, late flowering, a longer period from the appearance of the first square to first flowering and the production of the first sympodium at a relatively high node.

Soil requirements
Cotton tolerates a wide range of soils, from the acids sands of the Western Cotton Growing Area in Tanzania to the heavy, alkaline clays around Kibos in Kenya. Cotton growth may be restricted by inadequate drainage or by a soil pH less than

5·0–5·5 (depending on the texture). Liming has proved beneficial on acid soils in Uganda at the rate of ½ ton per acre on light soils and 2 tons per acre on heavy soils (c 1·2 and 5·0 t/ha respectively).

Varieties

During the first decade of this century many American Upland varieties were introduced into East Africa. These became physically mixed and eventually lost their identity owing to cross pollination; at the same time they underwent natural selection and adaptation to the East African environment. The most important of the resulting mixtures, which were of great genetic diversity, became known as Buganda Local, Mwanza Local and (in Nyasaland) Nyasaland Upland. The main varieties which were selected from these mixtures, and which were regularly improved by further selection, were BP52 and S47 in Uganda and the UK multiline seed issues in Tanzania. BP52 was selected at Bukalasa (BP stands for Bukalasa Pedigree); it was selected from Nyasaland Upland and was first released in 1943. It gave a longer, stronger lint than S47 but this was partly due to the fact that it was grown in a higher rainfall area nearer Lake Victoria. S47 was selected at Serere from BP50 which was a Buganda Local selection; it was first released in 1947. Slection from Mwanza Local started at Ukiriguru in the 1930s and led to the UK series of multiline seed issues; the prefix UK is followed by the year in which the lines were mixed prior to bulking up.

The most significant development in cotton breeding in East Africa has been the introduction, during the 1950s, of Albar; this is derived from Nigerian Allen (providing the letters Al) and is highly blackarm resistant (hence the last three letters). It was introduced because although previous selection programmes had imparted a certain degree of resistance to blackarm (not to mention great increases in lint yield and quality and virtual immunity to jassids), this degree of resistance was not considered satisfactory. In Uganda, selections from Albar produced the new varieties BPA and SATU. BPA stands for BP Albar; the letters BP were retained for trade purposes in order to indicate that the new variety has the same characteristics as BP52, i.e. long, strong lint. BPA was released in 1966 and is grown in the wetter areas near the lake, both in Uganda and Kenya, where these lint characteristics have the best chance of being expressed. SATU stands for Serere Albar Type Uganda; it was released in 1964.

At Ukiriguru, Albar was crossed with the local varieties and selections from the resulting populations gave lines which are designated by the prefix UKA. In Tanzania, UKA lines have been solely used for all seed issues after and including UK61; these seed issues are now widely distributed. A single line, UKA 59/240, has recently replaced UK51 in all Kenya's cotton growing areas east of the Rift Valley.

It has been estimated that without the programme of improvement outlined above, yields would be at least one-third less than they are now.

Seedbed preparation

There are several ways of preparing land for cotton; these range from ox-ploughing (in western Kenya, Teso and Lango Districts in Uganda and parts of Shinyanga Region in Tanzania) to drawing up ridges by tractor with no previous ploughing or harrowing (sometimes used on hill sands in Western Tanzania). One requirement is common to all systems: land preparation must be completed in time for sowing at the optimum period.

Ridging is almost universal in Tanzanian's Western Cotton Growing Area. Without ridging, water infiltration is very poor and erosion results, because the hill sands in this area form a surface cap when beaten by heavy rain. The ridges seldom run accurately along the contour so water conservation is not as efficient as it might be. It could be improved by tie-ridging, in fact experiments have shown that this measure raises yields by an average of 150 lb of seed cotton per acre (c. 170 kg/ha) when compared to ordinary ridging. Few farmers, however, have adopted tie-ridging.

Time of sowing

Sowing after the optimum period and poor pest control are the two most important reasons for poor cotton yields in East Africa. The severity of yield decline which results from late sowing varies from season to season, from place to place and with different standards of management; a general picture, however, is given by a survey of experimental results over several countries which gave average losses as follows:

Number of weeks delay in sowing after the optimum date	2	4	6	8
Percentage decrease in yield	14%	40%	50%	62%

Late sown crops usually produce a higher percentage of B grade cotton than early sown crops owing to increased damage by stainers. The quality of late sown crops is sometimes lower, e.g. in parts of the Western Cotton Growing Area in Tanzania where cold nights in the middle of the year may lead to reduced cellulose formation. Finally, excessively late sowing can lead to picking in a wet period; this happens with some crops in Western and Nyanza Provinces in Kenya which are ready for picking after the January/February dry season.

The reasons for the poor yields which result from untimely sowing are not fully understood. Fitting the moisture demands of the crop to the available water supply is doubtless important. For example, untimely sowing may bring the period of peak moisture demand into a period of inadequate rainfall. If sowing is too early or (in some instances) too late, picking may coincide with a wet period. The incidence of pests and, to a lesser extent, diseases tends to be higher in late sown crops. Other factors, including the direct effect of heavy rain on flowering, the amount of solar radiation received by the crop, night temperatures and various soil factors may also play a part.

Two types of incentive have been tried in order to persuade growers to sow at the correct time. In the Mwanza area in Tanzania the price paid for seed cotton sold in July was one cent higher than the price paid in later months. This scheme was

unsuccessful and was abandoned in the mid-1960s. In 1969, however, it was reintroduced. The premium was increased to 2 cents and the buying centres were opened in early June; an encouraging response has been reported. In Kenya free insecticide treatment was offered to Nyanza growers in 1960 and 1961 if they had more than one acre of cotton sown before 1st May and if it was well weeded and properly thinned. The ineffectiveness of the treatment at that time, i.e. dusting, which has since been abandoned, may explain the lack of success of this scheme.

It must be stressed that timely sowing is part of a 'package deal'. Good yields cannot be produced by this measure alone; it must be supported by proper spacing, thinning, weeding and, where necessary, fertilisers. Above all it must be supported by efficient methods of pest control.

Uganda

North of the 1°N line of latitude (which runs approximately from Mt. Elgon to the southern tip of Lake Albert) the recommended sowing period is from mid-April to mid-June; the best yields, however, usually come from crops sown in the first half of this period.

Fig. 66 shows the rainfall distribution at Serere; reference to this and to the moisture demand curve in Fig. 65 shows that when cotton is sown in the last half of April or the first half of May it receives good rainfall during its third or

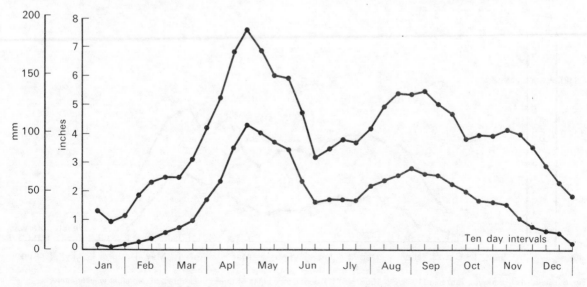

Fig. 66: Rainfall at Serere, Uganda (1921–1966); 1:1 confidence limits of twenty-day totals of rainfall at ten-day intervals.

fourth month and during the harvesting period the weather becomes progressively drier. South of the 1°N line the June/July drought is more pronounced and poses a threat to crops sown in April. The recommended sowing period is therefore from mid-May to July; crops sown at this time seldom suffer from the poor rainfall in June and July because their moisture needs are small at this time. In the extreme west of the country the drought in the middle of the year is usually so severe that farmers must either sow in dry soil or must wait for the onset of the rains in mid-July or August.

Tanzania
In the Western Cotton Growing Area the wet season is shorter and ends more abruptly than in Uganda (see Fig. 67 which shows the rainfall distribution for Ukiriguru). The optimum sowing period is therefore shorter; cotton should be sown in early December.

Kenya
In the Nyanza, Western and Coast Provinces the recommended time of sowing (if the crop is to be sprayed) is from mid-February to early March; if the rains are delayed, cotton should be dry sown at the end of March.

Eastern and Central Provinces present an unusual case because in most years neither the long rains (March to June) nor the short rains (October to December) are alone adequate to produce a

reasonable crop. The recommended sowing period is therefore in the short rains, preferably in the last half of October. If sown at this time cotton grows for about two months and is then checked, but not killed, during the dry season which is normally severe and lasts about two months. Few flowers are produced during the dry season so it is not worth giving the crop any chemical protection at this time; consequently any fruits from the short rains are readily attacked by bollworms, the plants usually become seriously tip bored by spiny bollworm and they become heavily infested with aphids. When the long rains break the aphids disappear, growth commences and there is a flush of flowering which later produces the bulk of the crop; from the onset of the long rains onwards the crop must be sprayed with insecticides.

Sowing

Almost all cotton in East Africa is sown by hand. Ox-drawn or hand-propelled seeders are sometimes used in Uganda. The number of seeds recommended per hole in Kenya and Uganda is five; in Tanzania it is six to ten. Free seed is issued in all areas except the Kenya coast, so it is common practice to sow many more seeds than necessary; this causes severe competition between the young seedlings. If the recommendations are followed, 15–40 lb of seed are needed per acre (c. 17–45 kg/ha) depending on the spacing. On most soil

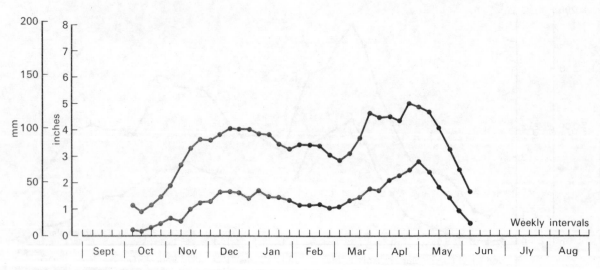

Fig. 67: Rainfall at Ukiriguru, Tanzania (1931/32–1958/59); 1:1 confidence limits of three-week totals of rainfall at weekly intervals. No significant rainfall occurs during July, August or September.

types the sowing depth should be no more than one inch. At Namulonge in Uganda the soil dries out quickly so deeper sowing is preferred.

Spacing

The spacing recommendations for the cotton growing areas of East Africa are summarised in the following table:

	Distance between rows	Distance between hills within the row	No. of plants per hill after thinning
N. and E. Uganda	2 ft (c. 0·6 m)	6 in (c. 15 cm)	1
Remainder of Uganda	3 ft (c. 0·9 m)	1 ft (c. 0·3 m)	2
Western Tanzania	3 ft (c. 0·9 m)	15 in (c. 38 cm)	2
Tanzania coast	3 ft (c. 0·9 m)	1 ft (c. 0·3 m)	1
Kenya coast	3 ft (c. 0·9 m)	1½ ft (c. 0·45 m)	2
Remainder of Kenya	3 ft (c. 0·9 m)	1 ft (c. 0·3 m)	2

In Western Tanzania almost all cotton is grown on 5 ft (c. 1·5 m) ridges; two rows, about 1½ ft (c. 0·45 m) apart, should be sown on each ridge; the hills should be 1½ ft apart within the row and there should be 2 plants per hill.

Thinning

Thinning must be done when the plants are 4–6 in (10–15 cm) high, i.e. when they have two true leaves. Postponement of thinning causes excessive competition between the plants but this may be unavoidable on heavy soils in dry spells when uprooting the unwanted seedlings would harm the others.

Fertilizers and manures

In Uganda and Tanzania the fertiliser recommendations for cotton are simple and uniform. Although they ignore differences in the nutrient contents of various soils, they represent average quantities of the fertilisers which have shown, in a great number of field trials, to produce economic increases in cotton yields.

Nitrogen

Cotton usually responds well to nitrogen. Responses can be as high as an additional 550 lb of seed cotton from an application of 50 lb of nitrogen per acre (c. 600 kg of seed cotton from 55 kg/ha). Nitrogenous fertilisers should be top-dressed about six weeks after sowing. In Uganda 1–2cwt per acre (c. 125–250 kg/ha) of ammonium sulphate are recommended in Northern and Eastern Regions. In Tanzania the recommendation is 1 cwt of ammonium sulphate or calcium ammonium nitrate. In Kenya no recommendations are available for the 1969–70 season, although a recent programme of fertiliser trials shows responses to nitrogen on most soils.

The more vigorous growth of the crop with nitrogen apparently presents a more attractive environment for insects, so efficient use of insecticides is especially important.

Phosphate

Phosphates increase cotton yields on most soils although responses may be small if the land has been recently cleared. Applications of 2 cwt per acre (c. 250 kg/ha) of single superphosphate are recommended. In Tanzania the recommendation is 1 cwt of single superphosphate annually or 2 cwt every other year. In Kenya recent trials indicate that phosphate responses are best on red soils and, at the coast, on Shimba grits.

Single superphosphate contains calcium and sulphur; cotton has a high requirement for both these elements. However, this fertiliser contains no magnesium and deficiencies of this element have been observed recently at Ukiriguru in Tanzania on plots which had received applications of superphosphate for several years.

Potassium

No recommendations for the use of potassium on cotton in East Africa have been formulated, but with increasing intensity of cultivation this nutrient could become deficient.

Boron

The only micro-nutrient suspected of being deficient in East African cotton areas is boron. Deficiency is thought to occur in parts of Tanzania, especially on Ukerewe Island. Correcting this is a complicated process, because a small excess of boron is toxic and can cause reductions in yield.

Mixed fertilisers

The 'fertile crescent' of Uganda, i.e. the 40–50 mile

wide (c. 65–80 km) strip of land which adjoins
Lake Victoria between Jinja and the Tanzania
border, has been a problem area as far as fertilisers
are concerned. Straight fertilisers have given no
consistent responses. Recent work, however,
suggests that mixed fertilisers, applied to the
seedbed before sowing, may be more satisfactory.
The following mixture has been successful although
it has not been widely tested: 2 cwt of calcium
ammonium nitrate, 2 cwt of single superphosphate
and 1 cwt of muriate of potash per acre (c. 250 kg
of both C.A.N. and single superphosphate and
125 kg of muriate of potash per ha).

Manure

Cattle manure, applied at 3–5 tons per acre (c. 7·5–
12·5 t/ha), gives good yield increases on cotton. It
is seldom applied, however, because little manure
is available and because the farmers are reluctant,
for a number of reasons, to use manure on annual
crops.

Weed control

Cotton is slow growing at first and can therefore
easily be overtaken by weeds. Several weedings
are usually necessary, the first coinciding with
thinning. Ridging reduces the weed problem by
burying all the plant material very thoroughly;
to get maximum benefit from this form of weed
control, cotton must be sown as soon as possible
after ridging.

Herbicides are only used on large scale cotton
growing schemes in East Africa. Diuron, fluome-
turon, trifluralin and prometryne are effective
pre-emergence herbicides and have given weed-
free conditions for 6–8 weeks after application.

Harvesting

Cotton harvesting, more commonly called picking,
is universally done by hand in East Africa (see Fig. 68).
Mechanical harvesting is only necessary when
insufficient labour is available, e.g. in the U.S.A.
where most of the crop is handled by machines.
Cotton picking is a very labour-consuming task
because it must be done at weekly intervals to
prevent discoloration of lint in the field. Hard-
working labourers, picking a heavy crop, harvest
more than 100 lb (c. 45 kg) of seed cotton a day
on experimental stations; smallholders, however,
average only about 20 lb (c. 9 kg) a day. Care must
be taken to avoid breaking off pieces of dried
plant material during picking because these can

Fig. 68: Picking cotton.

easily become mixed with the cotton, leading to
extra work during sorting. If heavy dew has fallen
during the night or if there has been rain, picking
must be delayed until the lint has dried.

After picking it is essential to sort the clean,
grade A cotton from the dirty and stained grade B
cotton which only fetches about half the price.
Average seed cotton has about 10% grade B but
this percentage can be much lower with careful
harvesting and with a low incidence of stainers.
Aprons with two large pockets are sometimes used
to enable a preliminary sorting to be done whilst
picking. Sorting is a laborious operation; the
combined operations of picking and sorting take so
much time that they limit the number of acres
which one family can handle, even if cultivation
and spraying are done mechanically.

Seed cotton is taken to ginneries in bags which
usually weigh 60–70 lb (c. 27–32 kg) when full.
Some people pack as much into a bag as possible
by compressing the contents with a stick; this is an
undesirable practice because it damages the lint.
Sisal bags must never be used for seed cotton, as
sisal fibres can become mixed with the lint and this
creates ginning and spinning problems.

Yields

Average yields of seed cotton per acre in East
Africa are 200–400 lb per acre (c. 220–450 kg/ha).
In the important growing areas 1 200–1 500 lb per
acre (c. 1 300–1 600 kg/ha) should be obtained in
most years with good husbandry and efficient pest

control. Well tended experimental plots sometimes produce 3 000 lb per acre (c. 3 400 kg/ha).

Ginning

A gin is a machine which removes the lint from the seeds. There are two types of gin: the roller gin and the saw gin.

All East African ginneries are equipped with roller gins. A roller gin consists of one or two rollers 8 in (c. 20 cm) in diameter covered with leather. Two knives, one stationary and one reciprocating, are situated along the length of each roller; they allow lint to pass between them and the rollers, but prevent the seed from doing so. The output of roller gins is much lower than that of saw gins, i.e. only 50–80 lb (c. 23–36 kg) of lint per hour; their action is more gentle than the action of a saw gin so they are essential for very long staple, fine cotton, e.g. *G. barbadense*.

A saw gin consists of a row of small circular saws, 8–12 in (c. 20–30 cm) in diameter revolving in a notched drum; there is no room for the seeds to be carried through the notches, but the lint adheres to the saws and is carried through, thus separating it from the seeds. Saw gins have a much higher output than roller gins: over 650 lb (c. 290 kg) of lint per hour. Their main disadvantage is that their action is too rough for the longest staple cottons and that if badly set they can even damage the lint of medium and short staple cottons. There is no reason why saw gins should not be used for East African cotton, provided they are well supervised.

Ginning is hindered by the presence of fuzz on the seed. Ginning outputs are increased by 50–100% with semi-naked or tufted varieties, e.g. SATU. Other advantages of these varieties are an increased ginning percentage, more rapid and more uniform germination, lighter and less bulky seed (thus facilitating storage and transport), easier oil extraction and machine planting and reduced seed coat nep (see section on lint quality). The only disadvantage of semi-naked seed is that seed dressings adhere less readily.

The ginning percentage of a sample of seed cotton is the weight of the lint divided by the weight of the seed cotton, expressed as a percentage. Ginning percentages in East Africa vary from 32–37%, depending largely on the variety. Constant improvements are being made by breeding.

After ginning, the lint is compressed into bales of approximately 400 lb (c. 180 kg); these are wrapped in hessian and bound with metal bands.

Lint quality

Lint quality depends on several characters; the most important are outlined below:

1 *Staple length*
   This is the length of the individual hairs. Long staple cottons are classified as being $1\frac{1}{8}$ in (c. 29 mm) or more. Medium long staple cottons are 1 in–$1\frac{3}{32}$ in (c. 25–28 mm). Medium staple cottons are $\frac{7}{8}$ in–$\frac{31}{32}$ in (c. 22–25 mm). Short staple cottons are under $\frac{7}{8}$ in. Most of East Africa's cotton falls into the medium long classification. Some of the best cotton around Mwanza in Tanzania and from the BPA cotton growing zone in Uganda is long staple. Staple length is the most important criterion in determining lint quality; the greater the length, the finer and stronger the yarn.

2 *Fineness*
   Fine hairs with a small diameter give stronger yarns for a given staple length.

3 *Maturity*
   Maturity is the degree of secondary cellulose thickening on the inner fibre wall; this affects both the strength of the individual fibres and the appearance of the lint.

4 *Fibre bundle strength*
   This is a measure of the strength of fibres before they are spun.

5 *Yarn strength*
   This is a measure of the ability of cotton to form a strong thread; many factors, including staple length, fineness and bundle strength, contribute towards yarn strength.

6 *Chalazal damage*
   Small pieces of seed coat may be broken off and included in the lint during ginning; these give rise to irregularities ('neps') in the yarn. Chalazal damage can be due to poor gin setting but it is also governed, to a certain extent, by the genetic constitution of the plant.

Pest control (individual pests)

The general principles of pest control are discussed in the next section after the following account of the most important pests.

Fig. 69: An American bollworm on a young cotton boll. Note the stripes. As in Fig. 70, the bracteoles are being held back.

American bollworm. *Heliothis armigera*

(American is a misnomer because *H. armigera* does not occur in America although the closely related *H. zea* does). The adults are small brown night flying moths. The larvae, which reach about 1½ in (c. 4 cm) long when fully grown, vary in colour and may be green, brown or yellow. The distinguishing features are the pale stripes on each side (see Fig. 69).

The moths are attracted to cotton fields when flowering begins. They lay their eggs, and these develop into very small larvae which are most attracted to the squares, the flowers and the very young bolls; occasionally they eat leaves, but only when there is no other food available. Early damage to squares and bolls causes shedding of these parts. As the larvae grow older they are attracted to more mature bolls; they seldom destroy these completely, although they may hollow out one or more loculi, making neat circular holes in the boll walls. Their characteristic feeding position is with the front half of the body inside the boll with the other half protruding outside; there is usually an accumulation of frass nearby, normally in the epicalyx. The pupae of American bollworms can go into diapause (a dormant stage) provided they are not near the equator. Diapause is therefore rare near Kampala but common in Tanzania.

DDT and endosulfan give good control of the young larvae but the fully grown instars are fairly resistant to these insecticides; for this reason it is important not to delay or interrupt the spraying programme. Spraying the tops of the plants must not be neglected because this is where many of the eggs are laid and where moulting takes place; recent work has shown that fewer eggs are laid on leaves which have a coating of insecticide, and larvae are much more susceptible to insecticides immediately after they have moulted. Due to the fact that the American bollworm can go into diapause, and that it can transfer to many alternate hosts (e.g. maize, tobacco, tomato, sorghum, millets, sunflower, pigeon peas, beans and many other pulses), it is impossible to control it by means of a close season. Trap crops of maize, planted as a border around cotton fields, have been tried in Kenya but have been found to be impracticable.

Of the many crops which can be attacked by

Fig. 70: Spiny bollworm.

American bollworm, cotton is far from being the most highly favoured. In Kenya, for instance, in a year of good rainfall the maize crop thrives and attracts most of the months, whilst in a year of poor rainfall cotton is the only available crop which is growing reasonably and it is therefore heavily attacked. Pigeon peas are preferred to cotton, and in areas where large quantities are grown they attract American bollworm from the cotton crop.

Spiny bollworm. *Earias spp.*

This pest seldom does as much damage as American bollworm. It cannot go into diapause and it has fewer alternate hosts; it can therefore be controlled to a certain extent by a close season.

The adult is an inconspicuous moth. The larvae characterised by short fleshy spines on most of their body segments (see Fig. 70); the colour is variable. They are smaller than *Heliothis* larvae, their maximum length being about $\frac{3}{4}$ in (c. 2 cm). In addition to eating bolls they may also eat the young shoots, causing death of the growing point. This damage causes stunting and the development of vigorous lateral growth. Spiny bollworms seldom hollow out the insides of bolls.

DDT does not control spiny bollworm efficiently; endosulfan or carbaryl should be used.

Pink bollworm. *Pectinophora (Platyedra) gossypiella*

This is smaller than the bollworms discussed above, being only $\frac{1}{2}$ in (c. 1·3 cm) long when mature. The larvae have a double pink bar on each segment of their bodies. The entry hole into each boll is minute, sometimes undetectable, and the larvae spend their life hollowing out the inside of one boll. They are often found inside seeds, characteristically in two seeds cemented together. Sometimes, shortly before they pupate, they bore a tunnel to the surface of the boll, leaving only a 'window' of epidermis. They then pupate inside the boll and the moths later emerge through these tunnels. Alternatively, and more commonly in warmer cotton areas, the larvae may drop to the ground and pupate in the soil, as is the case with *Heliothis*. Damaged bolls fail to open when they are thoroughly hollowed out. Pink bollworm larvae can only diapause if there are considerable changes in day length, so diapause is rare near the equator. In addition they have few alternate hosts, so they are usually of little importance if farmers have adhered to the regulations of the close season. In Central and Eastern Provinces in Kenya, where cotton is in

the ground almost all the year, pink bollworms have proved to be a serious pest. In the case of outbreaks, carbaryl gives fair control but DDT gives very little.

False codling moth. *Cryptophlebia leucotreta*

The larvae of the false codling moth are bollworms which occasionally cause serious damage in Uganda and the western parts of Kenya. They are about the same size as pink bollworms when mature but they are red above and have yellow underparts. They attack full size bolls. No satisfactory method of control is known.

Red bollworm. *Diparopsis castanea*

This bollworm spread from Mozambique to southern Tanzania in 1946. It has no alternate hosts other than cotton so a quarantine belt, approximately 150 miles wide and free of cotton, was immediately set up. This belt stretches from the southern boundary of Tanzania to a line which runs approximately from Kilwa on the coast to the southern end of Lake Rukwa. Permission was once given to grow cotton inside the quarantine belt, at Nachingwea, but there was soon an outbreak of this pest so permission was withdrawn.

Stainers. *Dysdercus spp.*

There are at least five species of stainer which occur in East Africa, differing slightly in the colouration of the abdomen, and differing markedly in their alternate hosts. Stainers are especially important throughout Kenya and in drier areas, e.g. the southern part of the Western Cotton Growing Area in Tanzania, where they survive the dry season on trees and herbs, e.g. baobab, kapok, *Hibiscus spp.* and *Abutilon*.

Stainers are sucking bugs which pierce the boll walls with their mouth parts and introduce a *Nematospora* fungus with their salivary juices as they suck the seeds. The fungus stains the lint and the grade of the seed cotton is consequently lowered. If stainers attack at an early stage they can cause shedding or mummification of the bolls. The adults are $\frac{1}{2}$–$\frac{3}{4}$ in (c. 1·3–1·9 cm) long with a characteristic red, brown and black marking on the upper side. They are easily observed by shaking cotton plants, whereupon they fall to the ground.

Stainers are best controlled by carbaryl. BHC gives some control but DDT and endosulfan are relatively ineffective.

Lygus. *Lygus (Taylorilygus) vosseleri*

This is a pest of the wetter areas, i.e. Uganda and

the lakeshore areas of Kenya. It is a small brown bug, about $\frac{1}{4}$ in (c. 0.6 cm) long, which sucks buds and, less frequently, young bolls. The leaves from damaged buds are tattered with many small holes and young bolls may be shed. Badly affected plants produce long internodes with short branches, giving a straggly appearance. Lygus attacks are especially severe where there are crops of sorghum nearby because this is an alternate host; when sorghum is harvested there are likely to be especially large outbreaks of Lygus on cotton. Lygus is efficiently controlled by DDT sprays.

Jassids. *Empoasca spp.*

Jassids are small, light green, sucking bugs about $\frac{1}{10}$ in (c. 0.25 cm) long, which fly away rapidly or hop sideways on the underside of a leaf if disturbed. The symptoms of jassid damage are curling down of the leaves and reddening of the leaf margins. Jassids used to cause extensive damage in East Africa but they have now lost in importance as the commonly grown varieties of cottons have been bred for leaf hairiness which discourages these insects. It is believed that leaf hairiness discourages egg laying. If hairless varieties are grown, DDT controls jassids effectively.

Aphids. *Aphis gossypii*

These black, yellow or green insects suck young shoots and the undersides of young leaves, causing the latter to curl downwards. Damage is seldom severe and yields are only affected if the lower leaves are shed. Outbreaks usually occur during dry spells: as soon as the rains resume aphids disappear. In the event of an exceptionally heavy outbreak of aphids, dimethoate or endosulfan should be used. A recent discovery is that the honey dew from aphids can cause damage to high speed spinning machinery if it falls onto the lint before picking.

Seed bugs. *Oxycarenus spp.*

These minute bugs are too small to get their mouthparts through the boll wall to suck the seeds, so they can only attack after the bolls have opened. Sucked seeds seldom germinate properly and seedbugs get amongst the seed cotton where their dead bodies cause discolouration. Chemical control against seed bugs is unnecessary but two cultural precautions can be taken: one is to pick at reasonably short intervals, thereby preventing the seed cotton from lying exposed in the field for too long; the other is to spread the seed cotton out in the sun after harvesting to allow any seed bugs to escape.

Calidea. *Calidea dregii*

This is a brightly coloured bug, mainly shiny blue, which feeds on bolls and which occurs in areas which are largely surrounded by bush. Carbaryl and BHC give fair control.

Red cotton mite (*Tetranychus telarius*) and Tea mite (*Hemitarsonemus latus*)

These mites are so small that they can barely be seen by the naked eye; they cause a yellow mottling of the leaves, then a reddening, followed by withering and leaf fall. As is often the case with mites, which are usually very immobile, attacks are normally confined to a single plant or to a small patch, so chemical control is rarely justified. Mite attacks are common after a series of DDT sprays.

Root knot nematodes

These are almost universal and are of special significance in the presence of *Fusarium* wilt (see section on diseases).

Pest control (general)

The wide range of pests described above is one of the most important problems facing cotton growers in East Africa. Cultural methods of control, although effective for some pests, cannot be relied upon; satisfactory yields can only be obtained if the crop is efficiently protected by insecticides.

Close season

A close season effectively controls the pests that have few alternative hosts and which cannot go into diapause. A strictly enforced close season can eliminate pink bollworm and can greatly reduce the incidence of spiny bollworm. All of the East African countries have legally enforced close seasons. In western Tanzania old plants must be uprooted and burned by 15th September. Uganda and Kenya have similar legislation, gazetting different dates for different areas each year.

Insecticidal dusts

Insecticidal dusts have been recommended in the past, especially in Kenya, when it was thought that sprays were impracticable because volumes as high as 20 gallons per acre (c. 225 litres/ha) were necessary. Dusts are much less efficient than sprays; they are easily dislodged by rain, they are difficult to apply evenly and they penetrate the crop much less effectively than sprays. They have not been recommended since it has been shown that insecticidal sprays are effective at considerably lower volume than 20 gallons per acre.

Yield increases from spraying

The benefit of a complete spraying programme can be great; the most dramatic responses are obtained in Kenya where a tenfold increase in yield can often be obtained by spraying, and where there can sometimes be a complete crop failure in the absence of insecticides. The figures below compare the yields of unsprayed cotton with those of cotton which received five applications of DDT; they are averages from 1963 to 1968 at Ukiriguru in Tanzania.

	Unsprayed	Sprayed	% increase from spraying
lb per acre	770	1140	48%
kg per hectare	863	1278	48%

Spraying often improves the quality of the seed cotton because damage by stainers may be reduced. It must always be stressed that economically worthwhile responses are only obtained if insecticides are applied to a crop which has been sown at the correct time and which has been properly thinned and weeded. Spraying badly grown cotton is a waste of both time and money.

Insecticides used

One of the main problems in cotton pest control is that no insecticide has been found which controls all the important pests. DDT controls American bollworms and Lygus but has little effect on spiny bollworms or stainers. Carbaryl controls stainers and spiny bollworms but has little effect on American bollworms. Endosulfan controls the bollworms but has little effect on stainers. At the time of writing, DDT and carbaryl are the most common insecticides; endosulfan has given good results in trials and may be used increasingly in the future.

Implements used

Almost all spraying is done by hand, using 2 gallon (c. 9 litres) knapsack sprayers and applying a spray volume of 8–12 gallons per acre (c. 90–135 litres/ha) depending on the availability of the water. An ultra low volume applicator was first tried in East Africa in 1969 and has given some promising results in Kenya, although it has not, as yet, been fully tested. It is battery operated and uses no water; the only liquid needed is $\frac{1}{2}-\frac{2}{3}$ gallon per acre (c. 5·5–7·5

litres/ha) of insecticide in the form of an emulsifiable concentrate. The operator walks down one inter-row in four or five (instead of every inter-row as with conventional hand spraying) allowing the wind to carry the insecticide across the rows. The most obvious advantage of using this machine is the reduction of time and physical effort involved in spraying. Tractor sprayers are seldom used even on large scale schemes; high clearance machines are needed to prevent damage to the crop and high rainfall, which usually coincides with the spraying period, can make the soil so muddy that tractor operations are impossible. Aerial spraying, which overcomes the two difficulties of tractor spraying, has been used on large scale schemes, e.g. at the irrigation scheme at Galole on the Tana River in Kenya, and on cotton blocks containing many individual adjoining plots in all three East African countries.

Timing of applications

The time when spraying should start depends on which pests are expected. In Uganda, where Lygus is one of the main pests, the first application should be made about six weeks after sowing because the buds can be damaged at this stage. In most other areas pests are attracted to the cotton crop at flowering, so applications should start about nine weeks after sowing. The recommended spraying interval depends largely on the persistence of the insecticide and the expected severity of the pest attack. DDT and carbaryl retain their effectiveness for only one or two weeks in the wet weather which usually occurs during the spraying period. In Kenya, where the pest complex is unusually damaging, the recommended interval is therefore 7–10 days. In Uganda and Tanzania it is two weeks. In Uganda four applications are recommended, although trials show economic benefits from eight or more. In Tanzania six applications are recommended whilst in Kenya, where a more flexible approach has been adopted (see below), as many as ten applications may be needed. It is important to apply the recommended number of sprays, especially if the more persistent and unselective DDT is being used, because there is a risk that an incorrect number of applications may kill not only the pests but also their insect predators and parasites; in this case the poor balance between parasites and pests could cause a rapid pest increase at a later date.

Variations in pest incidence

Cotton pests vary from place to place, from time to

time within the season, and from season to season. In Uganda and Western Province in Kenya, Lygus and American bollworms are the main pests. In the Western Cotton Growing Area of Tanzania, American and spiny bollworms and, towards the south, stainers do most damage. In Kenya, especially in Central and Eastern Provinces, the pest complex is more diverse, less predictable and usually more damaging than in other areas.

Within the season, Lygus often attacks before flowering. American bollworms are usually the first of the bollworms to appear after flowering, spiny bollworms usually appear later whilst stainers and mites usually attack towards the end of the season.

This variation within the season and the variation from year to year of both the nature and the intensity of the pest attack are severe problems. A farmer who follows a rigid spraying programme may be spraying when no pests are present or he may not be spraying when there is an unexpectedly early or late pest attack. Alternatively he may be applying an unsuitable insecticide. For these reasons more flexible programmes are being studied in East Africa and one has been introduced in Central and Eastern Provinces in Kenya; DDT is recommended if a certain number of American bollworms is observed per 100 plants, whilst carbaryl is recommended if there is a certain population of spiny bollworms or stainers. This is undoubtedly a more efficient programme but it requires a high degree of observational skill and very rigorous and regular inspection.

Diseases

Bacterial blight

This bacterial disease, caused by *Xanthomonas malvacearum,* may occur at any stage of the plant's life cycle; soft tissue of the leaves, stems or bolls may be affected. Seedling attack, usually called seedling blight, results in small water soaked patches on the cotyledons and young leaves. If the vascular tissues are affected cotyledons and leaves may turn brown and dry up. In older plants black lesions may occur on the stems and whole branches may break off; this phase is called blackarm. Bacterial boll rot may occur on young bolls, badly affected bolls failing to develop, whilst milder infection results in stained cotton in some of the loculi.

The disease is carried over from season to season partly by means of infected plant debris, provided that the soil is dry, but mostly by bacteria on the seed coats. A close season therefore gives a certain degree of control, but a copper fungicide seed dressing, as is applied by all of the Lint and Seed Marketing Boards, is more effective.* A bromine compound called Bronopol has given good results and may be used in the future. Mercury is a more efficient seed dressing but is very toxic to mammals. A copper seed dressing at 1:200 w/w should give a 95–100% healthy stand, compared with an approximately 30% healthy stand if susceptible varieties are grown and are not seed dressed. By introducing the variety Albar into the East African breeding programme, the present seed issues have been made resistant to bacterial blight. The combination of seed dressing and breeding has been successful in greatly reducing the importance of bacterial blight, although in wet conditions, especially after hail, heavy outbreaks can still occur. In such weather the bacteria travel rapidly from plant to plant by means of rain splash and once infection has occurred there is no cure.

Fusarium wilt

This fungal disease, caused by *Fusarium oxysporum,* causes stunting, yellowing of the leaves, brown dry patches which often occur between the veins of the leaves and brown staining of the wood which can be seen when the bark is peeled away. In the later stages of attack the leaves are shed. Resprouting from the base of the plant is common.

Fusarium wilt can be transmitted by infected seed (on the inner side of the seed coat), by infected plant residues or by infested soil. Seed dressings fail to kill the fungus because it is not situated on the surface of the seed; rotations are ineffective because infection can still occur after many years of resting a field from cotton. The only effective action against *Fusarium* is the use of resistant varieties; recent seed issues from Ukiriguru, e.g. UK61, are moderately wilt resistant. The disease has been slowly spreading ever since it was first observed near Geita, Tanzania, in 1953; at the time of writing it is still restricted, with a few scattered exceptions, to the lakeshore areas, where resistant varieties are now being issued. Nothing practical can be done to prevent spread by particles of soil on feet, implements, wheels etc. but precautions can be taken to prevent the widespread distribution of infected seed; seed husk is used for a number of purposes, particularly as packing material in lorries, and this can spread *Fusarium* wilt very rapidly.

*Only in Coast Province, Kenya, is cotton seed not dressed with fungicides.

Nematodes may be connected with the incidence of this disease (as is the case with Panama disease of bananas: *Fusarium oxysporum* f. sp. *cubense*) and it is assumed that these organisms wound the roots sufficiently to encourage the entry of the fungus.

Verticilium wilt
The symptoms of this fungal disease, which is caused by *Verticillium dahliae*, are similar to those of *Fusarium* wilt. *Verticillium* wilt is only spread by infested soil, so small groups of plants are infected rather than large patches, as with *Fusarium*. This disease occurs in the coastal regions of Tanzania, in parts of Kenya (especially near Kibos) and in Uganda. It is not important, largely because of the resistance shown by the East African varieties.

Bibliography

General
1   Brown, H. B. and Ware, J. P. (1958). *Cotton.* McGraw-Hill.

2   Christidis, B. G. and Harrison, G. J. (1955). *Cotton Growing Problems.* McGraw-Hill.

3   Ingram, W. R. (1965). *A survey of the cotton grown in Uganda in 1963.* Empire Cotton Gr. Rev., *42*:1.

4   Kibukamusoke, D. E. B. (1962). *Competitive effects of coffee on cotton production in Buganda.* Empire Cotton Gr. Rev., *39*:106.

5   Manning, H. L. (1962). *Increasing cotton production from Uganda.* Empire Cotton Gr. Rev., *39*:1.

6   Munro, J. M. (1966). *Cotton expansion in Kenya.* Empire Cotton Gr. Rev., *43*:140.

7   Smith, R. and Brown, R. A. (1963). *Improved methods of cotton growing in Eastern Region, Tanganyika.* Empire Cotton Gr. Rev., *40*:268.

Physiology, structure, etc.
8   Dale, J. E. (1962). *Fruit shedding and yield in cotton.* Empire Cotton Gr. Rev., *39*:170.

9   Lea, J. D. (1961). *Studies on the depth rate of root penetration of some annual tropical crops.* Trop. Agriculture, Trin., *38*:93.

10 Low, A. (1968). *Development toward tufted seed in varieties of cotton.* Cotton Gr. Rev., *45*:101.

11 Morris, D. A. (1964). *Variation in the boll maturation period of cotton.* Empire Cotton Gr. Rev., *41*:114.

Climate
12 Hutchinson, J. B. *et al.* (1958). *Crop water requirements of cotton.* J. Agric. Sci., *51*:177.

13 Jowett, D. and Eriaku, P. O. (1966). *The relationship between sunshine, rainfall and crop yields at Serere Research Station.* E. Afr. agric. for. J., *31*:439.

14 Russell, E. W. (1963). *Water requirements of the cotton crop.* Empire Cotton Gr. Rev., *40*:246.

15 Walton, P. D. (1962). *Estimates of water use by cotton crops at Serere, Uganda.* Empire Cotton Gr. Rev., *39*:241.

See also references 20, 22 and 29

Breeding
16 Arnold, M. H. *et al.* (1968). *BPA and SATU, Uganda's two new cotton varieties.* Cotton Gr. Rev., *45*:162.

17 Peat, J. E. and Brown, K. J. (1961). *A record of cotton breeding for the Lake Province of Tanganyika: seasons 1939–40 to 1957–8.* Emp. J. exp. Agric., *29*:119.

18 Spence, J. R. and Smithson, J. B. (1966). *Ukiriguru seed issues, their origins and characteristics.* Progress Report No. 5 of the Western Research Centre, Ukiriguru.

19 Spence, J. R. and Smithson, J. B. (1967). *The performance of Ukiriguru seed issues and their potential successors in the 1966–7 season.* Progress Report No. 13 of the Western Research Centre, Ukiriguru.

Ridging
20 Brown, K. J. (1963). *Rainfall, tie-ridging and crop yields in Sukumaland, Tanganyika.* Empire Cotton Gr. Rev., *40*:34.

21 Le Mare, P. H. (1954). *Tie ridging as a means of soil and water conservation and yield improvement.* Proc. 2nd Inter-African Soils Conf., Leopoldville, 1954, p. 595.

22 Peat, J. E. and Brown, K. J. (1960). *Effect of management on increasing crop yields in the Lake Province of Tanganyika.* E. Afr. agric. for. J., *26*:103.

23 **Walton, P. D.** (1962). *Cotton agronomy trials in N. and E. Provinces of Uganda, 1956 to 1961.* Empire Cotton Gr. Rev., *39*:114.

Time of sowing
24 **Reed, W.** (1964). *Problems posed by early sowing of cotton in Lake region, Tanganyika.* Empire Cotton Gr. Rev., *41*:255.

25 **Rijks, D. A.** (1967). *Optimum sowing date for yield; a review of work in the BP52 cotton area of Uganda.* Cotton Gr. Rev., *44*:247.

26 **Ruston, D. F.** (1962). *Effects of delay in sowing cotton.* Empire Cotton Gr. Rev., *39*:10.

See also references 3, 22 and 23

Spacing
27 **Grimes, R. C.** (1963). *Inter-cropping and alternate row cropping of cotton and maize.* E. Afr. agric. for J., *28*:161.

See also references 3, 15 and 23

Weed control
28 **Church, J. M. F.** (1969). *Evaluation of herbicides for use in cotton.* Cotton Gr. Rev., *46*:245.

Fertilisers and manures
29 **Brown, K. J.** (1962). *Effect of early rainfall on the response of cotton to nitrogen.* Empire Cotton Gr. Rev., *39*:177.

30 **Machado, V. I. L. S.** (1966). *External cotton*

Progress Report No. 7 of the Western Research Centre, Ukiriguru.

31 **Peat, J. E.** and **Brown, K. J.** (1962). *The yield responses of rain-grown cotton at Ukiriguru in the Lake Province of Tanganyika, I—The use of organic manure, inorganic fertilisers and cotton-seed ash.* Emp. J. exp. Agric., *30*:215.

32 **Peat, J. E.** and **Brown, K. J.** (1962). *The yield responses of rain-grown cotton at Ukiriguru in the Lake Province of Tanganyika, II—Land-resting and other rotational treatments contrasted with the use of organic manure and inorganic fertilisers.* Emp. J. exp. Agric., *30*:304.

33 **Scaife, A.** (1968). *The effects of a cassava 'fallow' and various manurial treatments on cotton at Ukiriguru, Tanzania.* E. Afr. agric. for. J., *33*:231.

34 **Stephens, D.** (1966). *Two experiments on the effects of heavy applications of triple superphosphate on maize and cotton in Buganda clay loam soil.* E. Afr. agric. for. J., *31*:283.

35 **Stephens, D.** (1967). *The effects of different nitrogen treatments and of potash, lime and trace elements on cotton, on Buganda clay loam soil.* E. Afr. agric. for. J., *32*:320.

36 **Stephens, D.** (1968). *Fertiliser trials on cotton and other annual crops on small farms in Uganda.* Expl Agric., *4*:49.

37 **Ingram, W. R.** and **Davies, J. C.** (1966). *Comparison of hand-operated spraying machines for cotton pest control in Uganda, II—Entomological assessment.* E. Afr. agric. for. J., *31*:416.

38 **Jones, T. R.** (1964). *The Uganda long boom.* E. Afr. agric. for. J., *29*:347.

39 **Jones, T. R.** (1966). *Comparison of hand-operated machines for cotton pest control in Uganda, I—Description of machines under test.* E. Afr. agric. for. J., *31*:409.

40 **Joy, J. L.** and **Lea, J. D.** (1960). *Some economic aspects of mechanised cotton farming at Namulonge, Uganda.* Emp. J. exp. Agric., *28*:223.

41 **Nutt, G. B.** (1964). *Harvesting cotton mechanically.* Empire Cotton Gr. Rev., *41*:163.

Pests and pest control
42 **Anon.** (1967). *Insecticides and the African cotton crop.* Cotton Gr. Rev., *44*:167.

43 **Coaker, T. H.** (1959). *Investigations on Heliothis armigera in Uganda.* Bull. ent. Res., *50*:487.

44 **Crowe, T. J.** (1967). *Cotton pests and their control.* Kenya Department of Agriculture pamphlet.

45 **Dale, J. E.** and **Coaker, T. H.** (1961). *Growth and yield of cotton sprayed with DDT in East and North Uganda.* Emp. J. exp. Agric., *29*:1.

46 **Davidson, A.** (1964). *Control of insect pests of cotton in E. Region, Kenya.* Empire Cotton Gr. Rev., *41*:276.

47 **Davies, J. C.** (1963). *Insecticides on cotton in Uganda.* Empire Cotton Gr. Rev., *40*:296.

48 **Davies, J. C.** (1964). *Experiments on cotton using DDT miscible liquids in Eastern Uganda in the period 1959–63.* E. Afr. agric. for. J., *29*:343.

49 **Davies, J. C.** (1967). *Spray interval studies on cotton in Eastern Uganda in 1961–5.* E. Afr. agric. for. J., *33*:37.

50 **Davies, J. C.** (1970). *Effects of spraying and cultural practices on cotton in Uganda.* Expl Agric., *6*:65.

51 **Davies, J. C.** and **Ingram, W. R.** (1965). *Insecticides on cotton in Uganda.* Empire Cotton Gr. Rev., *42*:300.

52 **Davies, J. C.** and **Kasule, F. K.** (1964). *A note on the relative importance of* Heteroptera *and bollworms as cotton pests in Eastern Uganda.* E. Afr. agric. for. J., *30*:69.

53 **Ingram, W. R.** (1967). *Insecticides on cotton in Uganda.* Cotton Gr. Rev., *44*:203.

54 **Ingram, W. R.** (1967). *An evaluation of five insecticides for the control of late-season cotton pests in Uganda.* E. Afr. agric. for. J., *33*:206.

55 **Ingram, W. R.** and **Davies, J. C.** (1964). *Insecticides on cotton in Uganda.* Empire Cotton Gr. Rev., *41*:124.

56 **Ingram, W. R.** and **Davies, J. C.** (1965). *Recent advances in pest control on cotton in Uganda.* E. Afr. agric. for. J., *31*:169.

57 **Kerridge, P.** et al. (1969). *Insecticide use on the 'sucking pest areas' of Western Tanzania.* E. Afr. agric. for. J., *35*:147.

58 **la Croix, E. A. S.** (1964). *Insecticide dusts on cotton in Coast Province, Kenya.* Empire Cotton Gr. Rev., *41*:137.

59 **la Croix, E. A. S.** (1966). *Stainer bugs in Coast Province, Kenya.* Empire Cotton Gr. Rev., *43*:41.

60 **McKinlay, K. S.** (1956). *Cotton pest control in Eastern Province, Tanganyika.* E. Afr. agric. J., *22*:20.

61 **Pearson, E. O.** (1958). *The insect pests of cotton in tropical Africa.* Empire Cotton Growing Corporation and the Commonwealth Institute of Entomology.

62 **Reed, W.** (1965). Heliothis armigera *in Western Tanganyika. I—Biology, with special reference to the pupal stage.* Bull. ent. Res., *56*:117.

63 **Reed, W.** (1965). Heliothis armigera *in Western Tanganyika. II—Ecology and natural and chemical control.* Bull. ent. Res., *56*:127.

64 **Reed, W.** (1966). *Recommendations for the protection of cotton shambas against bollworms in the Western zone of Tanzania.* Progress Report No. 1 of the Western Research Centre, Ukiriguru.

65 **Swaine, G.** (1955). *The cotton red bollworm problem in Southern Tanganyika.* E. Afr. agric. J., *20*:183.

See also references 3, 6, 37, 38 and 39

Diseases and disease control

66 **Arnold, M. H.** (1963). *The control of bacterial blight in rain-grown cotton, I—Breeding for resistance in African Upland varieties.* J. Agric. Sci., *60*:415.

67 **Arnold, M. H.** (1965). *The control of bacterial blight in rain-grown cotton, II—Some effects of infection on growth and yield.* J. Agric. Sci., *65*:29.

68 **Arnold, M. H.** and **Arnold, K. M.** (1961). *Bacterial blight of cotton; trash borne infection.* Empire Cotton Gr. Rev., *38*:258.

69 **Brown, A. G. P.** (1964). *Field trials of three fusarium wilt resistant cotton selections in Tanganyika.* Empire Cotton Gr. Rev., *41*:194.

70 **Brown, A. G. P.** (1966). *Fusarium wilt of cotton: the present situation.* Progress Report No. 4 of the Western Research Centre, Ukiriguru.

71 **Brown, A. G. P.** (1968). *Effects of fumigation to control nematodes on fusarium wilt of cotton.* Empire Cotton Gr. Rev., *45*:128.

72 **Brown, S. J.** (1969). *Bacterial blight of cotton: a note on trash borne infection in Uganda.* Cotton Gr. Rev., *46*:197.

73 **Perry, D. A.** (1962). *Fusarium wilt of cotton in the Lake Province of Tanganyika.* Empire Cotton Gr. Rev., *39*:14.

74 **Perry, D. A.** (1963). *Interaction of root knot and fusarium wilt of cotton.* Empire Cotton Gr. Rev., *40*:41.

75 **Wickens, G. M.** (1964). *Fusarium wilt of cotton; seed husk a potential means of dissemination.* Empire Cotton Gr. Rev., *41*:23.

15
Cowpeas

Vigna unguiculata (sinensis)

Introduction

Cowpeas are a dual leguminous crop. In most areas they are grown more for their leaves than for their seeds; this is especially true where rainfall is high and seed production, because of insect damage, is low.

In Uganda cowpeas are important in Teso and Lango Districts. In Kenya and Tanzania they are the most important legume at the coast and in Kakamega District (Kenya). They are also important in areas of marginal rainfall, e.g. Machakos District in Kenya and Central Province in Tanzania. They are a minor crop in almost all other parts of East Africa except the highlands of Kenya.

Plant characteristics

Cowpeas are an annual crop. Varieties may be spreading, semi-upright or erect in growth habit. Their flowers may be purple, pink, white, blue or yellow. The pods of most varieties hang downwards but in some varieties they point sideways or upwards. The seeds may be white, cream-coloured purple, brown, mottled brown or black. In most cases in East Africa the percentage of cross-pollination is only about 2%. The duration of the crop depends largely on the growth habit, the rainfall and local practice but it is seldom more than five or six months.

Ecology

In high rainfall areas, e.g. Teso and Lango Districts in Uganda and Kakamega District in Kenya, cowpeas are attacked by a wide range of insect pests and diseases; the former are particularly damaging and usually cause low yields of seed. Cowpeas need warm conditions and only give good yields of seed below about 5 000 ft (c. 1 500 m). At such altitudes rainfall is often marginal in Kenya and Tanzania, but cowpeas are drought resistant and can give better yields in these conditions than beans (*Phaseolus vulgaris*). They are intolerant of waterlogging and must therefore be grown on free draining soils.

Fig. 71: A cowpea plant.

Varieties

Cowpeas have undergone selection in Tanzania (first at Ilonga and later at Ukiriguru) since 1958. In Uganda further selections have been made at Makerere University College; this work started in 1966 as part of a programme on all aspects of growing cowpeas. Several high yielding, disease resistant varieties have been identified and are due for release during the 1970s. Varieties which yield well in one locality do not necessarily do well in others, so different varieties will be recommended for different areas.

Field operations

Most cowpeas are intersown with other food crops. Pure stands are sometimes established near the homesteads for their leaves. Experiments have shown that within reasonable limits spacing has no significant effect on seed yields; a spacing of 2 ft × 1 ft (c. 0·6 m × 0·3 m) strikes a reasonable balance between economising with seed and obtaining a sufficiently rapid cover to suppress weeds. In some areas, e.g. Kakamega District in Kenya, broadcasting is more common than dibbling. An experiment in Uganda has shown good responses to applications of single superphosphate.

Harvesting

To make a good spinach the leaves must be young and tender; the best are about the third and fourth from the apical ends of the shoots. Experiments in Uganda and Tanzania have shown that removing all the tender leaves three times at weekly intervals, starting five or seven weeks after sowing, has no adverse effect on grain yields although flowering may be delayed. Pods are usually removed individually as they ripen and are laid out in the homestead to dry.

Yields

Average yields are estimated to be 300–400 lb per acre (c. 340–450 kg/ha) but with good husbandry (without using insecticides) 600–800 lb per acre (c. 670–900 kg/ha) should be possible. In Uganda yields of about 2 000 lb per acre (c. 2 200 kg/ha) have been obtained by using insecticides on some of the better selections.

Pests

Pests are probably the most important factor limiting yields of seed, especially in the wetter areas. The most damaging are pod borers (*Maruca testulalis*), blossom beetles (*Coryna spp.*), thrips and a pod sucking insect (*Acanthomia horrida*). In an experiment in Uganda, DDT sprays, starting at flowering and continuing at weekly intervals, increased yields by about 40%. Seed quality was also improved. Insecticides which are toxic to humans must not be applied before the removal of leaves for spinach.

Diseases

Diseases can do much damage in the wetter areas but some of them also attack in the drier areas. The most important diseases are zonate leaf spots

(*Ascochyta phaseolorum* and *Dactuliophora tarii*) and pseudo rust (*Synchytrium dolichi*). Varietal resistance is the main weapon against these.

Utilisation

Leaves are usually crushed, fried and then boiled although they may simply be boiled. They are sometimes dried and ground into a powder which can be stored for later consumption.

The seeds may be boiled with maize or boiled in their pods. Alternatively their seed coats may be removed, after which they are boiled or fried to make a paste or sauce which can be eaten with 'ugali'.

16
Finger millet

Eleusine coracana

Introduction

Finger millet is a cereal crop which is important in Uganda and western Kenya. Its greatest advantage is that it can be stored for a longer period, without the use of insecticides, than any other cereal. As its seeds are small, they dry out quickly and insects cannot live inside them. Satisfactory storage for as long as ten years has been recorded although the viability of the seed drops after such a period. This advantage is of greatest significance in Uganda where the humidity usually prevents safe storage of the larger seeded cereals, notably maize. Because it stores so well, finger millet is sometimes termed a famine crop and during the colonial era each farmer in some of the northern and eastern parts of Uganda was required to grow a certain acreage.

Against its good storage properties, the following disadvantages must be considered: finger millet has a lower yield capacity than maize and requires more labour at all stages, particularly for seedbed preparation, weeding, bird scaring, harvesting and threshing. For these reasons it has declined in importance in Kenya and Tanzania whilst maize has become more important. Only in Uganda, where the climate is unfavourable for maize storage and where maize is usually considered less palatable, has finger millet retained its importance.

Uganda
The main finger millet growing districts, where it is the staple food, are Acholi, Lango and Teso. Large quantities are also grown in Bugisu, Busoga, Bukedi, West Nile, Ankole and Kigezi Districts. In 1968 an estimated 1 357 000 acres (c. 550 000 ha) were grown in Uganda.

Kenya
Finger millet is rarely grown east of the Rift Valley. In all well populated parts west of the Rift Valley, however, it is widely grown, especially in Kericho, Nandi, Kisii, Busia and Bungoma Districts.

Tanzania
North Mara District is the only place in Tanzania where finger millet is of any importance.

Plant characteristics

Finger millet seldom grows higher than four feet (c. 1·2 m). It tillers freely and has narrow grass-like leaves. The inflorescence consists of about six spikes each about four inches (c. 10 cm) long, and these are digitally arranged (see Fig. 73), giving the crop its name. The seeds are $\frac{1}{20}-\frac{1}{10}$ in (c. 1–2 mm) in diameter and the crop is almost entirely self-pollinated.

Ecology

Rainfall, and water requirements
Finger millet tolerates dry spells in the early stages of growth. In Uganda it is often sown in January or February, well before the main rains start and when rainfall is unreliable; the main object of

Fig. 72: Finger millet.

15 cm

1 in

Fig. 73: A finger millet head.

Station in Uganda in 1966. Several promising varieties have been selected and are due for release during the 1970s. They are high yielding and show resistance to lodging and blast (*Piricularia oryzae*).

Rotations

In Kenya and Tanzania finger millet is often grown as the first crop after clearing the land; at this stage of the arable break the weed population is at its lowest. In Uganda it is usually grown early in the arable break after cotton. Growing early in the arable break prevents it from growing in a soil which has been exhausted by previous cropping; growing it after cotton should provide it with a reasonably clean soil, provided the cotton was well weeded. Most finger millet is sown in the first rains in Uganda; these are the more suitable because they are the more reliable. Finger millet is better suited to the first rains than other cereals because it is not harmed when stored during the wet months which follow; other cereals do not dry quickly enough to store safely at this time and are better suited to the second rains.

Field operations

Seedbed preparation
Seedbed preparation should be thorough. The first reason for this is that the seed is small and is usually broadcast; an even distribution and coverage of the seed is only possible, therefore, if there is a fine seedbed. The second reason is that weed control is unusually difficult in finger millet; thorough seedbed preparation reduces weed competition in the early stages of the crop's growth.

When clearing the land and incorporating into the soil a large amount of organic matter which is low in nitrogen, soil nitrogen may be made unavailable and the following crop of finger millet may suffer. This effect appears to be recognised by most growers. In Kenya and Tanzania the vegetation is removed by digging it up; it is then piled into heaps, together with much of the topsoil, and burned. Brushwood from beyond the field is often added. The combination of ash and earth is later scattered.

Time of sowing
There is little experimental data to indicate the best time for sowing finger millet but it is usually assumed that the best yields are obtained by

sowing so early is to allow a second crop to be grown within the year. After the first month it requires a steady supply of soil moisture if good yields are to be obtained. It is usually grown in areas which receive an average annual rainfall of at least 35 in (c. 900 mm).

Altitude and temperature
Finger millet grows well from sea level up to 8 000 ft (c. 2 400 m) but owing to its reliance on a steady rainfall it is mostly grown above 3 000 ft (c. 900 m).

Soil requirements
Finger millet only yields well on fertile free draining soils.

Varieties

A breeding programme started at Serere Research

sowing as early as possible. As mentioned above, it is often sown in the 'grass rains' of January and February in Uganda; the earlier it is sown in these months the less it competes for labour with cotton, which is sown later in the year.

Spacing and sowing

Finger millet is almost always broadcast by hand; its seeds are so small and the optimum population is so high that dibbling would be difficult and labour-consuming. Sowing in rows is being encouraged in Uganda; 12–13 in (c. 30–33 cm) rows are recommended with the plants thinned to about 2 in (c. 5 cm) apart within the rows. This is best done by using ox-drawn seeders, which are becoming increasingly popular in Uganda. About 30 lb of seed per acre (c. 35 kg/ha) are needed for broadcasting whilst only 5–8 lb per acre (c. 5·5–9·0 kg/ha) are needed for sowing in rows. Wheat drills (see Fig. 149) have been used successfully for sowing finger millet on experimental stations and on a few large scale farms in Kenya.

About half of the finger millet in Uganda is sown in pure stands whilst most of the remainder is intersown with sorghum. In Kenya and Tanzania pure stands are more common.

Fertilisers

The current recommendation in Uganda is to broadcast 1 cwt of ammonium sulphate per acre (c. 125 kg/ha) when the plants are about 6 in (c. 15 cm) high. Experiments have shown that this usually gives yield increases of 500–700 lb of dried heads per acre (c. 550–800 kg/ha).

Weed control

Hand weeding is very labour consuming in finger millet. The first weeding, which is combined with thinning, must be done when the crop is only about 3 in (c. 7·5 cm) high; at this stage the plants are delicate and are difficult to distinguish from some grasses. Subsequent weedings require less precision but still require much labour because the crop is spaced so closely that a 'jembe' cannot be used. Thorough seedbed preparation and sowing in rows both reduce the labour needed for weeding. The most popular implements for hand weeding are pointed or notched sticks and pieces of metal, especially pieces of the bands used to secure cotton bales. When finger millet is sown in rows, ox-drawn weeders are usually used.

Eleusine africana and *E. indica* are common weeds. They often occur in large numbers because it is impossible for farmers to distinguish them from finger millet in the early stages of growth.

Harvesting

Finger millet straw is so strong that it is not easy to break the heads off by hand. Small hand knives are usually used to remove individual heads; the heads are stored without threshing. Combine harvesters have been used successfully on a field scale at experimental stations and on a few large scale farms in Kenya.

Yields

Yields are usually between 400 and 800 lb of dried grain per acre (c. 450–900 kg/ha) although 1 500 lb per acre (c. 1 650 kg/ha) can easily be obtained with good husbandry. Large scale farmers in Kenya have produced 3 000 lb per acre (c. 3 400 kg/ha) and work in Zambia has shown that 4 000 lb per acre (c. 4 500 kg/ha) can be achieved.

Pests and diseases

Finger millet is seldom damaged by pests other than birds which may be a problem when the grains are in the soft stage of development. The most damaging disease is head blast (*Piricularia oryzae*) which also attacks rice (see Chapter 26). Resistant varieties are being developed at Serere.

Utilisation

In Uganda and western Kenya finger millet is usually fried a little, before being mixed with dried cassava chips. The two are then ground into a flour which is used for making 'ugali' or 'uji'. When finger millet is used for brewing the seeds are germinated by being soaked in a sack for about a day and then left in the shade for two or three days. They are then dried in the sun and coarsely ground. Meanwhile, maize, sorghum or finger millet flour (which has been fermented in water for about a week) is fried. The ground germinated grains and the fried fermented flour are then mixed together and are placed in pots of water where they ferment further for two or five days. There are several variations of this method of brewing.

Finger millet straw can be used for thatching, for making the walls of small granaries or for making plates and various types of food containers.

17
Grams

Green gram: *Vigna aureus*
Black gram: *Vigna mungo*

Introduction

Grams are minor legumes in some of the lower parts of East Africa. In Kenya they are common in Central and South Nyanza Provinces, in Machakos and Kitui Districts and at the coast. In Tanzania they are grown in most low dry areas. In Uganda they are less common but are sometimes grown commercially in parts of Masaka District for sale to the Asian community. Green grams are more common than black grams throughout East Africa.

Plant characteristics

Green grams and black grams differ in the following respects: green grams have small round green seeds with a flat hilum whilst black grams have larger oblong dark seeds with a concave hilum. The pods of green grams are only slightly hairy, contain 10–15 seeds and are not beaked, whilst those of black grams are conspicuously hairy, contain 6–10 seeds and are beaked. Green gram pods point sideways or downwards whilst black gram pods point upwards. Green grams are erect whilst black grams are spreading in their growth habit.

Ecology

Grams grow best from sea level up to 5 000 ft (c. 1 500 m). They are drought resistant and give reasonable yields in areas which receive an average annual rainfall of only 25 in (c. 650 mm).

Field operations

There is almost no experimental data on grams in East Africa. A tentative spacing recommendation is 18 in × 6 in (c. 46 cm × 15 cm); this requires 10–15 lb of seed per acre (c. 11–17 kg/ha). Sowing in rows is uncommon; grams are usually intersown with cereals, root crops and other legumes or broadcast as a pure stand. Grams mature quickly, in 2½–3 months, and are therefore well suited to the shorter of the two rains in areas which have a bimodal rainfall distribution.

Harvesting and yields

Grams are usually harvested by removing individual pods as they ripen although the plants are sometimes uprooted with the pods still attached. Yields of dry seed are very variable, largely owing to the unpredictable attacks of pests and diseases. They are usually between 200 and 400 lb per acre (c. 200 and 450 kg/ha) when grown in a pure stand although they have occasionally been as high as 1 000 lb per acre (c. 1 100 kg/ha).

Pests

Pest damage is probably the most important factor limiting yields of grams. The most damaging pests are the bean fly (*Melanagromyza phaseoli*), American bollworm (*Heliothis armigera*) and the bean aphid (*Aphis fabae*).

Diseases

Important diseases include powdery mildew (probably *Erysiphe polygoni*), leaf spot and pod scab (*Asochyta phaseolorum*), target leaf spot (*Dactuliophora tarii*) and a bright yellow mosaic mottle disease caused by a virus.

Utilisation

Grams can be used as a subsistence pulse crop, their seeds being prepared in the same way as cowpeas. Alternatively, they can be sold as a cash crop to the Asian community. Asians use green grams as a split pulse or a sprouting bean. They use black grams, because of their glutenous proteins, for grinding into flour.

18
Groundnuts

Arachis hypogea

Introduction

Most groundnuts in East Africa are grown as a subsistence pulse crop. Some, however, are sold to local oil mills; groundnut oil, which constitutes 40–50% of the kernels, can be used for cooking oil, margarine, salad oil and other edible products. A small proportion of East Africa's groundnuts is exported for the confectionery trade; the kernels must be of good quality (see especially the problem of aflatoxins in the section on diseases) and large (40–50 per oz: 140–180 per 100 g).

Uganda
Groundnuts are most widely grown in Busoga, Bukedi, Teso and Mengo Districts. They are also grown in all other areas except the highest and driest. An estimated 444 000 acres (c. 180 000 ha) were sown in Uganda in 1968.

Tanzania
Groundnuts are grown in all areas below 5 000 ft (c. 1 500 m) which receive a reasonable rainfall, e.g. at the coast, around the lakes and in the south.

In the late 1940s the British Government made an abortive attempt to grow groundnuts on a very large scale in Tanganyika. A total of 3¼ million acres (c. 1 300 000 ha) was planned, based on Kongwa, Urambo and Nachingwea. A number of factors led to the scheme's downfall; these included the unforseen problems of large scale mechanisation, insufficient rainfall data, unsuitable varieties and administrative difficulties. Kongwa later became a beef ranch. Urambo has become an important smallholder tobacco area whilst at Nachingwea, during the 1950's, a scheme was implemented for growing maize, sorghum, groundnuts and soya beans on a large scale; its acreage and production targets were much lower than originally planned.

Kenya
Groundnuts are relatively unimportant in Kenya. They are seldom grown east of the Rift Valley because of heavy attacks by leaf hopper. They are grown for home consumption in Western and Central Nyanza Provinces and are grown around Homa Bay in South Nyanza as a cash crop for export.

Plant characteristics

The groundnut plant is a low growing annual legume. The roots are almost always well nodulated. The two main groups of varieties exhibit, respectively, sequential or alternate branching. Plants with sequential branching have an upright main stem and five or six branches carried at an acute angle to it; secondary branching does not normally occur. These plants therefore have an erect bunch growth habit. Plants with alternate branching have a main stem whose more widely angled branches produce first a pair of secondary branches, then a pair of inflorescences, then another pair of secondary branches and so on. These plants therefore have a more spreading growth habit; the more spreading are known as runner types and the less spreading as spreading bunch types. The leaves of groundnuts, unlike those of most other pulse crops which are trifoliate, are pinnate.

Each inflorescence bears several yellow flowers; these appear between one and two months after sowing. Pollination usually occurs before the flowers open; cross-pollination is therefore rare. After pollination the yellow floral parts are shed, the meristem at the base of the ovary becomes active and grows to form a peg. The peg grows downwards bearing the ovary at its tip; the tip of the peg is lignified, thus protecting the ovary when it penetrates the soil. The growth of the peg puts the ovary in a horizontal position a few inches below the soil surface.

The fruit is an elongated pod containing one to six seeds; these are covered by a thick fibrous shell. The seeds have a testa which is paper-like when dry and which may be white, pink, red, purple, brown or variegated.

The time from sowing to maturity depends mostly on the variety and on the altitude. In

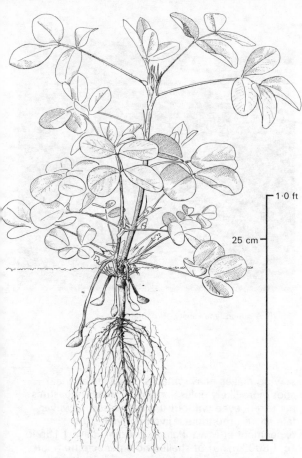

Fig. 74: A groundnut plant of a bunch variety with pods in an early stage of development.

Uganda most erect bunch varieties mature in 90–110 days whilst spreading bunch varieties mature in 120–130 days; runners may take even longer.

Ecology

Rainfall, and water requirements
Good distribution of rainfall is essential for satisfactory yields and ease of field operations. Until harvesting the soil must be moist; droughts can cause poor vegetative growth, poor peg penetration if the soil is hard and the development of small shrivelled nuts. Wet seasons of three months' duration, as are common in areas with a bimodal rainfall distribution, are usually sufficient for growing the quick growing erect bunch

varieties. Spreading bunch and runner varieties, however, need a longer period of rainfall and are better suited, in Uganda, to the areas in the north which have a single long wet season.

Dry conditions are needed for harvesting and drying, but the soil must not be too dry, otherwise many nuts remain in the soil after pulling. In wet weather soil adheres to the nuts, the seeds of non-dormant varieties may germinate and aflatoxins (see below) may develop.

Altitude and temperature
Groundnuts need a warm climate and are only grown in East Africa below 5 000 ft (c. 1 500 m).

Soil requirements
Groundnuts grow well in all reasonably fertile, light soils. Heavy soils present problems because it is difficult to remove the nuts from the ground and because pieces of soil adhere to them.

Varieties

The broad classification of the sequential branching, Spanish-Valencia section and the alternate branching, Virigina section is given below.

SEQUENTIAL BRANCHING: (Spanish-Valencia section)
erect bunch growth habit, light green foliage, no seed dormancy, many different seed colours, not resistant to leaf spot, about 90–100 days to maturity.
I—Spanish-Natal group. 2 seeded.
II—Valencia group. 2–4 seeds, no beaks on pods.
III—Manyema group. Similar to Valencia but pods are beaked.

ALTERNATE BRANCHING: (Virginia section)
runner or spreading bunch habit, dark green foliage, dormancy of several months, seeds usually sepia brown, moderately resistant to leaf spot, 120–150 days to maturity.

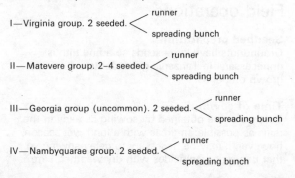

I—Virginia group. 2 seeded. ⟨ runner / spreading bunch

II—Matevere group. 2–4 seeded. ⟨ runner / spreading bunch

III—Georgia group (uncommon). 2 seeded. ⟨ runner / spreading bunch

IV—Nambyquarae group. 2 seeded. ⟨ runner / spreading bunch

Mani Pintar

This is a runner variety of the Nambyquarae group. The seeds are red and white variegated and their size and shape is irregular; these characteristics make it unsuitable for confectionery. It is suitable for other purposes, however, and consistently outyields other varieties in trials in East Africa provided the soils are fertile, the growing season is long enough and the seeds are old enough to have broken dormancy.

Asirya Mwitunde

This Virginia spreading bunch variety is moderately tolerant of rosette and has outyielded other varieties when the incidence of this disease was high. Its seeds are too small for the confectionery trade.

Homa Bay

This runner variety has large pink seeds and is one of the Matevere group. It is grown in Kenya for the confectionery trade.

Bukene

This variety, of the Manyema group, has light brown seeds. It is preferred in Uganda for the confectionery trade because the light colour of its skin makes damage to the kernels less obvious.

B1

This red-seeded, multiline variety, also called Red Beauty, belongs to the Valencia group. It is recommended in Uganda and is suitable for the confectionery trade.

Makulu Red

This variety has been introduced recently from Zambia to Uganda. It is a red skinned selection from Mani Pintar which has a high oil content and which usually outyields Mani Pintar.

Field operations

Seedbed preparation

Groundnuts have large seeds so a fine tilth is unnecessary. In Tanzania groundnuts are usually grown on ridges.

Time of sowing

Best yields are obtained by sowing as early in the rains as possible. In areas with a long wet season, however, it may be necessary to delay sowing so that harvesting coincides with dry weather. Late

Fig. 75: A bunch type variety of groundnuts.

sowing makes heavy attacks of rosette and leaf spot more likely unless chemical control measures are taken; even without these diseases, however, late sown groundnuts give poor yields. An experiment at Mwanhala in Tanzania gave 1 155 lb (c. 1 300 kg/ha) of shelled nuts per acre from late December sowing, 224 lb per acre (c. 250 kg/ha) from late January sowing and only 9 lb per acre (c. 10 kg/ha) from late February sowing. These results are typical of 28 experiments carried out in Tanzania between 1956 and 1962.

Seed preparation

Groundnuts can be sown in their shells but this gives slow and uneven germination and prevents any exclusion of diseased or damaged seeds. This practice is therefore rare and is not recommended. A combined insecticidal and fungicidal seed dressing is recommended. Seed dressings with toxic chemicals such as mercury and dieldrin should be avoided; thiram/BHC seed dressings are preferred. When varieties of the Virginia group are being grown the seed must be old enough for its dormancy to have broken. Seed must be sown as soon as possible after shelling because viability drops quickly as soon as the seeds lose the protection of their shells.

Sowing

The seed should be sown between two and four inches deep (c. 5–10 cm); this is usually done by dibbling although broadcasting and then covering the seed with a 'jembe' is common in Uganda. Tractor-drawn maize planters, fitted with suitable plates, have been used successfully.

Spacing

Spacing has an important effect on the incidence of groundnut rosette, which is a virus disease transmitted by aphids. When there is a close spacing, with a complete soil cover, winged aphids are discouraged from landing and rosette incidence is consequently low. When there is a wide spacing, with a discontinuous soil cover, aphids land more readily and unless they are controlled by using insecticides the incidence of rosette is likely to be high. The reasons for aphids' preference for wide spacing are not clear, despite several hypotheses.

The recommended spacing for erect bunch varieties is 2 ft or 1½ ft rows with the plants four inches apart within the row (c. 0·6 m or 0·45 m × 10 cm); this gives plant populations of 65 000–87 000 plants per acre (c. 160 000–210 000 plants per hectare) and requires 80–100 lb of seed per acre (c. 35–45 kg/ha). For spreading bunch and runner varieties a spacing of 2 ft × 6 in (c. 0·6 × 15 cm) is recommended, i.e. a plant population of about 44,000 plants per acre (c. 108 000 plants per hectare) requiring about 35 lb of seed per acre (c. 40 kg/ha). Smallholders are reluctant to adopt these spacing recommendations; the high cost of the seed and the labour involved, which may be as much as 20 man days per acre (c. 50 per hectare), presents an expense they can seldom afford. When aphids are controlled by using insecticides, optimum plant populations are probably lower than those mentioned above, but recommendations have not yet been formulated.

Fertilisers and manures

Phosphatic fertilisers usually give large increases in groundnut yields. In Uganda and Tanzania 1–2 cwt of single superphosphate per acre (c. 125–250 kg/ha) are recommended. Farmyard manure usually gives very good responses.

Weed control

Groundnuts should be clean weeded during the early stages of growth. Weeding after flowering should be avoided or should be done by hand pulling, otherwise it may interfere with the growth of the pegs. In *Western* Region in Tanzania a common practice is to bury the base of the plant during the early stages of growth so that flowers grow underground; many consider this to be a beneficial practice but no trials have been carried out to evaluate its effects. Ox-drawn implements are often used for weeding in eastern Uganda.

Harvesting

Lifting

Groundnuts should be lifted when most of them are mature; 60–70% is sometimes quoted as the proportion that should be mature at lifting. Maturity is indicated by a darkening of the veins on the inner surface of the shells; by this stage most of the leaves are yellow and many are shed. Postponing harvesting is undesirable because varieties of the Spanish–Valencia group are liable to germinate and many of the nuts may be detached from their plants during lifting. In East Africa groundnuts are usually harvested by pulling; this may be done by hand alone if the soil is sandy but a 'jembe' is usually used to help ease the plants out of the soil with as many nuts as possible still attached. In Uganda an ox-plough with the share removed is sometimes used for lifting. Mechanical harvesters have been developed in other parts of the world and one has been used on experimental stations and seed multiplication farms in Uganda; it lifts the plants and removes the nuts from them in one operation.

Drying

The usual practice after lifting is to turn the plants upside down and to leave them to dry in the field for a few days; the nuts are then removed by hand. Drying on stooks has been tried with success in Tanzania but is not adopted by smallholders even though it ensures quicker drying.

Shelling

Hand shelling, which is the universal practice in East Africa, is a laborious task; one man can produce only about 30 lb (c. 14 kg) of shelled nuts in a day. Hand operated machines are available and these can produce about 200 lb (c. 90 kg) of shelled nuts in a day; very few are used in East Africa, however. The shelling percentage of groundnuts is 65–75%.

Yields

Actual yields of shelled nuts are usually between 400 and 600 lb per acre (c. 450–670 kg/ha). With good husbandry, but without using insecticides, yields of about 1 200 lb per acre (c. 1 300 kg/ha) should be achieved. Using insecticides, yields of 2 000 lb (c. 2 200 kg/ha) have been achieved on a field scale in Uganda.

Pests

There are few pests which damage groundnuts directly. Thrips, millipedes, ants, termites and the *Systates* weevil occasionally pose problems. The aphid vector of the groundnut rosette virus is discussed in the next section; recent work suggests that it is also a direct pest since it sometimes migrates below the ground and feeds on the pegs or the developing pods. In Central and Eastern Provinces of Kenya groundnuts cannot be grown successfully owing to attacks by the groundnut hopper (*Hilda patruellis*). No effective control measures are known for this pest.

Diseases

Rosette
This disease is caused by a complex of at least five viruses, some of which are found throughout East Africa whilst others are restricted in their distribution. The symptoms vary, according to which viruses are present; they include yellowing, mottling and mosaic symptoms of the leaves and stunting and distortion of the shoots. If the plants are infected when they are young they may produce no nuts. The viruses are transmitted by aphids, *Aphis craccivora* and, possibly, *A. gossypii*, which mainly feed on the undersides of the leaves and on the flowers.

Sowing as early as possible in the rains and using a close spacing reduce rosette damage but it can only be completely checked by using insecticides. Menazon, endosulfan, phosphamidon, dimethoate and dicrotophos are recommended; when applied four times at ten day intervals, starting ten days after emergence, they have given yield increases of over 600 lb per acre (c. 650 kg/ha) of shelled nuts in rainfed crops and over 900 lb per acre (c. 1 000 kg/ha) under irrigation. One of the difficulties is that spraying must be done before disease symptoms appear and before aphids can be found easily; it is difficult to persuade a farmer to spray a healthy and apparently uninfested crop. Some varieties, e.g. Asirya Mwitunde, are reported to be tolerant of rosette but the general experience in East Africa has been that varieties which are resistant in one area often succumb in another.

Leaf spot diseases
The most common of these is caused by the fungus *Cercospora personata*. It causes dark brown, almost black spots on the upper and lower surfaces of the leaves; the spots are well defined and have no halo. In wet weather the stems, petioles and pegs may show symptoms. *C. arachidicola* is less important; it produces larger, less distinct, pale brown spots, and these are surrounded by a yellow halo. Both diseases are spread by airborne spores, by infected trash in the soil and, to a lesser extent, by infected seed. The incidence of both is higher in conditions of high humidity. Leaf spot damage can be reduced by early planting, seed dressing, crop rotation and the destruction of infected debris. Organo-tin fungicides give good control and might be economic on high yielding crops which had been sprayed with insecticide.

Bacterial wilt
This disease, which is caused by *Pseudomonas solonacearum*, occurs in Uganda but is not yet widespread. Plants develop a wilt, which is often asymmetric, and then die.

Groundnut blight
This is caused by the fungus *Sclerotium rolfsii*. Wilt occurs in patches in the field and a white mycelium can be found on the roots. Sunken brown lesions occur on the stems. The incidence of blight is greatest in wet weather and when late weeding or heaping weeds around the plants causes high humidity around the base of the stems.

Aflatoxins
In humid conditions a fungus, *Aspergillus flavus*, may grow inside the pods and infect the kernels. This fungus can produce toxic substances called aflatoxins. These were responsible for the deaths of large numbers of turkeys which had been fed with groundnut cake in Europe; they are also suspected of being carcinogenic substances which can cause liver cancer. Countries which import groundnuts are becoming increasingly conscious of this problem and many have introduced inspection and testing with a very low tolerance of aflatoxins.

Rapid drying to a moisture content of about 10% is the only means of preventing infection by

Aspergillus flavus. Damage to the nuts during harvesting should be minimised because the fungus can easily enter the shell if this is broken. It can also enter through holes in the shells made by termites; these insects may attack groundnuts that are left in the soil for too long.

Utilisation

When groundnuts are used as a subsistence crop they are usually shelled, roasted, pounded into a flour and made into a sauce or relish, sometimes with other vegetables, for eating with 'ugali'. Alternatively they may be boiled in their shells, boiled with maize or roasted after shelling and eaten whole.

Bibliography

1 **Akehurst, B. C.** and **Sreedharan, A.** (1965). *Time of planting—a brief review of experimental work in Tanganyika, 1956–62*. E. Afr. agric. for J., *30*:189.

2 **Anderson, G. D.** (1970). *Fertility studies on a sandy loam in semi-arid Tanzania, II—Effects of phosphorus, potassium and lime on yields of groundnuts*. Exp. Agric., *6*:213.

3 **Bunting, A. H.** (1955). *A classification of cultivated groundnuts*. Emp. J. exp. Agric., *23*:158.

4 **Bunting, A. H.** (1958). *A further note on the classification of cultivated groundnuts*. Emp. J. exp. Agric., *26*:254.

5 **Bunting, A. H.** and **Lea, J. D.** (1955). *The choice of groundnut varieties for production in three areas of Tanganyika*. Emp. J. exp. Agric., *23*:29.

6 **Clinton, P. K. S.** (1957). *A note on a wilt of groundnuts due to* Sclerotium rolfsii. E. Afr. agric. J., *22*:137.

7 **Davies, J. C.** and **Kasule, F. K.** (1964). *The control of groundnut rosette disease in Uganda*. Trop. Agriculture, Trin., *41*:303.

8 **Evans, A. C.** (1954). *Groundnut rosette disease in Tanganyika, I—Field studies*. Ann. appl. Biol., *41*:189.

9 **Evans, A. C.** (1954). *Rosette disease of groundnuts*. Nature, Lond., *173*:1242.

10 **Evans, A. C.** (1956). *A study of a rosette-resistant groundnut variety, Asirya Mwitunde*. E. Afr. agric. J., *22*:27.

11 **Evans, A. C.** (1960). *Studies of intercropping, I—Maize or sorghum with groundnuts*. E. Afr. agric. for. J., *26*:1.

12 **Evans, A. C.** and **Sreedharan, A.** (1962). *Studies of intercropping, II—Castor-bean with groundnuts or soya-bean*. E. Afr. agric. for. J., *28*:7.

13 **Goldson, J. R.** (1967). *Weeding requirements of groundnuts in Western Kenya*. E. Afr. agric. for. J., *32*:246.

14 **Hemingway, J. S.** (1954). Cercospora *leafspots of groundnuts in Tanganyika*. E. Afr. agric. J., *19*:263.

15 **Hemingway, J. S.** (1957). *The resistance of groundnuts to* Cercospora *leafspots*. Emp. J. exp. Agric., *25*:60.

16 **Hill, A. G.** (1947). *Oil plants in East Africa*. E. Afr. agric. J., *12*:140.

17 **Hull, R.** (1964). *Spread of groundnut rosette virus by* Aphis craccivora. Nature, Lond., *202*:213.

18 **Oram, P. A.** (1958). *Recent developments in groundnut production, with special reference to Africa*. Field Crop Abstr., *11*:1 and 75.

19
Maize

Zea mays

Introduction

Maize is the most important cereal crop in East Africa. It was first introduced in the sixteenth or seventeenth century by Portuguese traders but the varieties came from the Caribbean and were only suited to the coastal strip. Its proliferation in the East African highlands and in the medium altitude areas, with a consequent decline in most places of the indigenous cereals, is a comparatively recent development; it was largely due to the introduction by European settlers of varieties from South Africa

Fig. 76: Hybrid maize.

which were better suited to the inland environment.

The main reasons for the popularity of maize are as follows: it has a higher yielding potential than the indigenous cereals in areas with satisfactory rainfall and free draining soil; it is seldom seriously damaged by pests or diseases in the field and is virtually untouched by birds, which can cause a complete crop loss in some of the indigenous cereals; land preparation, weed control and harvesting all required little labour (when done manually) compared with some of the indigenous cereals, and no threshing or winnowing is required. Some people would quote the fact that maize is more palatable as an additional advantage but this appears to be a recent and local adaptation of taste; in Northern and Eastern Uganda and on Ukara Island in Lake Victoria the majority still prefer millet.

Kenya

Maize is the staple food throughout the arable areas of Kenya. Approximately 2 500 000 acres (c. 1 000 000 ha) are grown annually, very largely on smallholdings. The area produced on large-scale farms in 1969 was only 100 000 acres (c. 40 000 ha). This area is important because it produces a large proportion of the grain that is marketed instead of being used for subsistence. The most important large-scale maize growing area is Trans Nzoia District, where the climate is comparatively warm and humid compared with the other large-scale farming areas. Kenya's maize production varied between 980 000 tons and 1 340 000 tons during the latter half of the 1960s; this variation was mostly due to variation in the rainfall. In the past Kenya's maize production has fluctuated between years of surplus, when exports were possible, and years of shortage, when imports were necessary. The last import was in 1965. It is now hoped that, as in the late 1960s, expanded production will not only cater for the country's needs but will also allow for the accumulation of a reserve for years of poor production, for considerable exports and for alternative uses for maize within the country. The use of grain for a livestock feed and for industrial processing (e.g.. into starch or oil) are being investigated.

Tanzania
Maize is a staple food in the highland areas, i.e. the wetter parts of Bukoba, Ngara, Kibondo, Mara, Moshi, Arusha, Pare, Lushoto, Morogoro, Iringa, Rungwe and Songea Districts and in many of the medium altitude areas, especially in the cultivated areas of Western Region. The total area has been estimated as 1 500 000 acres (c. 600 000 ha).

Uganda
Maize is relatively unimportant in Uganda where it surpasses neither bananas in the south nor finger millet in the north as a staple food. In the south there are storage problems owing to the high humidity and in the north maize has been officially discouraged because it harbours cotton pests, especially American bollworm. The total area has been estimated as 700 000 acreas (c. 280 000 ha) in 1968. Busoga was the district with the largest acreage but there was also considerable production in all districts bordering on Lake Victoria and in Bugisu and Kigezi Districts.

Plant characteristics

Maize is an annual crop which can grow to a height of 13–14 ft (c. 4·0–4·3 m). The first roots to emerge after germination are the seminal roots which arise from the seed; there are usually four of them. A few weeks later adventitous roots develop from nodes above the seed; these become increasingly important until at maturity they supply all the water and nutrients. Roots often grow from some of the lower nodes above ground level; these are usually called 'prop roots' and may lend some support. The depth of rooting depends on a number of factors, especially the soil type and the rainfall. In deep, well drained, fertile soils and with reasonable rainfall, roots may extend to a depth of 12 ft (c. 3·6 m). The tassel, i.e. the male inflorescence, emerges at the top of the plant and sheds its pollen over a period of about a week. The silks, i.e. the stigmas, emerge from the ears, i.e. the female inflorescences, towards the end of pollen shedding and remain receptive for a period of about three weeks in suitable conditions; the percentage of self-pollination is therefore low and is usually about 5%. There is usually only one ear per plant but with high-yielding varieties and good growing conditions there may be two or three. Four ears per

plant are very seldom seen. Tillers are occasionally produced and they may bear hermaphrodite inflorescences. These produce very little useful grain; their removal is not considered worthwhile.

Life cycle
The period to flowering and to maturity varies greatly, according to variety and altitude. At low altitudes, short-maturing varieties flower in two months and mature in four. At Ol Joro Orok in Kenya, at 7 800 ft (c. 2 400 m), the longer term varieties, which are the only ones that can be grown at this altitude, flower in $6\frac{1}{2}$ months and take more than a year to mature.

The Kitale hybrids (prefixed by the number 6) flower in about $3\frac{1}{2}$ months at an altitude of 6 000 ft (c. 1 800 m). At 1 000 ft (c. 300 m) above or below this the period to flowering is lengthened or shortened, respectively, by approximately 17 days. The period to physiological maturity, i.e. grain maturity, for the Kitale hybrids is about 6 months at 5 000 ft (c. 1 500 m), about 7 months at 6 000 ft (c. 1 800 m) and about 8 months at 7 000 ft (c. 2 100 m). In practice maize is usually left in the field to dry for several weeks after maturity.

Ecology

Rainfall, and water requirements
The young maize plant is moderately drought resistant but is usually susceptible to unfavourable soil air/moisture relationships during the first four or five weeks of its life. During this time, when it relies primarily on its seminal roots, and its growing point is below the soil surface, it requires a very well aerated soil; ideally there should be just enough moisture to maintain uninterrupted growth. Any more moisture than this prevents optimum growth and any waterlogging provides a check from which the plant never recovers; in both instances yields are reduced. Recent experiments at Kitale indicate that every inch (c. 25 mm) of rain that falls in January, February and March *before* sowing in mid-March reduces yields by about 200 lb of grain per acre (c. 220 kg/ha) by creating less favourable soil air/moisture relationships. The optimum rainfall during the five weeks after sowing is 8 in (c. 200 mm); seedling growth and yields were reduced at Kitale when there was more rain than this. These facts have a profound effect on the optimum time of sowing (see below).

From five weeks onwards the maize plant is less drought resistant. The most critical period is at silking when a small degree of wilting can cause incomplete pollination, and when a severe drought can cause a crop loss. The experiments at Kitale showed that, within the limits of rainfall experienced, the more rain that fell after five weeks the higher the yields became. Dry conditions are needed towards harvesting in order to reduce the moisture content of the grain and to prevent the incidence of ear rots.

Maize is grown in a wide range of rainfall regimes in East Africa. In Machakos District in Kenya satisfactory yields are usually obtained by sowing early maturing varieties at or before the beginning of the long rains; these average about 12 in (c. 300 mm) and have an average duration of only 2 months. In Kakamega District in Kenya, on the other hand, the rainfall pattern allows longer maturing varieties, with a higher yielding potential, to be grown; the average annual rainfall is about 70 in (c. 1 800 mm) with reliable rainfall in most months of the year.

Temperature and altitude
Different types of maize are suited for different temperatures at altitudes from sea level to 8 000 ft (see section on varieties below). On the East African coast day temperatures are seldom far from the optimum for maize growth, i.e. 86°F (30°C). The high night temperatures, however, lead to high rates of respiration, thus limiting yields. The highest yields in experiments in Kenya have been obtained at an altitude of about 7 000 ft (c. 2 100 m) where night temperatures are considerably lower although seldom below the critical temperature of 50°F (10°C) below which little growth occurs. The cool conditions at high altitudes lengthen the life cycle and above 8 000 ft (c. 2 400 m) severely limit yields. Nevertheless, maize can be seen on roadsides and in smallholdings between 8 000 and 9 500 ft (c. 2 900 m) in Kenya. At these altitudes potatoes would be a very much more productive crop. Maize is killed by light frosts.

Soil requirements
Maize needs well drained soil and a good supply of nutrients. It cannot tolerate the slightest degree of waterlogging; it can be killed if it stands in water for as long as a day.

Varieties

Caribbean Flint
This is the type of maize, genetically very diverse, which was introduced to the East African coast three or four centuries ago. It is only suited to lowland areas.

Kenya Flat White Complex
Many varieties were introduced by European settlers in the early part of this century. Most of these, e.g. Hickory King and Natal White Horsetooth, come from South Africa. They have now become genetically mixed and are called the Kenya Flat White Complex. This type of maize is suited to altitudes between 3 000 and 7 500 ft (c. 900–2 300 m) and is grown by the majority of smallholders in the highland areas of Kenya. It has been one of the main constituents of the improved varieties bred in Kenya and Tanzania.

Cuzco
Cuzco types, which have very large, soft, white grains, were introduced by missionaries in the early part of this century. They came from high altitude areas in Peru and are only suitable for altitudes of about 8 000 ft (c. 2 400 m) in East Africa. They are still grown by smallholders.

Kitale synthetics
Synthetic varieties, which are produced by open pollination when several inbred lines are grown together, were the first product of a maize breeding programme which started at Kitale in Kenya in 1955. The first release, Kitale Synthetic II, was first grown commercially in 1961. Synthetics are still available although they are seldom grown because hybrids have a higher yielding potential. They have been used as parents in some of the varietal hybrids.

Kitale hybrids
Hybrids are bred by crossing inbred lines or varieties under conditions of controlled pollination. Fresh seed must be purchased each year; if a grower uses second generation seed the resulting population is very variable, owing to genetic segregation, and yields are poor. All hybrids bred in Kenya are identified by three numbers. The first indicates the approximate altitude at which the crop has been bred: 6 for Kitale at 6 000 ft and 5 for Embu at 5 000 ft. The second number indicates the type of hybrid: 1 for a varietal hybrid (when a variety is used as one of the parents); 2 and 3 for classical hybrids (when inbred lines are used as the parents). 2 is used for double crosses, e.g. (G × D) × (A × F), and 3 is used for three-way crosses, e.g. (F. × A) × G. The last number is a series

number; it is sometimes followed by a letter which also denotes the series.

The Kitale hybrids are long term varieties which are bred for areas with a long wet season of at least five months. Their life cycle has been described in a previous section. When grown in a suitable area, with the expected rainfall and with good standards of husbandry, they are capable of yielding 30–50% more than local varieties. They give greater responses than local varieties to correct time of sowing, correct plant population (when compared to late sowing and low plant population) and to fertiliser applications. As a corollary, it is a waste of money to grow hybrids without high standards of husbandry. Long term hybrids may even yield less than local varieties without good standards of husbandry; if they are planted late in an area with a limited duration of rainfall, e.g. Central Province in Kenya, they can suffer more than faster maturing local varieties. In 1970 approximately 350 000 acres (c. 140 000 ha) of Kitale hybrids were grown in Kenya; an estimated 250 000 acres (c. 100 000 ha) were grown on smallholdings.

The classical hybrids are bred by crossing inbred lines which have been derived from the Kenya Flat White Complex. The first, 621, was released in 1964. The classical hybrids are best suited to medium altitudes and give best yields between 4 000 and 5 000 ft (c. 1 200–1 500 m). They have also given excellent results in Tanzania in areas between 3 000 and 4 000 ft (c. 900–1 200 m) with a good rainfall, where they outyield local varieties by at least 25%.

The varietal hybrids are composed by crossing varieties rather than inbred lines. The most successful parent for crossing with varieties or inbreds from the Kenya Flat White Complex is Ecuador 573; it was used as a parent in all the Kitale varietal hybrids released in the 1960s. Having a different centre of origin, i.e. South America, the genetic differences between it and the Kenya Flat White Complex are very great, thus creating a high degree of hybrid vigour when the two are crossed. The varietal hybrids are best suited to altitudes about 5 000 ft (c. 1 500 m).

Embu hybrids
A varietal hybrid, 511, has been bred at Embu in Kenya for areas with a medium duration rainfall of $3\frac{1}{2}$ to 4 months. At 5 000 ft (c. 1 500 m), 511 flowers in approximately 80 days and matures in about 5 months. This variety is intended primarily for Central Province in Kenya although the Kitale hybrids can also be grown successfully in this area

provided that they are grown at a suitable altitude and sown at the very beginning of the rains.

Katumani Composites
Composite varieties are bred by growing a number of varieties of diverse genetic composition together and allowing them to interpollinate freely; continual improvement is possible by the process of recurrent selection. An advantage of composite varieties and of synthetics is that farmers can retain the seed for successive generations. They can even improve it by selection. The Katumani Composites (which were preceded by Katumani Synthetics) are short maturing varieties bred primarily for Machakos and Kitui Districts in Kenya where the long rains seldom last for longer than two months. They have also given good yields in parts of Tanzania where the duration of the rainfall is limited, e.g. Mara District and Kongwa. Between 4 000 and 5 000 ft (c. 1 200–1 500 m) Katumani Composites flower in about two months and mature in about four.

Ukiriguru Composites
Maize breeding started at Ukiriguru in Tanzania in 1964. At the time of writing Ukiriguru Composite A (UCA) is available for areas with a reasonable rainfall between 3 000 and 5 000 ft (c. 900–1 500 m). It yields 25% more than the local varieties but 5% less than Kitale hybrid 622. Ukiriguru Composite B, which is planned for release in the early 1970s, will be higher yielding than UCA. UCA matures more slowly than the local varieties, flowering in 73 days compared to 65.

Ilonga Composites
Composites have been bred for inland areas below 3 000 ft (c. 900 m) at Ilonga in Tanzania and have given higher yields than local maize and varieties bred for higher altitudes.

Coast Composites
Composites are being bred for coastal conditions at Muheza in Tanzania and at Mtwapa in Kenya but at the time of writing they have not been released. At Mtwapa the variety called Pp (*Puccinia polysora* resistant) is being used as a parent in the composites; it was bred in the 1950s but was never widely distributed.

Field operations

Land preparation
When the land is prepared by hand a very rough

Fig. 77: Ox-ploughing for maize.

seedbed is usually formed. This is sufficient, provided that the weeds are killed, because each seed can be placed at the correct depth by hand however rought the seedbed. A rough seedbed may be an advantage because it encourages water infiltration and resists erosion to a greater extent than a fine seedbed.

The usual practice when using mechanical cultivations is to use the disc plough and the disc harrow. Some farmers prefer to use a large number of operations, possibly three ploughings and four or five harrowings, in order to prepare a very fine seedbed; this has the advantage of encouraging rapid and even germination and creating a relatively weed-free environment; there is, however, a risk both of soil erosion if the field is on a slope and of soil compaction and poor aeration. Other farmers prefer to reduce their costs by cultivating as little as possible; the rough seedbed which results may not be so free of weeds and germination may be uneven but there is less risk of soil erosion and poor aeration. Methods of mechanical cultivation for maize, which have been studied very little in East Africa, are now under investigation. The mouldboard plough, unlike the disc plough, helps to improve the soil structure and aeration by incorporating organic matter such as maize stover, and leaves a relatively smooth surface which requires fewer operations to break down to a suitable tilth; it is becoming increasingly popular in Trans Nzoia District. Tined harrows, which pulverise the soil less than disc harrows, have also been tried; one of their problems is that they require a fairly moist soil and in dry conditions, which prevail if the land is being

prepared for sowing before the beginning of the rains, are unable to break up large clods of soil.

Maize stover is almost invariably burned before cultivating. There are problems involved in its incorporation into the soil but, as with those of wheat straw (see chapter 36), these are not insuperable.

Time of sowing

Experiments in East Africa, and in particular a country-wide series of experiments in Kenya which started in 1966, have shown that time of sowing is the most critical factor affecting maize yields. As a general rule maize should be sown as near the beginning of the rains as possible. In areas where the duration of rainfall is limited, e.g. Machakos and Kitui Districts in Kenya, it is safest to sow in dry soil before the onset of the rains. An exception to this rule occurs in the Mwanza area in Tanzania, where early sowing in November or early December would bring flowering into January or February when rainfall is relatively uncertain; the recommended time of sowing is therefore early January: this leads to flowering in the wetter month of March. Losses that can be expected from late sowing in Rift Valley Province in Kenya are 50–100 lb of grain per acre (c. 55–110 kg/ha) per day delayed after the onset of the rains. In Central and Eastern Provinces, where the rains are shorter, losses are as high as 150 lb per acre (c. 170 kg/ha) per day delayed. Timely sowing, which costs the farmer little or nothing, is the cheapest and most effective step towards ensuring satisfactory maize yields.

Several reasons have been suggested for the decline of maize yields which accompanies late sowing. The first reason is that early sown maize benefits from the flush of nitrogen that occurs when a dry soil is wetted; this can now be dismissed because it has been found that late sown crops still yield less than early sown crops even when they are generously supplied with nitrogen, in fact they respond less to nitrogen than do early sown crops. The second reason put forward is that early sown maize suffers less from fungal diseases; this can also be dismissed because late sown crops yield less than early sown crops even when they are protected by fungicides and, as with nitrogen, early sown crops respond more to applications of fungicide than late sown crops.

It is now considered that the soil air/moisture relationship is the most important factor in the time of sowing effect; it may affect yields in the early stages of growth and later in the cropping period. If the soil air/moisture relationship is ideal for seedling growth, i.e. if the soil pores are only partially filled with water, maize will grow well in the early stages and will consequently yield well; these soil conditions only occur at the very beginnning of the rains, hence the importance of sowing at this time. After the onset of the rains, which is usually characterised by several light showers, the soil air/moisture relationship steadily deteriorates, the soil pores becoming increasingly filled with water; as noted above in the section on ecology, these conditions limit the growth of maize seedlings and the crop never fully recovers. As regards the soil air/moisture relationship later in the life cycle, late sown crops may mature or even flower after the end of the rains; this is a serious risk in the short rainfall areas although it is less of a risk in the long rainfall areas such as western Kenya and Trans Nzoia District. In the short rainfall areas late sown maize may suffer both in the early stages, when the soil is too wet, and later, when the soil is too dry, thus causing dramatic yield reductions of up to 6% per day delayed after the onset of the rains.

Sowing

Maize seed is sown manually by dibbling and mechanically by maize planters. When sown by hand, two seeds should be placed in each hole or, alternatively, two should be placed in one hole and one should be placed in the next. The seedlings should be thinned to one plant per hole when they are about 6 in (c. 15 cm) high. In moist soil the seed should be placed 1–2 in (c. 2·5–5·0 cm)

deep but in dry soil it should be placed about 4 in (c. 10 cm) deep to prevent it germinating as a result of only a light shower. A fungicidal and insecticidal seed dressing should be used to prevent infection by soil-borne fungi and insects in the seedling stage.

Spacing

Low plant populations are one of the important reasons for the low average maize yields in East Africa. In eleven experiments in western Kenya a population of 8 000 plants per acre (c. 20 000 plants per hectare), which is near the population usually found in smallholdings, yielded an average of 920 lb per acre (c. 1 030 kg/ha) less than a population of 16 000 plants per acre (c. 40 000 plants per hectare). The recommended population depends on the variety and the reliability of the rainfall. For the Kitale hybrids, which should be grown in areas of reliable rainfall, it is 18 000–19 000 plants per acre (c. 44 000–47 000 plants per hectare); a spacing of 3 ft × 9 in (c. 0·9 m × 23 cm) is suitable. In drier areas, e.g. western Tanzania, a population of 14 500 plants per acre (c. 36 000 plants per hectare) is recommended, i.e. a spacing of 3 ft × 1 ft (c. 0·9 m × 0·3 m). The Embu hybrids, which are smaller, yield best at a higher population, i.e. 30,000 plants per acre (c. 74 000 plants per hectare), provided that there is unlimited rainfall. Because of the risk of dry periods in the area where the Embu hybrids are grown a lower plant population is recommended, i.e. 21 000–22 000 plants per acre (c. 52 000–54 000 plants per hectare); a spacing of 2 ft × 1 ft (c. 0·6 m × 0·3 m) is suitable. The Katumani Composites are smaller than either the Kitale or the Embu hybrids and with unlimited rainfall yield best at about 40 000 plants per acre (c. 99 000 plants per hectare). Owing to the risk of severe moisture stress in the areas for which they have been bred, however, they should be grown at a population of about 14 500 plants per acre (c. 36 000 plants per hectare); a spacing of 3 ft × 1 ft (c. 0·9 × 0·3 m) is suitable.

Apart from the risk of moisture stress during droughts, the main risk of too high a population is that of lodging, which makes harvesting more difficult and which increases the likelihood of damage to the ears by fungi, animals and termites.

Fertilisers and manures

The two main conclusions provided on fertilisers by the widespread husbandry trials in Kenya during the last half of the 1960s were, firstly, that they give smaller and less predictable responses than the less expensive improvements in husbandry, i.e. early sowing, correct spacing, efficient weed control and use of an improved variety, and, secondly, that they only give worthwhile responses when standards of husbandry are high. To restate the latter point in another way, a farmer who does not sow at the optimum time, who does not achieve an adequate plant population and who controls his weeds inefficiently, is wasting money by applying fertilisers.

Fertiliser experiments have been carried out throughout the maize growing areas of Kenya and western Tanzania. The responses varied greatly, depending primarily on the soil type, the rainfall and the cropping history of the soil. The recommendations, which vary according to these factors and according to the prices of both fertilisers and maize, would occupy too much space if mentioned in detail here; they are available at the National Agricultural Research Station, Kitale, in Kenya, and the Research and Training Institute, Ukiriguru, in Tanzania.

Nitrogen should be applied as a top-dressing when maize is about knee high. Recommendations are as high as 170 lb of nitrogen per acre (c. 190 kg/ha) on the sandy clay soils of Trans Nzoia District where such an application can increase yields by over 2 000 lb of grain per acre (c. 2 200 kg/ha) provided that there is no other limiting factor. In Central Province and Kisii District in Kenya worthwhile responses from nitrogen are seldom obtained. In western Tanzania, where the standard of husbandry and the variety grown almost always limit yields, the recommendations vary between 0 and 70 lb of nitrogen per acre (c. 0–80 kg/ha). Contrary to the general experience with wheat, nitrogen applications reduce lodging rather than increase it.

Phosphate should be incorporated into the soil at the time of sowing. When maize is sown by hand it should be stirred into the bottom of the planting holes; when maize is sown mechanically it should be placed in a band about 2 in (c. 5 cm) away from the seed, either at the same level or slightly below it. Recommendations are as high as 107 lb of P_2O_5 per acre (c. 120 kg/ha) on the sandy clay soils of Trans Nzoia District. In this area maize without fertilisers seldom produces more than 3 000 lb per acre (c. 3 350 kg/ha) but with the recommended applications of nitrogen and phosphate it can yield 7 000 lb per acre (c. 7 850 kg/ha). The recommendations in Tanzania range from 0–50 lb per acre (c. 0–55 kg/ha).

Potassium has given no positive responses in Kenya or western Tanzania, even when large amounts of nitrogen and phosphate were applied. It may become limiting in the future, however, if a succession of high yielding crops exhausts the potassium reserves of the soil.

Sulphur deficiency was first observed in Trans Nzoia District in the late 1960s. It only occurs when grassland has been ploughed late, thus allowing insufficient time for decomposition and sulphur release, and when fertilisers with little or no sulphur have been used. The symptoms, yellow stripes on the younger, upper leaves in the centre of the plants and small, dull, white grain, can be avoided by using ammonium sulphate, ammonium sulphate nitrate or single superphosphate, all of which contain more than 10% sulphur.

Farmyard manure gives good responses on maize but is very seldom applied.

Weed control

Inefficient weed control is one of the main factors causing the low average yields of maize in East Africa. The husbandry trials in Kenya showed that three weedings, when the crop was 3, 18 and 36 in high (c. 7·5, 45 and 90 cm high) gave 1 040 lb more grain per acre (c. 1 170 kg/ha) than one weeding when the crop was 18 in high. If maize growth is checked by weeds in its early stages of growth it never recovers fully, however well weeds are controlled subsequently, and yields are reduced. Weeding may therefore be necessary when the crop is only about 3 in (c. 7·5 cm) high.

In properly spaced maize in areas of reliable rainfall weed-free conditions need only be maintained until the crop is about 18 in (c. 45 cm) high; after this the crop itself suppresses weeds and further cultivations are of no benefit. In areas of less reliable rainfall, e.g. western Tanzania and Machakos and Kitui Districts in Kenya, weed-free conditions should be maintained until flowering to minimise the risk of moisture stress at this critical stage.

Small scale farmers rely solely on hand cultivation for controlling weed growth. Large scale farmers can reduce weed competition in the early stages of growth by careful seedbed preparation (see above), by a light harrowing no later than five days after sowing and by the use of herbicides. 2, 4-D amine is the most commonly used herbicide; it kills most of the broad-leaved weeds and should be applied when the maize is about 4 in (c. 10 cm) high. Atrazine, applied as a pre-emergence herbicide, has also proved successful.

Harvesting

Maize grain is physiologically mature, i.e. it accumulates no more dry matter, when the moisture content is about 35%. It is usually left in the field to dry until the moisture content is about 19–20%; at this stage it can safely be stored on the cob, after the husks have been removed, in cribs, i.e. stores with walls of wire netting. Most of East Africa's maize is harvested by hand. On large scale farms the plants are usually cut and stooked several weeks before the removal of the cobs. The main advantages of stooking are that damage by fungi, animals and termites, which often attack lodged plants, is reduced, that the land can be prepared for the next crop by cultivating between the stooks and that thieves, who can hide in a standing crop of maize, are discouraged.

Fig. 78: Combine harvesting.

Maize combine harvesting was introduced into Kenya in the mid-1960s and has since become increasingly popular in the large scale farming sector. It involves much less labour and grain can be harvested when the moisture content is as high as 25–28%, thus reducing losses from lodging, diseases and theft, provided that it is put through a drier immediately after harvesting. Supplementary benefits of early combining are that ploughing can be done earlier, before the soil has completely dried out, and that weed growth, which can be prolific when maize leaves wither, can be checked more quickly.

The shelling percentage of maize, i.e. the proportion of the de-husked cobs made up by grain, is usually between 70% and 80%.

Yields

The average yield of maize in Kenya is 1 000–1 200 lb per acre (c. 1 100–1 350 kg/ha). In Tanzania it is thought to be as low as 600 lb per acre (c. 670 kg/ha). These yields reflect a very low average standard of husbandry for even in experimental plots in Kenya where unimproved varieties were sown one month after the optimum date, at populations of only half the optimum, with no fertilisers and with only one weeding, yields averaged 1 760 lb per acre (c. 1 970 kg/ha). The same series of experiments showed that yields could be raised to 4 360 lb per acre (c. 4 890 kg/ha) simply by sowing at the correct time, using the correct spacing and controlling weeds effectively. When these measures were combined with the use of hybrid seed and fertilisers, the yield rose to 7 160 lb per acre (c. 8 030 kg/ha).

Yields achieved with good husbandry on a field scale vary according to the length of the growing season. In Machakos District in Kenya, where the growing season is short, 3 000 lb per acre (c. 3 350 kg/ha) is the most that has been obtained. On the large scale farms around Kitale the growing season is longer; the average yield is three times the national average and yields of 5 000 lb per acre (c. 5 600 kg/ha) are regularly obtained by good farmers. The highest yield obtained in that area on a field scale is 7 600 lb per acre (c. 8 520 kg/ha). The highest yield obtained in Kenya on an experimental plot is 12 400 lb per acre (c. 13 900 kg/ha). Yields are usually quoted in bags per acre: one bag contains 200 lb (c. 90 kg) of grain.

Pests

Stalk borers

The most important stalk borer is *Busseola fusca*. Below 3 000 ft (c. 900 m) *Chilo zonellus* (*partellus*), *C. argyrolepia* and *Sesamia calamistis* occur and *Busseola* is seldom seen. The adult moths lay eggs on the foliage and the larvae, after eating small patches of the upper leaves and making small 'windows', penetrate to the centre of the plant where they feed on and near the growing point. Later generations may feed in the developing ears. DDT, applied to the funnels of the young plants either as a dust or a liquid, usually prevents serious damage. One application about three weeks after emergence is usually sufficient but one or more additional applications may be necessary, e.g. in western Tanzania. Late sown crops are likely to be more severely attacked than early sown crops.

Armyworm. *Spodoptera exempta*
Armyworms attack a wide range of crops but maize is one of the more seriously damaged. Larvae usually occur in very large numbers and may eat all the maize leaves until only the midribs remain. The worst attacks seem to occur in years when the onset of the rains is delayed. DDT sprays are most commonly used against armyworms although several other insecticides are equally effective.

Diseases

White leaf blight
This disease is caused by the fungus *Helminthosporium turcicum* (*Trichometasphaeria turcica*). The symptoms are oval, grey, papery lesions on the leaves. Damage is seldom severe because there appears to be a degree of resistance in the local varieties. Blight resistance is a secondary objective of the maize breeding programmes in East Africa.

Maize streak
This disease, which is caused by a virus, can cause damage in a few isolated areas in East Africa, e.g. around Geita, Biharamulo and Ifakara in Tanzania and around Embu in Kenya. In other areas it seldom causes much damage. It is spread in the field by a leaf hopper, *Cicadulina mbila*. Early sown crops are rarely attacked.

Rust
Two fungal species cause rust on maize: *Puccinia sorghi* occurs from sea level to the higher maize growing areas, whilst *P. polysora* seldom occurs higher than 4 000 ft (c. 1 200 m). The symptoms are red or brown pustules on the leaves. *P. polysora* did much damage in West Africa but since its first appearance in East Africa in 1952 it has caused less harm than was expected. A maize breeding programme, carried out on the Kenya coast in the 1950s, produced a variety called Pp, but its distribution appears to have been very limited.

Utilisation

The majority of maize is dried and ground into flour for making 'ugali'. Sometimes, however, the dried grains are cooked whole; they may be eaten with beans cooked in the same way or may be incorporated into boiled and mashed potatoes. 'Green maize', i.e. maize which has not fully dried out, is popular; it is roasted and eaten on the cob. Maize is the most popular silage crop in the large scale farming areas in Kenya; it is usually ensiled shortly after flowering when the grains are still in the milky stage. Maize stover is often fed to cattle if it is not too dry; otherwise it can be used for bedding.

Bibliography

1 **Akehurst, B. C.** and **Sreedharan, A.** (1965). *Time of planting—a brief review of experimental work in Tanganyika, 1956–62.* E. Afr. agric. for. J., *30*:189.

2 **Allen, A. Y.** (1968). *Maize population and spacing recommendations for 1968.* Kenya Weekly News, 2nd February.

3 **Allen, A. Y.** (1968). *Better husbandry in maize growing.* Kenya Weekly News, 23rd February.

4 **Allen, A. Y.** (1968). *Maize diamonds—some valuable results from the District Husbandry Trials in 1966.* Kenya Farmer, January.

5 **Allen, A. Y.** (1968). *The control of weeds in maize.* Kenya Farmer, August and October.

6 **Anderson, G. D.** (1969). *Responses of maize to application of compound fertilisers on farmers' fields in ten districts of Tanzania.* E. Afr. agric. for. J., *34*:382.

7 **Anderson, G. D.** (1970). *Fertility studies on a sandy loam in semi-arid Tanzania, I—Effects of nitrogen, phosphorus and potassium fertilisers on the yields of maize and the soil nutrient status.* Exp. Agric., *6*:1.

8 **Bolton, A.** and **Scaife, M. A.** (1969). *Maize variety trials in Western Tanzania, 1963–1966.* E. Afr. agric. for. J., *35*:11.

9 **Coaker, T. H.** (1956). *An experiment on stem borer control on maize.* E. Afr. agric. J., *21*:220.

10 **Coaker, T. H.** (1959). *'Insack' treatment of maize with insecticide for protection against storage pests in Uganda.* E. Afr. agric. J., *24*:244.

11 **Coaker, T. H.** and **Davies, J. C.** (1958). *Palatability of maize under storage in Uganda.* E. Afr. agric. J., *24*:57.

12 **Davies, J. C.** (1960). *Experiments on the crib storage of maize in Uganda.* E. Afr. agric. for. J., *26*:71.

13 **Doggett, H.** and **Jowett, D.** (1966). *Yields of maize, sorghum varieties and sorghum hybrids in the East African lowlands.* J. Agric. Sci., *67*:31.

14 **Dowker, B. D.** (1963). *Rainfall reliability and maize yields in Machakos District.* E. Afr. agric. for. J., *28*:134.

15 **Dowker, B. D.** (1964). *New cereal varieties—1963: Katumani Synthetic No. 2.* E. Afr. agric. for. J., *30*:31.

16 **Dowker, B. D.** (1964). *A note on the reduction in yield of Taboran maize by late planting.* E. Afr. agric. for. J., *30*:33.

17 **Drysdale, V. M.** (1965). *The removal of nutrients by the maize crop.* E. Afr. agric. for. J., *31*:189.

18 **Evans, A. C.** (1960). *Studies of inter-cropping, I—Maize or sorghum with groundnuts.* E. Afr. agric. for. J., *26*:1.

19 **Glover, J.** (1948). *Water demands by maize and sorghum.* E. Afr. agric. J., *13*:171.

20 **Glover, J.** (1957). *The relationship between total seasonal rainfall and yield of maize in the Kenya highlands.* J. Agric. Sci., *49*:285.

21 **Glover, J.** (1959). *The apparent behaviour of maize and sorghum stomata during and after drought.* J. Agric. Sci., *53*:412.

22 **Goldson, J. R.** (1963). *The effect of time of planting on maize yields.* E. Afr. agric. for. J., *29*:160.

23 **Gray, R. W.** (1970). *The effect on yield of the time of planting of maize in south-west Kenya.* E. Afr. agric. for. J., *35*:291.

24 **Grimes, R. C.** (1963). *Intercropping and alternate row cropping of cotton and maize.* E. Afr. agric. for. J., *28*:161.

25 **Hemingway, J. S.** (1955). *Effects of* Puccinia polysora *rust on yield of maize.* E. Afr. agric. J., *20*:191.

26 **Hemingway, J. S.** (1957). *Effects of population density on yield of maize.* E. Afr. agric. J., *22*:199.

27 **Kockum, S.** (1953). *Protection of cob maize stored in cribs.* E. Afr. agric. J., *19*:69.

28 **Kockum, S.** (1958). *Control of insects attacking maize on the cob in crib stores.* E. Afr. agric. J., *23*:275.

29 **Kockum, S.** (1965). *Crib storage of maize. A trial with pyrethrin and lindane formulations.* E. Afr. agric. for. J., *31*:8.

30 **Kockum, S.** and **Graham, W. M.** (1962). *Prevention of insect reinfestation of bagged maize.* Trop. Agriculture, Trin., *39*:231.

31 **la Croix, E. A. S.** (1967). *Maize stalk borers in the Coast Province of Kenya.* E. Afr. agric. for. J., *33*:49.

32 **Nattrass, R. M.** (1952). *Preliminary notice of the occurrence in Kenya of a rust on maize.* E. Afr. agric. J., *18*:39.

33 **Nattrass, R. M.** (1954). *Note on* Puccinia polysora *rust of maize in Kenya.* E. Afr. agric. J., *19*:260.

34 **Scaife, M. A.** (1968). *Maize fertiliser experiments in Western Tanzania.* J. Agric. Sci., *70*:209.

35 **Semb, G.** and **Garberg, P. K.** (1969). *Some effects of planting date and nitrogen fertiliser in maize.* E. Afr. agric. for. J., *34*:371.

36 **Stephens, D.** (1966). *Two experiments on the effects of heavy applications of triple super-phosphate on maize and cotton in Buganda clay loam soil.* E. Afr. agric. for. J., *31*:283.

37 **Storey, H. H.** *et al.* (1958). *East African work on breeding maize resistant to the tropical American rust.* Emp. J. exp. Agric., *26*:1.

38 **Storey, H. H.** and **Howland, A. K.** (1967). *Transfer of resistance to the maize streak virus into East African maize.* E. Afr. agric. for. J., *33*:131.

39 **Turner, D. J.** (1966). *An investigation into the causes of low yield in late-planted maize.* E. Afr. agric. for. J., *31*:249.

40 **Walker, P. T.** (1960). *A survey of the use of maize stalk borer control methods in East Africa.* E. Afr. agric. J., *25*:165.

41 **Weiss, E. A.** (1967). *Fertiliser trials on maize and wheat.* E. Afr. agric. for. J., *32*:326.

42 **Wheatley, P. E.** (1961). *The insect pests of agriculture in the Coast Province of Kenya, V—Maize and sorghum.* E. Afr. agric. for. J., *27*:105.

20
Mangoes

Mangifera indica

Introduction

Mangoes are a tropical fruit grown in all the lower altitude areas in East Africa except the driest. In Kenya they are a minor export crop.

Fig. 79: Mangoes on the East African coast.

Plant characteristics

The mango tree is an evergreen perennial. Its height, when mature, depends on the variety; some of the smaller fruited varieties grow to a height of 60 ft (c. 18 m) or more, whilst other varieties grow to only half this height. The foliage is produced in flushes; at first it is yellow, pale green or red, but it becomes dark green when mature. One part of a tree may be producing young flush growth whilst the rest of the foliage is mature. The inflorescences are also produced in flushes; these occur at different times of the year in different parts of East Africa but they generally occur after the main periods of vegetative growth. Each inflorescence consists of as many as 6 000 flowers, most of which are male, the remainder being hermaphrodite. The hermaphrodite flowers are insect pollinated but as few as 0·1% set fruit. Each inflorescence produces very few fruit; in the larger fruited varieties only one is usually produced but in varieties with smaller fruits there may be three to five. The fruit consists of the following: the skin, or exocarp, whose colour may be green, yellow, red or purple according to the variety; the edible flesh, or mesocarp; the woody husk, or endocarp, which usually has fibres extending into the flesh, and the seed, which is embedded in the husk (see Fig. 80). The period from flowering to fruit maturity is about five months. Most of the East African mango varieties exhibit polyembryony, i.e. each seed produces, in addition to the sexual seedling, one to five nucellar seedlings which are genetically identical to the parent plant. Monoembryonic varieties produce only one embryo per seed; as this embryo is of sexual origin such varieties do not breed true to type. Uniformity can only be achieved in monoembryonic varieties by vegetative propagation.

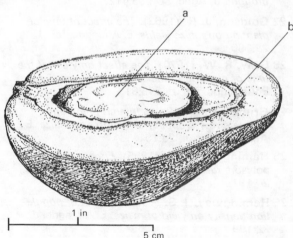

Fig. 80: A mango fruit which has been cut in half. Note the seed (a) and the husk (b).

Ecology

Mangoes grow well in areas which receive poor and erratic rainfall, e.g. the central part of Tanzania where an average annual rainfall of only 25 in (c. 650 mm) is common. They grow best at a higher rainfall than this but a dry period is essential at flowering and fruiting; rain at flowering can cause a high incidence of fungal infections on the inflorescences. Mangoes are limited to the lower altitude areas and are seldom seen above 5 000 ft (c. 1 500 m). Two Kenya selections, Sabre and the Harries mango, can be grown as high as 6 000 ft (c. 1 800 m).

Mangoes do not require soils with a high nutrient content but they must be free draining and deep.

Varieties

The best quality varieties in Kenya are Ngowe, Boribo, Batawi and Apple. The highest yielding varieties are considered to be Boribo and Dodo although no figures on relative yields are available. Dodo produces somewhat fibrous fruits.

Propagation

In East Africa mangoes are propagated by seed although vegetative propagation is possible and is practised in some other parts of the world. Almost all trees are the results of discarded seeds. If mangoes are to be grown deliberately seed must be obtained from fresh fruit; the husk must be removed immediately to reveal any weevil larvae and the seed must be sown as soon as possible. As soon as the shoots are a few inches high the seedlings should be transplanted into pots. Unless this is done the tap root grows so strongly and with so few laterals that it is almost inevitably broken during transplanting, seriously reducing the chances of survival in the field. Transplanting into pots allows an early selection of the most vigorous seedlings. During transplanting surplus seedlings should be separated from the selected one if the variety is polyembryonic. Seedlings should be shaded and kept well watered. They are ready for transplanting after about four months in the nursery.

Field establishment

The spacing should be 35 ft × 35 ft or 40 ft × 40 ft (c. 10·5 m × 10·5 m or 12 m × 12 m). Large planting holes should be dug and should be supplied with a generous amount of organic manure. The young plants should be mulched and shaded and a weed-free circle should be maintained around each. Capping the seedlings at a height of 3 ft (c. 0·9 m) helps to produce a spreading framework of branches.

Field maintenance

Very little work is involved in field maintenance other than preventing excessive growth of vegetation beneath the trees. Nitrogen fertilisers are beneficial and should be applied when the trees are bearing a heavy crop of fruit in order to stimulate the vegetative growth necessary for the next crop. Pruning, to remove dead wood and low hanging branches and to 'open up' the tree, is also beneficial.

Harvesting

Fruits are usually allowed to fall to the ground if they are out of reach whilst on the tree. If they are to be exported, however, they must be picked with care to avoid bruising.

Yields

Mango trees start to bear four or five years after planting. They give economic returns after about eight years and by the time they are twenty years old each tree should be producing from 200 to 500 fruits a year. Dodo and Boribo trees can produce as many as 1 000 fruits a year.

Some mango varieties bear fairly evenly but most are erratic; they may yield well every other year, once a year instead of the normal twice a year or on only part of the tree at any one time.

Pests

The most important pest is the mango weevil, *Sternocochetus mangiferae*. The larvae enter the fruit, leaving no external sign of their entry, and attack the seeds. There is often a hard white area in the flesh of mangoes which may be caused by the entry of mango weevil larvae. Fruits usually fall if they are attacked at an early stage of growth. The flesh of mature fruits may rot in storage or in transit. Field hygiene, i.e. disposal of fallen fruit, is the best method of controlling the mango weevil.

Scale insects can be found on the leaves and the fruit; they excrete honey dew and a sooty mould grows on this. The honey dew and the mould can be washed off but the latter may reduce photosynthesis. Thrips, mites and fruit fly may also be troublesome.

Diseases

Mangoes are fairly free of diseases. The main problem occurs in wet weather when a powdery mildew, *Oidium mangiferae,* attacks the inflorescences. Mangoes are very susceptible to anthracnose (*Colletotrichum gloeosporioides*) which causes discolouration of the fruits.

Bibliography

1 **Sethi, W. R.** (1949). *Mangoes on the East African coast.* E. Afr. agric. J., *15*:98.

2 **Vickers, M. E. H.** (1964). *An experiment on the cold storage of mangoes on the Kenya Coast.* E. Afr. agric. for. J., *30*:46.

3 **Wheatley, S. M.** (1956). *The mangoes of the Kenya coast.* E. Afr. agric. J., *22*:46.

4 **Wheatley, P. E.** (1960). *Insect pests in the Coast Province of Kenya, I—Mango.* E. Afr. agric. for. J., *26*:129.

21
Pawpaw

Carica papaya

Introduction

In East Africa almost all pawpaws are grown
semi-wild from discarded seeds. A few trees are
found near most homesteads in all areas except the
highest and the driest. Pawpaws are mostly grown
for their fruit but in Tanzania there are estates in
the Rift wall area to the south of Lake Manyara and
on the southern slopes of Mt. Meru where they
are grown for papain. Papain production has also
been tried in parts of Kenya although there is no
commercial production at the time of writing.

Fig. 81: A young pawpaw tree.

Papain is an enzyme; it is extracted from the latex
which is obtained by tapping pawpaw fruits. Dried
latex is exported mainly to the U.S.A., western
Europe and Japan, where the papain is extracted.
Papain is a protein digestive; it can be used as a
meat tenderiser, for indigestion cures, for clarifying
beer or even for reducing warts. A recent development, common in Europe, is the injection of papain
into beef animals' bloodstreams shortly before
slaughter.

Plant characteristics

The pawpaw plant is a short-living perennial tree,
seldom growing higher than 30 ft (c. 9 m). It has
a crown of large, palmately lobed leaves; these have
long petioles.

Pawpaw varieties are either dioecious or
hermaphrodite. Seeds from dioecious trees produce
50% male plants and 50% female plants, provided
the pollen came from a male tree. Male plants can
be recognised as soon as they flower because the
flowers are borne on long stalks. Female flowers
(see Fig. 82) have very short stalks and are
therefore borne close to the stem. When a dioecious
variety is grown a small proportion of male trees
must be allowed to grow to ensure pollination.

If hermaphrodite flowers are self-pollinated the
seeds produce 67% hermaphrodite plants, which are
genetically uniform and identical to their mother
plant, and 33% female plants. Hermaphrodite
flowers externally resemble female flowers but
anthers can be found if the petals are removed (see
Fig. 83). To keep a hermaphrodite variety true to
type, self pollination is necessary. To keep a
dioecious variety true to type, sib-pollination is
necessary.

Ecology

Pawpaw trees are found in areas with a wide
range of rainfall regimes and from sea level up to
7 000 ft (c. 2 100 m). They grow best, however,
with a good supply of water and at the lower
altitudes. For papain production the moisture and

137

Fig. 82: A female pawpaw flower with the two nearest petals removed.

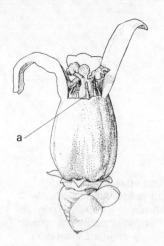

Fig. 83: A hermaphrodite pawpaw flower with the two nearest petals removed. Note the anthers (a).

altitude requirements are more precise. There must be a constant supply of soil moisture; for this reason irrigation is necessary in the dry season in most areas. The flow of latex is best at altitudes below 3 000 ft (c. 900 m); as altitudes increase above this level the flow becomes progressively worse. The soil must be free draining and there must be a good supply of nutrients.

Varieties

Solo is a hermaphrodite variety, with good quality fruit, that has been tried with success in East Africa. The plants normally found in smallholdings are dioecious.

Field establishment and maintenance for papain production

Pawpaws are either raised in nurseries or are sown at stake, i.e. directly in the field with no transplanting. The spacing should be between 7 ft × 7 ft and 9 ft × 9 ft (c. 2·1 m × 2·1 m and 2·7 m × 2·7 m). If sown at stake, about ten seeds should be sown ½ in (c. 1·5 cm) deep. The seedlings emerge about three weeks after sowing; five (the most vigorous) should be allowed to grow until their sex can be determined at six months after germination when they should be thinned, to allow about one male tree for every 20–30 females. Fertilisers are often applied but on volcanic soils high in phosphate and potassium, fertilisers containing these elements may cause damage. According to an Australian

recommendation an 8:12:6 NPK mixture should be applied at the rate of 4 oz (c. 0·11 kg) per hole at sowing, 12 oz (c. 0·34 kg) at two months, 6 oz (c. 0·17 kg) at six months and then 2–3 lb (c. 0·9–1·4 kg) per tree divided into three applications per annum.

Tapping

Small fruits, about 4–5 in (c. 10–13 cm) in diameter should develop about a year after germination. These can be tapped by making a longitudinal cut in each. A razor blade is used; it is embedded in a rubber holder which is tied to a lance. Only about ⅛ in (c. 3 mm) of the blade should protrude; a smaller length fails to penetrate the latex bearing vessels whilst a greater length makes deep cuts which become blocked up with latex and can become infected. Tapping should not be done more frequently than once every week. It should not be done in hot sunny weather when the latex flows slowly; early morning tapping should therefore be practised. Cuts on any one fruit should be as far as possible from previous cuts. By the time a fruit begins to turn yellow and to dry up it has a series of parallel scars about ½ in (c. 1·5 cm) apart. Although only one cut can be made at each tapping when the fruits are small, two or even three can be made on full sized fruits. Tapping should be stopped during the driest months of the year if there is no provision for irrigation.

As the latex drips from the fruit it is caught in 'umbrellas' which are clamped to the trunks. 'Umbrellas' have wooden frames with canvas or polythene stretched on each side to receive the latex. The latex coagulates on the canvas or polythene and is scraped into the box; this provides the first grade coagulum. Second grade coagulum accumulates on the fruits and may be removed subsequently. Impurities must be removed before drying.

Drying

The coagulated latex must be dried immediately after collection to prevent discolouration. Flue drying is universal; the coagulum is laid on wire mesh over a 3 ft (c. 0·9 m) flue. The ideal temperature is between 130° and 140°F (c. 55°–60°C). Over-heating causes an undesirable brown colour. Drying should be completed in 5–7 hours; at this stage the coagulum should be crumbly and not sticky.

Yields

Tapping usually continues for only 2 years. If it were continued for longer the tops would become too high and yields would fall to an uneconomic level. In ideal ecological conditions yields of dried latex should average 125–150 lb per acre per annum (c. 140–170 kg/ha/annum). In Tanzania, where the estates are at a somewhat higher altitude than the optimum, average yields are only 40–50 lb per acre per annum (c. 45–55 kg/ha/annum).

Tapped fruits cannot be sold for eating because of their poor appearance but they are as tasty as untapped fruit. Yields of 4–6 tons of fruit per acre (c. 10–15 t/ha) should be obtained annually if standards of husbandry are high.

Pests and diseases

Pest problems are negligible but disease damage can be considerable and is one of the main reasons for the short life of pawpaw trees. Mildew, root rot, foot rot, damping off, leaf spots, fruit spots, anthracnose and some viruses attack pawpaw.

Bibliography

1 **Greenway, P. J.** (1948). *The pawpaw or papaya.* E. Afr. agric. J., *13*:228.

2 **Kulkarni, H. Y.** and **Sheffield, F. M. L.** (1968). *Interim report on virus diseases of pawpaw in East Africa.* E. Afr. agric. for. J., *33*:323.

3 **Peregrine, W. T. H.** (1968). *A survey of pawpaw debility in Tanzania.* E. Afr. agric. for J., *33*:316.

4 **Wallace, G. B.** (1948). *The establishment and running of a pawpaw plantation.* E. Afr. agric. J., *13*:234.

5 **Wallace, G. B.** and **Wallace, M. M.** (1948). *Diseases of pawpaw and their control.* E. Afr. agric. J., *13*:240.

22
Pigeon peas

Cajanus cajan

Introduction

Pigeon peas are an important pulse crop in many of the lower altitude areas of East Africa. In Uganda an estimated 220 000 acres (c. 90 000 ha) were grown in 1968; approximately 180 000 of these (c. 73 000 ha) were in Lango and Acholi Districts. In Kenya pigeon peas are widely grown in the lower parts of Eastern and Central Provinces. In Tanzania they are common in all the lower altitude areas, especially in the southern part of the country.

Plant characteristics

Pigeon peas are short lived perennial legumes. They have deep woody tap roots. Their stems, which also become woody, may grow as high as 12 ft (c. 3·5 m). The leaves are trifoliate with narrow leaflets. The flowers are usually yellow but are sometimes red, orange or a mixture of these colours. The pods contain up to eight seeds but 4–6 are more common; there are marked depressions in the pods between each seed. Pigeon peas are mostly self-pollinated but there can be as much as 40% cross-pollination. The first pods are ready for harvesting about five months after sowing if the local varieties are grown; some of the introduced varieties produce their first pods more quickly. The life of the crop may be two or three years; in Kenya, where the incidence of pests and diseases is high, they seldom last for longer than a year.

Ecology

Pigeon peas are drought resistant and are therefore well suited to areas of low and uncertain rainfall such as central Tanzania and the lower altitude areas in Machakos and Kitui Districts in Kenya. They grow between sea level and 5 000 ft (c. 1 500 m) but are sometimes found as high as 7 000 ft (c. 2 100 m). Any free draining soil is suitable.

Varieties

At the time of writing no improved varieties of pigeon peas are available in East Africa. A breeding

Fig. 84: Pigeon peas.

programme started in 1968 at Makerere University College in Uganda, however, and high yielding varieties should be released during the 1970s.

Field operations

In Uganda pigeon peas are usually intersown with finger millet; they remain in the field after the millet has been harvested and simsim or cowpeas are often sown amongst them in the following year. In Kenya they are usually intersown with maize and beans at the beginning of the long rains; they remain as a pure stand after the maize and beans have been harvested. In Tanzania they are almost always intercropped with cereals, root crops and other pulses.

The spacing recommended for pure stands is 5 ft × 4 ft (c. 1·5 m × 1·2 m) or 6 ft × 6 ft (c. 1·8 × 1·8 m). Short varieties, which may be introduced in the future, will need a closer spacing.

It is likely that spacing is not a critical factor in pigeon pea production because experiments outside East Africa have shown that there is no variation in yield between plant populations of 1 000 plants per acre (c. 2 500 plants per hectare) and 14 000 plants per acre (c. 35 000 plants per hectare).

Harvesting

Harvesting can begin about five months after sowing and may continue for a further six months or for several years. In Kenya, pigeon peas are usually grown for only one year owing to the high incidence of *Fusarium* wilt and mealy bug in old crops; the plants are seldom left for longer than two years. Harvesting is done either by picking individual pods or by cutting off the bearing branches. Sometimes the branches are dried in the field; sometimes they are taken to the homestead for drying.

Yields

Yields are usually 400–600 lb of dried seeds per acre per annum (c. 450–670 kg/ha/annum) but with good husbandry they should be at least 1 000 lb per acre per annum (c. 1 100 kg/ha/annum). The potential is much higher; 4 000 lb per acre per annum (c. 4 400 kg/ha/annum) have been produced in the West Indies.

Pests

A root mealy bug occurs in Kenya and may build up in the soil to such an extent that fields have to be rested from pigeon peas for several years. American bollworm (*Heliothis armigera*) and spotted pod borer (*Maruca testulalis*) can cause considerable damage.

Diseases

The yields of pigeon peas are seldom seriously reduced by diseases. An exception is the incidence of *Fusarium* wilt, (a fungal disease caused by *Fusarium udum*) which is common in Kenya. The symptoms of this soil borne, fungal disease are blackening of the base of the stem and of the internal tissues of the roots. Crop rotation is the only method of avoiding *Fusarium* wilt.

Utilisation

The dried seeds are usually either boiled with maize seeds or fried and eaten as a vegetable with 'ugali'. The green seeds are occasionally harvested and eaten before they are fully ripe; there is a potential canning market for this product. Pigeon peas are often grown as boundary plants, hedges or windbreaks. The woody stems can be used for firewood.

23
Pineapple

Ananas comosus

Introduction

Pineapple is grown widely throughout the warmer parts of East Africa although it is usually produced on a small scale in order to meet home needs and supply local markets. An exception occurs near Thika in Kenya where an important canning industry has been established and where estates and, to a declining extent, smallholders, grow a considerable acreage of pineapple. There is also a canning factory at Machakos which caters for a smaller acreage. Near Nairobi the fresh fruit market is also important; the fruit is sold either for local consumption or for air freighting to Europe.

Smallholders

Smallholders were encouraged to grow pineapple during the 1950s and 1960s, largely to maintain a good supply of fruit to the Thika and Machakos factories. Such production has never lived up to its expectations, however; the main reason for this is the reluctance of smallholders to invest in a crop which only yields well when there is a high level of

Fig. 85: Mature pineapple almost ready for harvesting.

capital input, e.g. for soil fumigation, black polythene mulch and heavy applications of nitrogenous fertilisers. Expansion of smallholder pineapple is not anticipated during the 1970s.

Estates

A large international canning company produces most of the estate fruit; its dominance will almost certainly increase during the 1970s. This company has mainly adopted Hawaiian techniques (as have the other estates) but it is engaged in a research programme to find improvements more suited to Kenya conditions. Some of these techniques are so sophisticated that they could only be recommended to growers with considerable pineapple experience; in the company's interest, information on some other field aspects cannot be released. For these reasons and because of the declining role that will be played by outgrowers, much detail has been intentionally omitted from this chapter.

Plant characteristics

The pineapple plant has a fibrous root system which seldom penetrates more than 2 ft (c. 0·6 m) below the soil surface; most of the roots are found in the top foot. The main stem, which only grows to a height of about 2 ft, produces suckers from the axils of the leaves (see Fig. 87); suckers are leafy shoots and can be removed and used as planting material. Suckers start to appear at the same time as the inflorescence, either from below or above the soil surface; there are usually one or two on each plant but there are sometimes as many as five. At the tip of the stem one inflorescence is produced; this appears as a red bud 12–18 months after planting, depending on the type of planting material used. The fruit is ready for harvesting about six months after the bud first appears; it developes asexually, i.e. without any previous pollination. Leafy shoots also grow at the base of the fruit; these are called slips. They only develop, however, on vigorous, healthy plants. There are usually two or three slips on each plant although as many as six are sometimes seen. They can be removed and used as planting material two or three months after harvesting the fruit. After harvesting the first crop, ratoon crops can be taken by allowing suckers to develop; the general practice in Kenya is to allow one sucker to grow on each plant and to take only one ratoon crop but in some other parts of the world more than one ratoon crop is taken. The ratoon crop is ready for harvesting about 14 months after harvesting the plant crop.

Ecology

Rainfall, and water requirements
The pineapple plant is very hardy and can withstand considerable droughts. For high yields, good quality and rapid growth, however, a regular supply of soil moisture is essential. A well distributed average annual rainfall of 40 in (c. 1 000 mm) is about the minimum requirement for optimum growth. Most of the estates near Thika receive an average of under 40 in and usually suffer from a severe dry spell at the beginning of the year. Irrigation is being increasingly used to prevent damage to the crop during such times.

Altitude and temperature
In East Africa pineapple is grown from sea level up to 7 000 ft (c. 2 100 m) although at the highest altitudes temperatures are so low that growth is slow and the fruits are acid. The altitude limits for pineapple for canning are 4 500 and 5 700 ft (c. 1 370–1 750 m); above the upper limit the fruits contain too much acid and not enough sugar; below the lower limit they have too little fibre, leading to a 'mushy' product, and too high a content of sugar compared to acid, leading to a 'bland' taste.

Soil requirements
Pineapple is grown on a wide variety of soils. Free draining soils are essential. On heavy soils, e.g. near Thika, deep cultivation is necessary in order to prevent waterlogging.

Varieties

Smooth Cayenne is the only variety grown in Kenya. Some other varieties are grown on Uganda smallholdings. The leaf margins of Smooth Cayenne are smooth except for the tips which have a few marginal spines. The margins of most other varieties are finely serrated.

Fig. 86: Pineapple recently planted through black polythene mulch.

Field operations

Land preparation
Deep cultivation is essential for good pineapple yields. Pineapple is a weak-rooted crop and without a deep seedbed the root range is severely limited with consequent bad effects on the growth of the crop. Soil fumigation to reduce the damage done by nematodes is a standard practice on some estates.

Mulching
Mulching with black polythene (see Fig. 86) has proved to raise yields in the plant crop by 50% and to give very substantial improvement in the size of the fruits. No experimental results for ratoon crops are yet available. Black polythene mulch improves the growth of pineapple not only by conserving moisture and suppressing weeds but also by raising the temperature of the soil. Organic mulches, which reduce the temperature of the soil, tend to depress pineapple growth. Polythene mulch is laid in strips prior to planting; one strip covers one bed (see section on spacing).

Planting material
Crowns, from the tops of the fruits, slips and suckers can all be used for planting material. Crowns take longer from planting to harvesting than slips or suckers (i.e. about 24 months); their

main advantage is that they grow more evenly in the field. Suckers mature fastest and the plant crop can be harvested about 18 months after planting; even if they are carefully graded into different sizes, however, they tend to show uneven growth in the field. Suckers should be removed when they weigh 8–10 oz (c. 230–280 gm); if they weigh more than 1 lb (c. 450 gm) many produce a minute fruit within a few months after planting. Slips are intermediate between crowns and suckers both in their uniformity and speed of growth; they mature within 20–22 months after planting.

Whatever type of planting material is used must be graded into at least three different groups according to size, and each group must be planted separately in order to get as much uniformity in the field as possible. Planting material should be cured after removal from the mother plants; this involves spreading it out in the sun for three days to two weeks, depending on the weather, thus allowing the wounds to cork over and to become more resistant to infection after planting. Finally, the planting material should be dipped in an insecticidal solution as a precaution against mealy bug.

Spacing

A double row spacing is most convenient with pineapple as it allows easy access between the rows. The double row beds are normally 5 ft (c. 1·5 cm) apart from centre to centre; the two rows within a bed are 20 in (c. 0·5 m) apart with the plants 11 in (c. 28 cm) apart within the row. This gives a plant population of almost exactly 19 000 per acre (c. 47 000 per hectare). The risk of plant populations higher than this is that they cause too high a proportion of small low grade fruit which is unsuitable for canning.

Depth of planting

Shallow planting causes weak root development. Deep planting causes the crowns to be so low that they become filled with soil and later rot. If the bases of the plants are placed 3–4 in (c. 7·5–10·0 cm) deep the crowns are well away from the soil and the plants are deep enough for a healthy root system to develop.

Fertilisers

Pineapple needs heavy applications of nitrogen if it is to give high yields. In a recent experiment pineapple without nitrogen yielded about 12 tons of fruit per acre (c. 30 t/ha) whilst regular urea sprays produced 20 tons per acre (c. 50 t/ha) of considerably higher grade fruit. Nitrogen can be

1·0 ft

0·5 m

Fig. 87: A mature pineapple plant. For clarity, several of the nearer leaves have been removed. Note the slip, the lower sucker, which would be unsuitable for selection for the ratoon crop, and the upper sucker, which originates from just below the slip, pointing to the left of the figure, and which would be suitable for ratoon crop selection.

applied either as a foliar spray or as a soil applied fertiliser. The optimum rates are still being investigated but common rates of application are 100 lb of nitrogen per acre (c. 110 kg/ha) in both the long and the short rains or monthly sprays of 75 lb of urea per acre (c. 85 kg/ha).

Phosphate gives good responses and is best applied about four months after planting in the axils of the lower leaves.

Weed control

Weeds are usually controlled by a combination of mulch, herbicides and hand cultivation. The most popular herbicide is bromacil which is applied to the soil and suppresses weeds for 6–12 months.

Stripping

Stripping involves the removal of all slips and all unwanted suckers. Only one sucker should remain as the basis of the ratoon crop and it should not arise from below the ground; suckers which arise from below the ground are invariably weak because they are heavily shaded by the foliage above them. Stripping should start at around the time of harvesting; if correct sized suckers and slips are needed for planting material, two or three operations at monthly intervals are usually needed.

Forcing

Forcing consists of spraying the crop with an auxin (plant hormone) in order to encourage uniform flowering, thus controlling the time of harvesting. This technique is being tried at Thika on a limited scale.

Prevention of sun scorch

When fruits remain in an upright position they are reasonably protected by their crowns from sun scorch. If they lean over to one side, however, the upper part is unprotected and in cloudless weather can become sun scorched; the affected areas are discoloured and soft. It is often necessary to protect some of the fruits by shading them with dried grass or by tying the leaves of the plant over the top of the fruit.

Harvesting

There is a popular fallacy that a pineapple is ripe when the small leaves in the centre of the crown can be pulled out easily. The best test for ripeness is to tap the fruit; a ripe fruit gives a dull sound but this can only be recognised after considerable experience. When ripe, a fruit is easily snapped off its stem by hand.

Yields

With adequate cultivation, soil fumigation, mulching and nitrogen applications, plant crops should yield about 25 tons of fruit per acre (c. 63 t/ha) whilst ratoon crops should yield about 20 tons per acre (c. 50 t/ha).

Pests

Mealy bug. *Dysmicoccus brevipes*

Mealy bug is undoubtedly the most damaging insect pest of pineapple in Kenya. The insects are found on the roots, in the leaf bases or on the fruit. They carry a virus which causes a characteristic red or yellow colouration on the leaves, stunting and in severe cases, die-back from the tips of the leaves. Mealy bugs are difficult to control; the planting material may be dipped in an insecticidal solution, chlordane or heptachlor may be spread around the soil to control the attendant ants or the crop may be sprayed with parathion, but none of these give long term control.

Nematodes

Pineapple roots are almost always heavily infested with nematodes unless the fields have been fumigated before planting. The extent to which nematodes reduce yields is currently under investigation, but is generally considered to be very great.

Scales

Various species of scale insects are found on the fruits and leaves but they usually occur in isolated patches and seldom cause serious damage.

Diseases

Apart from the virus disease discussed above, various fruit diseases cause considerable losses, both to canning and fresh market fruit. No effective control has yet been found.

24
Potatoes

Solanum tuberosum

Introduction

Potatoes, sometimes called European, Irish, English or Solanum potatoes, to distinguish them from sweet potatoes, are an important root crop in some of the East African highlands. At the higher altitudes they have a greater yielding potential than other food crops and are a more suitable subsistence crop than maize, which may take more than a year to mature. Most potatoes are consumed as a subsistence crop although some are marketed internally, often reaching towns far from the growing areas, such as Kampala, Mombasa and Dar es Salaam.

Kenya

Approximately 700 000 acres (c. 28 000 ha) of potatoes are grown annually in Kenya. In 1964 it was estimated that 5 000 acres (c. 2 000 ha) were grown on large scale farms but this area has declined and at the time of writing is probably below 1 000 acres (c. 400 ha). The higher altitude parts of Central Province are the most important potato growing areas. Potatoes are also widely grown in the higher parts of Meru, Machakos, Kericho and Nandi Districts and in most of the forest reserves, where forestry labourers are allowed to grow crops between newly planted trees during the first few years of establishment.

Tanzania

Potatoes are occasionally grown in the Northern Highlands, e.g. near Moshi, Arusha and Lushoto. They are an important crop in the Southern Highlands, especially in Njombe and Mbeya Districts.

Uganda

Potatoes are only grown in the higher altitude areas of Bugisu and Kigezi Districts.

Plant characteristics

A potato tuber (see Fig. 88) is a swollen underground stem. On its surface are a number of

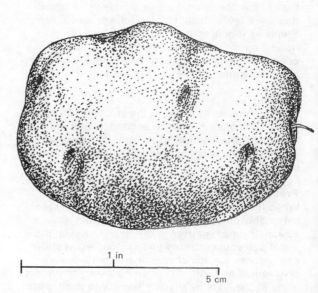

1 in

5 cm

Fig. 88: A potato tuber. Note the eyes and the point of attachment to the parent plant.

'eyes', at each of which is a bud in the axil of a scale leaf. After undergoing a period of dormancy, which is usually two to three months, some of these buds sprout and produce stems. One 'eye' can produce more than one stem owing to branching at the base of the original stem. The nodes of the stems produce roots and, later, short stolons whose ends swell into tubers (see Fig. 89). The potato plant produces flowers; the petals are white, pink or blue or purple according to the variety. As a general rule varieties with white skinned tubers have flowers with white petals whilst varieties with coloured tubers have coloured petals. In East Africa, although not in Europe, seedlings often grow in the field from true seed dropped by the previous crop. Despite the fact that potato flowers are largely self-pollinated, seedlings differ considerably from their parent plant in genetic constitution and can therefore lead to the genetic impurity of a variety.

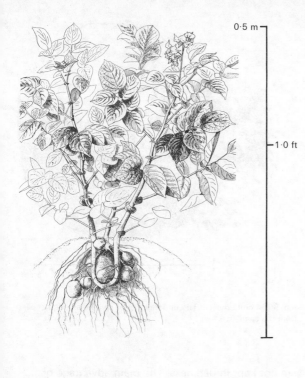

0·5 m

1·0 ft

Fig. 89: A two month old potato plant. Note the flowers, the parent tuber and the developing tubers, which are borne on short stolons arising from the stems.

Life cycle

The tubers of most varieties in temperate countries stop growing at about the time of flowering. In East Africa, however, tuber growth often continues after flowering, provided that there is adequate soil moisture. In the absence of diseases the productive life of the potato plant is therefore governed primarily by the duration of the rains. Between 4 000 and 9 500 ft (c. 1 200–2 900 m), altitude has little effect on the rate of growth.

Ecology

Rainfall, and water requirements

A steady rainfall of about 1 in (c. 25 cm) a week is enough to maintain optimum growth of potatoes on most soils in the East African highlands. A good yield should be obtained if the rains continue for $3\frac{1}{2}$ months, provided that damage by diseases is not serious.

Altitude and temperature

In warm conditions the potato varieties now available in East Africa give poor tuber growth. They are best suited to the cooler conditions above 6 000 ft (c. 1 800 m) although they are occasionally seen on smallholdings between 5 000 ft and 6 000 ft (c. 1 500–1 800 m). They can be grown successfully as high as 9 500 ft (c. 2 900 m).

Soil requirements

Soils must be free draining. Heavy soils restrict tuber expansion and make harvesting difficult. Potatoes only give good yields when they have a good supply of nutrients, either from a naturally fertile soil or when fertilisers or manures are applied.

Varieties

Almost all potato varieties grown in East Africa have been introduced from Europe. A breeding programme started at the National Agricultural Laboratories at Nairobi in 1962; it is still in progress and at the time of writing is being considerably expanded. Its main aims are to breed varieties resistant to blight and bacterial wilt, using the type of resistance that operates equally against all races of the pathogen and that does not break down when new races appear. Many varieties have been introduced in the past with the type of resistance that breaks down too quickly to be of agricultural value.

Dutch Robijn

This variety has a red skin and yellow flesh and is known as 'Ngorobu' in Central Province in Kenya. It was introduced from Holland in 1946 and rapidly became popular with smallholders; at one time it was said to contribute more than half of Kenya's crop. It has moderate resistance to blight but it is highly susceptible to bacterial wilt. Without fungicidal spraying its yields are generally low and it cannot be grown on land heavily infested with bacterial wilt. Nevertheless there is a high consumer preference for this variety.

Roslin Eburru

Frequently and incorrectly called B/53. This variety was introduced from Roslin in Scotland in 1953. It has white skin and white flesh. Its blight resistance was originally high but has recently broken down with the appearance of new races of the pathogen. Roslin Eburru rapidly displaced Dutch Robijn in many parts of Central Province, but is now, in its turn, expected to be replaced by Kenya Akiba.

Kerr's Pink

This variety was introduced in the 1930s and is still by far the most popular variety in Meru District in Kenya. It has red and white skin and white flesh. It is very susceptible to both blight and bacterial wilt. However, blight is less severe in Meru District and bacterial wilt is relatively uncommon. This variety has been grown for about 70 vegetative generations and is often quoted as an example of the unimportance of 'degeneration' of stocks due to viruses in Kenya.

Kenya Akiba

This is the first release of the Kenya potato breeding programme. It has white skin and white flesh and longer stolons than most varieties. It has a high level of resistance to blight and can safely be grown without spraying against this disease. The type of resistance is such that new races of blight will not be able to attack the crop. Kenya Akiba has a moderate resistance to bacterial wilt and matures more slowly than most varieties. Further releases of disease resistant varieties in the 'Kenya' series are planned for the early 1970s. Each new variety will be given a Swahili name in alphabetical order of release.

Atzimba

This variety has been introduced from Mexico where it has recently been bred for blight resistance. Its resistance to this disease is the highest known in the world. It is under observation in Kenya but has not yet been released.

Propagation

Small tubers, called 'seed' or 'setts', are used for propagation. They should be between $1\frac{1}{4}$ and $2\frac{1}{2}$ in (c. 3–6 cm) in diameter. About $\frac{3}{4}$ ton of seed is needed to plant an acre (c. 1·85 tons per hectare).

Chitting, sometimes called sprouting, is strongly recommended. It involves spreading the seed in a layer no more than two or three tubers deep in diffuse light. This encourages the development of short, green, healthy sprouts (see Fig. 90). Complete darkness must be avoided because this causes the development of long, white, thin sprouts which are easily broken before or during planting. When subjected to light, buds take several weeks (depending on the variety and the temperature of storage) to produce sprouts about $\frac{1}{2}$ in (c. 1·3 cm) long; the seed can be stored in this form for a further two or three months during which the sprouts grow little, if at all, provided that they

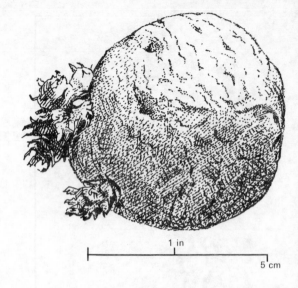

Fig. 90: A chitted seed. The upper group of sprouts all originate from the same eye.

are not kept in darkness. The main advantage of chitting is that stem growth commences immediately the seed is planted, thus making maximum use of the available rainfall and causing rapid and even emergence. Crops grown from chitted seed yield more than those from un-chitted seed.

Cutting seed tubers into two or more pieces is sometimes practised in temperate countries, especially in the U.S.A. It should be strongly discouraged in East Africa because it spreads bacterial wilt from infected tubers to clean ones, by means of the implement used for cutting.

Field operations

Land preparation

Planting on ridges which are $2\frac{1}{2}$ ft (c. 0·75 m) apart is recommended; it conserves soil and water and gives the ideal conditions for tuber expansion. Planting on ridges is the general rule on large scale farms but is very seldom practised by small scale farmers although they often heap soil around the stems whilst weeding. Heaping soil around the stems should be greatly encouraged because only when the lower nodes are covered with soil can they produce tubers; the more nodes that are covered, the higher are the yields.

Time of planting

Potatoes should be planted as near the beginning of the rains as possible. Dry planting has proved very successful. Delayed planting reduces the period of tuber growth because the onset of dry weather occurs earlier in the crop's life.

Planting

All planting in East Africa is done by hand. The seed is usually placed about 4 in (c. 10 cm) deep. When potatoes are planted on the flat it is recommended that they should be planted at the bottom of holes about 6 in (c. 15 cm) deep and covered with a shallow layer of soil; they can then be earthed up easily during normal weeding operations, thus encouraging tuber production at many of the lower nodes. When using ridges large scale farmers usually place the seed, together with the fertiliser, in the furrows; they then cover the seed by harrowing or by splitting the ridges with ridging bodies.

Spacing

The recommended spacing for the varieties currently grown is $2\frac{1}{2}$ ft (c. 0·75 m) rows with plants 9–12 in (c. 23–30 cm) apart within the row. With the introduction of varieties with greater vegetative growth and with a wider spread of tubers, such as Kenya Akiba, a spacing of $1\frac{1}{2}$ ft (c. 0·45 m) within the row may be necessary. In practice small scale farmers almost always use a considerably wider spacing than that recommended, even when they plant a pure stand. Interplanting with crops such as maize and beans is common at the lower altitudes.

Fertilisers and manures

Potatoes respond well to fertilisers and manures in most parts of the world and Kenya is no exception. Economic responses have been obtained from applications of 20–40 lb of nitrogen per acre (c. 22–45 kg/ha) and 40–60 lb of P_2O_5 per acre (c. 45–65 kg/ha). Some large scale farmers apply as much as 100 lb of nitrogen per acre and 200 lb of P_2O_5 per acre (c. 110 kg N and 225 kg P_2O_5/ha). No responses to potassium have been shown. Potatoes give greater responses to farmyard manure than most other crops. The use of manure by smallholders, however, is very limited. The reason for this is possibly that potatoes usually suffer from black scurf when the seed is placed near organic manure; for successful results it is essential that manure is dug deeply into the soil. Fertilisers and manures should only be applied when high yields are confidently expected. It is almost certainly a waste of time and money to apply them when blight susceptible varieties are grown without fungicidal protection; in these conditions crop losses are almost always high.

Weed control

It potatoes are planted at the correct spacing they rapidly cover the ground and suppress weeds. Weeding should not be necessary after the first six weeks if there is a good stand. On large scale farms ridging bodies are usually used for weeding; they heap soil around the stems at the same time as killing weeds.

Harvesting

All potatoes in East Africa are harvested by hand. No harvesters of the elevator type have been imported and spinners, which lift the tubers and place them on the surface of the soil, have proved to be unsuitable for East African soil conditions. Blunt sticks or 'jembes' are the implements usually used for harvesting. Tubers must be exposed to the minimum of direct sunlight during harvesting, otherwise they turn green.

If the tubers are to be transported after harvesting it is highly advisable to cut or pull the tops off two or three weeks before lifting. The effect of this is to harden the skins of the tubers by preventing further growth. Hard skins are less likely to be bruised during transport.

Potatoes cannot be stored for long in East Africa because the high temperatures encourage sprouting. They are most conveniently stored in the soil during the dry season although there is a risk of nematode damage if they are left in dry soil for more than 4–6 weeks.

Yields

The national average yield in Kenya is considered to be between two and three tons per acre (c. 5·0–7·5 t/ha). This could probably be doubled if blight was controlled, because with good husbandry and the use of either blight resistant varieties or fungicides, yields of 6–8 tons per acre (c. 15–20 t/ha) can be expected. The highest commercial yield achieved on a field scale in Kenya is 16 tons per acre (c. 40 t/ha).

Pests

Potato aphid. *Aulacorthum solani*

Aphids may occur on the leaves and stems in the

field and on the sprouts of chitting seed. They do little damage themselves but may carry a number of virus diseases. Chitting seed can be protected by spraying with dimethoate but spraying in the field is not considered necessary.

Potato tuber moth. *Phthorimaea operculella*
This pest is seldom damaging unless potatoes are grown out of season. In the field larvae bore into the vines and from there into the tubers; in stores they bore straight into the tubers. During storage tubers can be protected by dusting with aldrin if they are to be used for seed or with pyrethrum if they are to be eaten.

Nematodes
Several species of nematode occur on potatoes in East Africa. They are entirely different from the cyst-forming nematodes of temperate regions and should not be confused with these very serious pests. It is unlikely that any potato growing area is free from nematodes in East Africa. They normally do very little damage, however, and no measures need to be taken against them. They are occasionally damaging in badly eroded soils and when tubers have been left in the ground too long.

Fig. 91: Potato blight. A healthy leaf is shown on the right for comparison.

Diseases

Blight
This disease is caused by the fungus *Phytophthora infestans*. Blight epidemics are more severe in East Africa than they are in temperate countries and are the most important factor limiting potato yields in this part of the world. The first reason for the severity of blight epidemics is the absence of either a winter or a prolonged dry period to check the disease; it thrives throughout the year not only on potato crops, which are planted in many months of the year, but also on volunteer potatoes and wild *Solanum* species. The second reason is climatic: the climatic requirements of both the fungus and the crop are identical and are met in most months of the year in the East African highlands.

The first symptoms of blight are irregular, brown, necrotic patches on the leaves (see Fig. 91). These spread rapidly, especially if the weather is overcast, wet and humid, and all the vegetative parts may finally be destroyed. Diseases of the tubers, i.e. discolouration and rotting, are seldom seen in East Africa although they are common in temperate countries.

Blight can be spread by infected seed and possibly by infected debris in the soil. The most important method of spread, however, is by spores which are blown in the air or which are splashed from one leaf to another.

The most effective method of preventing blight is growing resistant varieties, especially those such as Kenya Akiba and Atzimba which have a high level of the type of resistance that does not break down. Varieties such as these do not need to be sprayed with fungicides. Their release and distribution in Kenya during the 1970s should make potatoes a very much more popular subsistence crop.

Fungicidal control is expensive and time consuming and is beyond the means of a smallholder growing potatoes for subsistence. It is sometimes worthwhile, however, when blight susceptible varieties are grown as a cash crop. Copper, zineb, maneb or a combination of the last two should be applied every five days during weather favourable for the spread of blight.

Bacterial wilt
This disease is caused by the bacteria *Pseudomonas solanacearum*. It was first observed in Kenya at Embu in 1940 and spread throughout the main potato growing areas during the late 1950s and the

1960s. Once soil is contaminated with bacterial wilt it remains contaminated indefinitely. For this reason potato growing is impossible, using existing varieties, in about 15% of Central Province and almost all of Embu District. It also occurs in Tanzania and in most other potato growing areas in Kenya.

The external symptom is a wilting of the vegetative parts in spite of a moist soil. A white bacterial mass oozes from the vascular tissue when the base of the stem or a tuber is cut (see Fig. 92).

The main method of spread is by diseased seed tubers. The bacteria are spread very effectively by furrow irrigation water. Once the bacteria is in the soil it remains there almost indefiniately both because it can survive saprophytically and also because it parasitises a number of very common weeds.

Bacterial wilt does not occur in the potato breeding countries of Europe; potato selection has been in progress in these countries for 400 years and all resistance to bacterial wilt has been lost. It follows that any European potato variety is exceptionally susceptible to wilt and this is one of the main reasons that potatoes have remained a minor crop in East Africa. Resistant varieties are being bred in Kenya and are expected to be released in the early 1970s. These will be the only means of combating bacterial wilt. Until their release potatoes cannot be grown on infested soil. An important precaution when growing susceptible varieties on clean land is to use clean seed. The use of bare fallowing during the dry season is under investigation; this practice is thought to reduce the amount of inoculum by desiccation but it seems that it cannot eliminate it entirely. Soil fumigation is effective but is prohibitively expensive.

Virus diseases
No detailed study has yet been made of potato virus diseases in East Africa. Potato virus X and leaf roll virus occur, together with a virus that appears to be potato virus Y. The symptoms of virus X and virus Y vary with variety but usually include leaf mottling or mosaic. When affected by leaf roll the leaves not only roll up but become starchy and rattle together when shaken. The yellow dwarf virus and the spindle tuber virus, which can cause serious damage in Europe, do not occur. Virus diseases seldom do much damage in East Africa and varietal degeneration due to virus build-up appears to be slower than in other parts of the world. The reason for this is not yet known.

Black scurf
This disease is caused by the fungus *Rhizoctonia solani*. White encrustations occur on the tubers and stems. Black scurf seldom occurs unless potatoes have been planted in close contact with organic manure.

Utilisation

Africans, like Europeans, cook potatoes in a number of ways; boiling, frying and roasting are all common. One of the most important foods in Central Province in Kenya consists of mashed potato with whole maize and legume grains; the maize and legumes are included with the water and the potatoes towards the end of the boiling period.

Bibliography

1 **Linton, R. D.** (1947). *Observations on European potatoes in Tanganyika.* E. Afr. agric. J., *13*:19.

2 **Nattrass, R. M.** (1946). *Note on the bacterial wilt disease of the potato in Kenya.* E. Afr. agric. J., *12*:30.

3 **Rayner, R. W.** (1945). *Notes on the effect of day length on potato yields.* E. Afr. agric. J., *11*:25.

4 **Robinson, R. A.** and **Ramos, A. H.** (1964). *Bacterial wilt of potatoes in Kenya.* E. Afr. agric. for. J., *30*:59.

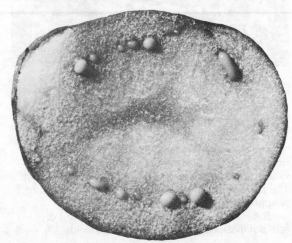

Fig. 92: The cut surface of a potato tuber which is infected with bacterial wilt. The exudate only appears when considerable pressure is applied.

25
Pyrethrum

Chrysanthemum cinerariaefolium

Introduction

Pyrethrum is a cash crop that is grown in the
highlands of Kenya and Tanzania. The growers sell
dried flowers from which an insecticide is extracted.
The insecticidal properties of pyrethrum are
discussed at the end of this chapter.

Kenya

Pyrethrum was first introduced into Kenya in 1931.
During the late 1960s about 90% of Kenya's crop
was produced by smallholders; in 1968 they
numbered approximately 70,000. Kisii and
Nyandarua Districts are the most important
smallholder areas; Kiambu, Nyeri, Meru, Murang'a,
Embu and Elgeyo-Marakwet Districts (in declining
order of importance) have smaller acreages. The
number of large scale growers, most of whom farm
on the western side of the Rift Valley between, and
including, Mau Narok and Londiani, has declined
since 1962 when there were about 1 000 of them.
During the late 1960s the area of pyrethrum in
Kenya was estimated as approximately 100 000
acres (c. 40 000 ha).

Tanzania

Tanzania produces only about half as much
pyrethrum as Kenya. Most is grown by smallholders
in the Southern Highlands. There is some large
scale production on the western slopes of Mt.
Kilimanjaro.

Plant characteristics

Pyrethrum is an annual herb. Plants produce long
deeply lobed leaves from their central crowns and
seldom grow more than 2 ft (c. 0·6 m) high. The
root system is fibrous and most of the roots are
found in the top foot (c. 0·3 m) of the soil profile.
Daisy-like flowers are borne on long stalks which
arise from the central crown. Each flower is about
1½ in (c. 4 cm) in diameter but in triploid plants,
which are usually larger in all respects, they may be
over 2 in (c. 5 cm) in diameter. The base of the
flower, i.e. the receptacle, is surrounded by many

Fig. 93: Picking pyrethrum.

green bracts. The circumference of the receptacle
bears one row of ray florets with white petals; the
remainder of the receptacle bears yellow disc
florets. The outer disc florets are the first to open,
whilst those in the centre of the flower are the last.
The period from bud emergence to flower opening
is approximately six weeks. Pyrethrum is
predominantly cross-pollinated; the amount of
self-pollination has been estimated as only 1%.

Pyrethrins content

Pyrethrins are esters of two acids, chrysanthemic

acid and chrysanthemum dicarboxylic acid, and are the insecticidal constituents of the pyrethrum plant. There are six esters: Pyrethrin I, Pyrethrin II, Cinerin I, Cinerin II, Jasmolin I and Jasmolin II. The ratios of the different esters vary considerably from one clone to another and can therefore be used for clone identification. The following factors affect the pyrethrins content:

1 *Part of the plant*
All parts of the pyrethrum plant contain pyrethrins but their concentration is negligible in all parts except the flowers. 90% of the pyrethrins in the flowers occur in the achenes (developing fruits) of the disc florets.

2 *Stage of flower development*
Pyrethrins are only produced in the achenes in useful quantities after the flowers have opened. The peak pyrethrins content occurs when two or three rows of disc florets have opened (see Fig. 99).

3 *Genetic constitution of the plant*
Pyrethrum seedlings are highly variable in their genetic constitution owing to the high percentage of out-crossing. They therefore vary greatly in their ability to produce pyrethrins. A breeding programme was carried out in Kenya during the 1950s and early 1960s, resulting in the issue of so-called 'high toxic varieties'; these varieties, however, always produced a large proportion of seedlings with an undesirably low pyrethrins content. For this reason a programme of clone selection and distribution started in 1966 in Kenya. (All plants belonging to a clone are genetically identical because they are vegetatively propagated).

In seedling populations there is a general inverse correlation between flower yield and pyrethrins content; this means that most plants selected for good performance in one characteristic perform poorly in the other. This inverse correlation does not appear to be very marked, for in one of the most successful clone selection programmes in Kenya plants were first selected for number and size of flowers; chemical analysis showed that 15% of these had high pyrethrins contents. One of the factors which must be considered during clone selection is the ability of the plants to be propagated vegetatively; the clone with the highest pyrethrins content yet recorded (3·25%) unfortunately failed in this respect. Any clone

whose dried flowers have a pyrethrins content of over 2% is considered very good. During the late 1960s, farmers who produced high content flowers in Kenya were paid a premium price whilst producers of low content flowers were penalised by lower prices.

4 *Temperature*
This factor is discussed in the following section on ecology.

5 *Length of time after planting*
The pyrethrins content tends to be a little higher in the year of planting than in subsequent years.

6 *Care during picking, drying and despatch*
Lack of care during these operations may lead to fermentation and overheating with a consequent drop in the pyrethrins content. The necessary precautions are discussed in later sections.

Ecology

Rainfall, and water requirements
Flushes of flowers are only produced when the rainfall is adequate. 4 in (c. 100 mm) per month is ideal, giving a total of about 50 in (c. 1 250 mm) per annum. Pyrethrum gives reasonable yields, however, in areas that receive a lower rainfall than this; a well distributed average annual rainfall of 40 in (c. 1 000 mm) is usually considered the minimum. Ol Joro Orok in Kenya receives only 37–38 in (c. 940–965 mm) per annum; the reason that pyrethrum can yield well in this area is that cloudy or misty conditions reduce evapo-transpiration for much of the year. A short dry season is an advantage because it gives a wintering effect and the following flush of flowers is considered to be greater than the production that would have occurred if there had been no break in the rains. However, the long dry season in southern Tanzania, which usually lasts for five or six months, undoubtedly prevents optimum yields.

Altitude and temperature
There is an inverse correlation between temperature and pyrethrins content. Therefore the higher pyrethrum is grown, the higher is the pyrethrins content; a rise of 1 000 ft (c. 300 m) is estimated to cause a rise of 0·08%, up to 8 300 ft (c. 2 530 m). The pyrethrins content is also affected by the changes of temperature throughout the year; it

rises about five weeks after the fall in mean temperatures which accompanies the onset of the rains.

Bud initiation is also affected by temperature and is directly correlated with the number of hours when the temperature is below 60°F (15·5°C). Flowering therefore varies at different times of the year, being low during warm periods, and at different altitudes. The optimum altitudes for flowering are 8 000 ft (c. 2 400 m) or more. The recommended lower limit for pyrethrum in Kenya is 6 500 ft (c. 2 000 m) outside the Rift Valley and 7 000 ft (c. 2 100 m) within the Rift Valley. Vegetative growth is greatest in warm conditions: at 6 000 ft (c. 1 800 m) in Kenya pyrethrum grows rapidly but produces few flowers. In Tanzania 5 000 ft (c. 1 500 m) is the lower limit for commercial pyrethrum. Presumably this figure is lower than that for Kenya because there is a cooler 'winter' in Tanzania.

Mild frosts cause no reductions in yields or pyrethrins content although they may cause a browning of the extremities.

Soil requirements

Fairly rich soils with a good structure are ideal for pyrethrum. A good structure is essential to ensure water infiltration and to prevent erosion. During the three years that pyrethrum is in the ground the soil suffers so much from repeated weeding and trampling that unless the structure is fairly good in the first place it soon breaks down, causing poor water infiltration and, in the absence of ridging, soil erosion. Very rich soils, as are found when land is first cleared from bamboo forest or where there is an old hut site, often cause excessive vegetative growth and an increased incidence of root rot. Pyrethrum does not tolerate waterlogging although a large acreage is grown on ridges on the poorly drained soils of the Kinangop Plateau in Kenya. The lower limit for the soil pH is 5·6.

Varieties

Until the 1960s pyrethrum was almost universally propagated by seed. Seed propagation has a few advantages over vegetative propagation: seed is easier to transport to remote areas; seedlings are less trouble in the nursery and it is fairly easy to time the sowing of a seedling nursery so that seedlings are ready for transplanting into the field at exactly the correct date. (It is not so easy to have splits available in the correct quantities at the correct time). Owing to their great variability, especially in pyrethrins content, existing seed varieties have

fallen out of favour in Kenya and vegetatively propagated clones are now recommended as part of a nation-wide campaign to raise both pyrethrins content and yields. Research is in progress, however, to produce seed of more uniform genetic background and several such new varieties are being subjected to trial. Names of individual clones are not given here because it is likely that current issues will soon be superseded.

Propagation

Nurseries are needed whenever rapid multiplication and large numbers of splits are needed for vegetative propagation. They have have established by the Pyrethrum Marketing Board in Kenya and by large scale farmers specialising in the sale of splits.

If possible nurseries should be sited at fairly low altitudes to ensure rapid vegetative growth. The rainfall is usually unreliable at such altitudes so it may be necessary to provide irrigation. The main aim must be to maintain the plants in a young, vigorous, vegetative condition and to prevent the accumulation of woody material in the centre of the plants (see Fig. 94). Old woody plants give few splits and the survival rate may be poor after splitting; young vegetative plants, however, are easily divided into many splits with little wounding of the roots.

Nursery beds should be about 5 ft (c. 1·5 m) wide and should be dug deeply to ensure good root growth. The beds should be separated by paths 2 ft (c. 0·6 m) wide. Nitrogenous and phosphatic fertilisers should be incorporated into the beds. The best spacings are 9 in × 9 in or 12 in × 6 in (c. 23 cm × 23 cm or 30 cm × 15 cm); the latter allows easier earthing up. Holes should be dug about 6 in (c. 15 cm) deep; it is important that they should have one vertical side.

The preparation of splits is an important operation; many failures may be caused by careless work at this stage. If the plants to be multiplied are old and woody the stems are cut off at the level of the top of the leaves. If on the other hand, the plants are young and have previously been in a nursery, there should be no flowers or stems. A young vegetative plant breaks up easily into separate splits with little damage to the roots and with only a little finger pressure needed (see Fig. 95). Provided that young living roots are present, a split can be very small with only one or two leaves but if there are very few leaves it is important to check that there are no woody parts or areas with extensive root breakage. After

Fig. 94: A pyrethrum plant taken from the nursery.

splitting, the roots are trimmed to 4–6 in (c. 10–15 cm) long; roots longer than this would be bent upwards in the hole and splits do not grow well when planted in this manner. The splits must be planted as soon as possible, preferably on the same day as uprooting and splitting; many plant deaths are caused by allowing the splits to dry out. If it is necessary to leave them overnight before planting the following day they should be spread out rather than left in a heap in which they would heat up. The splits should be dipped in a maneb-zineb fungicidal solution before planting.

When planting, the split should be held against the vertical wall of the hole with one hand, making sure that it is neither too high nor too low; it must be in the same relationship with the soil as it was previously. Deep planting causes the crown to rot whilst shallow planting causes root exposure and death. The other hand is used to fill in the hole after making sure that the roots hang downwards and are not bent. Filling must be done bit by bit, firming the soil around the roots at least three times. When the operation is complete it should

Fig. 95: These splits were obtained from the plant in fig. 94.

not be possible to remove the split from the soil with a gentle pull.

During the first month in the nursery the soil should be kept permanently moist; irrigation is necessary if there is no rain. Shade may be necessary if the sun shines for most of the day. The beds must be carefully hand weeded during this month; pulling the weeds ensures the least possible damage to the roots of the young splits. Flower buds must be removed as soon as they are visible; this is essential to make the splits grow vegetatively. De-budding is a very labour-consuming operation, especially at high altitudes where pyrethrum produces flowers profusely, but it must not be ignored at any stage in the nursery. About two months after planting the splits are earthed up, using dry earth, to a depth of about two inches (c. 5 cm). Both shade and watering should be reduced well before earthing up and should be kept to a minimum afterwards. Fungicide sprays may be necessary if leaf rot is noticed; a maneb-zineb mixture has proved to be the most effective. If all the recommendations are followed plant deaths should be reduced to about 5%.

Nursery plants should be ready for uprooting and re-splitting three or four months after planting. If they are to be replanted in a nursery it should be possible to get ten or twelve small splits from each plant. If they are to be planted in the field, however, they need to be larger to withstand the harsher conditions, and only four or five splits should be taken from each plant.

A nursery system has been developed with the plants grown under polythene. They are carefully enclosed and this reduces the work of watering.

Rotations

Pyrethrum should remain in the ground for no longer than three years; after this yields drop below an economic level and the soil structure deteriorates, causing poor infiltration and, possibly, soil erosion. In addition, pyrethrum tends to suffer from root rot and nematodes if left for too long. The only practical way to restore soil structure and to reduce soil pests and diseases is to rest the land under grass or a bush fallow for at least three years. Most smallholders ignore this recommendation; some fields near Limuru in Kenya have now been under pyrethrum for as long as ten years. Small-holders sometimes plough up a whole field and replant after three years but more commonly they simply split up plants to fill in nearby gaps whenever these appear.

A cleaning crop such as maize or potatoes is beneficial after breaking land from grass and before planting pyrethrum; this helps to get rid of weeds, especially perennial grasses.

Field establishment

Land preparation
The most important aspect of land preparation is the eradication of perennial grasses. Failure to eradicate these means that deep and thorough weeding is necessary during the life of the crop; this is very harmful and can cause die-back of the above-ground parts of pyrethrum.

During land preparation the land should not be worked excessively; this would tend to destroy the soil structure.

Ridging
Many experiments have shown that ridging gives higher yields than flat cultivation. Damage from waterlogging is reduced; trampling is confined to the bottom of the furrows; young plants are not smothered during weeding and, if the ridges are built on the contour, soil and water conservation are improved. $2\frac{1}{2}$ ft–3 ft (c. 0·75–0·90 m) ridges are ideal. Most large growers plant pyrethrum on ridges. Most smallholders, however, find ridge preparation too laborious and therefore prefer flat cultivation. An exception occurs on the Kinangop Plateau in Kenya where drainage is usually so poor that pyrethrum cannot be grown without ridges.

Time of planting
Pyrethrum should be planted as early in the main rains as possible; planting later leads to poor establishment and poor yields in the first year although subsequent yields are unaffected. An exception to this rule occurs in the Kisii Highlands in Kenya where the rainfall is so well distributed that any month except January is suitable for planting.

Planting
Splits for planting in the field, whether they are from a nursery or from the farmer's own fields, must be a reasonable size in order to withstand short dry periods. As with nursery planting, the stems must be removed, the roots must not be bent in the bottom of the planting hole and the plants must be re-established with as little delay as possible. A mechanical planter was tried in Kenya but its use was not recommended owing to its cost and the fact that it bent the roots during planting.

Spacing

A plant population between 14 500 and 22 000 plants per acre (c. 36 000–54 000 plants per hectare) is satisfactory but it should not fall outside these limits. The usual spacing between plants within the row is 1 ft (c. 0·3 m) whilst the spacing between the rows depends on the needs of the grower. Rows 2 ft (c. 0·6 m) apart are adopted by most smallholders; they give higher yields in the first year but the extra value of these is no more than the extra cost of planting. Large scale growers, who usually rely on tractor cultivation between the rows, prefer rows 3 ft (c. 0·9 m) apart.

Fertilisers

Phosphatic fertilisers are the only ones that show predictable, positive responses. The recommended rate is 300–400 lb of single superphosphate per acre (c. 330–440 kg/ha) which, on red soils at reasonably high altitudes, usually gives an additional 300 lb of dried flowers per acre for each of the three years that the crop is in the ground. The superphosphate should be mixed into the soil at the bottom of the planting holes.

Organic manures and other fertilisers have been tried in Kenya, both in the seedbed and as top dressing, but responses are seldom positive and are sometimes negative. Uwemba in the Southern Highlands of Tanzania is the only place where there are regular, economic responses to nitrogen; there is also a positive inter-action between nitrogen and phosphate in this area.

Replanting

A certain proportion of plant deaths is inevitable in the early stages; they may amount to 15–20% of the total population. Gaps must be filled in but this operation is only economic if done in the first year; after this the remaining plants will have grown into the gaps.

Field maintenance

Weed control

Apart from picking, weeding is the most labour-consuming operation in pyrethrum growing. Weeds grow vigorously because pyrethrum does not shade the ground as much as many other crops, e.g. maize or potatoes, and because pyrethrum is grown in high rainfall areas. The weeds are relatively easily removed from the inter-rows but their removal from between the plants within the rows is an arduous task. During the wetter parts of the year weeding should be done about once a month, although this is not an inflexible guide. 'Jembes' and 'pangas' must never be used for weeding pyrethrum; they can easily damage the roots of the crop, causing die-back of the aerial parts. Small 'forked jembes' should be used and during weeding the soil should be drawn up around the plants. Mechanised weeding is best done with a ridger; this also places soil around the base of the plants.

There are very few herbicides suitable for pyrethrum owing to its susceptibility to the usual range of hormone weed killers. Some herbicides were found to kill weeds and to do no harm to the crop but they had to be withdrawn because they reduced the pyrethrins content. Paraquat is the only herbicide used with any degree of success. It can only be applied by using a flood jet between the rows because it scorches the crop very rapidly if it is allowed to come into contact with pyrethrum foliage.

Cutting back

Cutting back old stems is an essential operation during the dry season at the end of each cropping year. It not only leads to an earlier and increased flush of flowers in the following rains but it also reduces the damage done by bud disease. The operation consists of cutting the old stems back to the level of the top of the foliage. Some stems may have grown sideways or may have lodged; these must also be cut back to the foliage. Sickles or shears are the most effective implements although 'pangas' are more commonly used. A reciprocating mower can also be used but hand cutting is necessary afterwards to remove all the stems that point outwards. Some growers are reluctant to cut back in the first dry season after planting because a large number of flowers are being produced at that time, but cutting back should still be done because it stimulates an even greater flush of flowers at a later date.

Harvesting

Picking is extremely laborious; it must be done every two or three weeks during months when a flush of flowers is being produced.

The weight of pyrethrins in a flower increases from the time that it opens until two or three rows of disc florets are open (see Figs. 96–101). From that time onwards the weight of pyrethrins in a

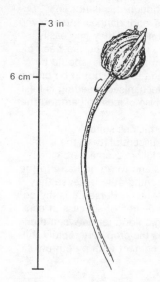

3 in

6 cm

Fig. 96: A pyrethrum flower bud.

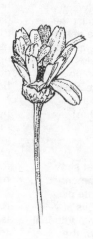

Fig. 97: A pyrethrum flower recently opened with most of its petals still vertical.

Fig. 98: A pyrethrum flower with its petals horizontal but with all disc florets closed.

flower increases only very slightly. The pyrethrins content, expressed as a percentage of the dry weight, also increases from the time that a flower opens until two or three rows of disc florets have opened, but from that time onwards the percentage falls; this is due to the fact that the dry weight of the plant, mostly carbohydrates, is steadily increasing whilst the weight of pyrethrins remains almost constant. When all the disc florets have opened, the flower is 'overblown'.

It has long been recognised that it is a harmful practice to allow overblown flowers to remain on the plant because at this stage nutrients and carbohydrates are being transported to the developing seeds at the expense of new buds; yields are thus reduced. The picking policy until the mid-1950s, therefore, was to pick only the flowers that had reached the maximum pyrethrins content as shown by the number of rows of opened disc florets. This practice is called selective picking. It has serious disadvantages: picking must be done very frequently to make sure that the flowers are at exactly the right stage; pickers must inspect each flower very carefully, thus preventing rapid work, and the pickers must be carefully supervised to ensure that they are picking suitable flowers. It was

later found that strip picking, i.e. picking all the flowers regardless of the number of rows of open disc florets, gave even higher yields of pyrethrins per acre and could be done less frequently than selective picking.

The current strip picking recommendation is to pick all flowers with horizontal petals and to pick at two or three week intervals. The aim must be to adjust the picking interval so that no more than 10% of overblown flowers are picked. The interval depends on the season and the area; it is likely to be longer, for instance, at higher altitudes where the flowers take longer to open.

It might be asked why strip picking is recommended, because large numbers of flowers are harvested soon after their petals have reached a horizontal position but well before they have reached the maximum weight or percentage of pyrethrins. The reason is that early removal of flowers stimulates an even greater production of new flowers so although the pyrethrins content of the flowers may be slightly reduced there is a greater yield of flowers per acre and this more than compensates for this reduction.

Picking should be done with forefinger and thumb, rolling the flower over so that it breaks off

Fig. 99: The stage of optimum pyrethrins content. Two rows of disc florets are open.

Fig. 100: An overblown pyrethrum flower. All disc florets are open.

Fig. 101: A completely overblown pyrethrum flower.

with no adhering stem. An upward pull usually removes a bit of stalk with the flower; it has been shown that if each flower has only one inch (c. 2·5 cm) of stalk on it the pyrethrins content of the sample is 6% lower than if no stalks are included. A further disadvantage of including stalks is that these parts take longer to dry than any others; drying costs consequently rise. An undesirable practice is interlocking the fingers of both hands around a cluster of stems and pulling upwards to remove many flowers at once; this should be strongly discouraged.

Buds have an extremely low pyrethrins content and their inclusion can reduce the pyrethrins content of a sample more than the inclusion of stems. If all the buds are harvested at any one picking the pyrethrins content can be reduced by 25%. If both stems and buds are picked, a potential content of 1·5% may be reduced to 1·05%.

Picking wet flowers is undesirable because they may heat up and ferment before they are dry. If rain or dew has dampened the flowers, morning picking should be delayed until they have dried in the sun. Flowers must be picked into open weave baskets; these allow air to circulate between the flowers, thus preventing fermentation; metal or polythene containers must never be used. If flowers cannot be dried soon after picking they should be spread out under cover in a layer not more than one inch (c. 2·5 cm) thick as a precaution against fermentation. Fermentation must be carefully avoided because even a small amount can cause a rapid reduction in the pyrethrins content.

Women do most of the picking and on average pick 25–30 lb (c. 11–14 kg) of wet flowers a day. Skilled workers, however, can pick anything between 60 and 100 lb a day (c. 27–45 kg).

Drying

Efficient drying is an essential part of pyrethrum production; mistakes at this stage can cause serious reductions in the pyrethrins content. Drying is necessary not only to prevent fermentation and the associated loss of pyrethrins but also to enable the flowers to be ground into a fine powder at the factory.

Sun drying

Almost all smallholders dry their flowers in the sun. This method involves very little capital expenditure and no running costs. The great disadvantage of

159

sun drying is that the growers are entirely at the mercy of the weather; drying may not be possible at certain times of the year. In the Kisii Highlands in Kenya growers are fortunate because the mornings are sunny for most of the year. In Central Province, however, dull misty conditions continue for many weeks during July and August and at such times it is impossible to dry flowers properly.

Sun drying should be done on trays which are raised about 2 ft (c. 0·6 m) above the ground; this allows good air circulation. The floors of the trays should be made of coffee tray wire, hessian or sisal cloth. Split bamboo or Kavirondo matting are unsuitable because they do not allow an even passage of air. Too often smallholders dry their flowers on bare earth or on earth that has been smeared with cow dung and allowed to dry; this is a bad practice because air cannot pass freely between the flowers and much dirt is picked up. To prevent fermentation and to ensure even drying flowers must be spread in a layer no more than one inch (c. 2·5 cm) thick and must be turned at least three times a day. At night and when rain falls the trays must be covered or taken indoors. Drying takes six to fourteen days depending on the weather.

Whether drying is done in the sun or in artificial driers, the aim is to get the moisture content of the flowers down to 10–12%. This involves a weight loss of approximately 75%, i.e. about four pounds of wet flowers give one pound of dry flowers. The most common test to discover whether drying is complete is to crush a flower between finger and thumb; if the floral parts adhere closely to each other or if considerable pressure is needed to break the flower up, drying is incomplete. If, on the other hand, the flower crumbles easily between the fingers the correct moisture content has been reached. In sun drying, which is normally a small scale operation, five flowers should be taken at random from each tray. If four of them pass the finger and thumb test drying is considered complete. With artificial driers, better sampling is needed and 20 flowers should be selected at random of which 16 should pass the test. The reason for not waiting until every flower passes the test is that if every flower was dry some would be overdry and this would cause a reduction in the pyrethrins content.

The Henderson Drier

This is a cheap construction which includes both sun and artificial drying. Trays are arranged on a series of rails so that they can be moved out when the sun is shining or can be rapidly covered under a roof at night or when it rains. There is provision for a charcoal burner at the bottom of the covered portion. Henderson driers would be ideal for smallholders in areas which commonly have prolonged wet weather; unfortunately, few have been constructed.

Artificial drying

Most large scale pyrethrum growers, and consequently most cooperative societies on Settlement Schemes in Kenya, use kerosene or charcoal fired driers. These vary in size from small units to large batch driers which hold as much as 3 000 lb (c. 1 400 kg) of wet flowers.

The three points that need careful attention during artificial drying are checking for complete drying (discussed above), constant stirring and temperature control. The flowers can be left for the first one or two hours but must be turned over at least once every half hour towards the end of the drying period. An effort should be made to turn those at the edges into the centre and those at the top to the bottom. The operation must be done gently, using hands or wooden paddles, otherwise flowers that are nearly dry may break up.

Temperature control is very important because overheating can cause a marked decline in the pyrethrins content, the flowers becoming scorched and brown. In addition overheating can cause case hardening, i.e. a rapid loss of moisture from the outer parts of the flowers, creating an impassable barrier for moisture from the inner parts. To get rid of the moisture efficiently, drying must be a gradual process. The temperature, as measured in the air flow immediately before it reaches the flowers, should never exceed 185°F (85°C) and this can only be permitted if there is a forced flow of air through the flowers. In driers which have no fans, and therefore no forced air flow, 140°F (60°C) must be the maximum temperature. There is no harm in temperatures lower than these except that drying is slower, and drying costs therefore rise. If dried at about 185°F, flowers should be ready in about ten hours. A word of warning is needed on fast drying: if the temperature is near the maximum permitted, flowers very rapidly pass from dry to over-dry; it is safest to reduce the temperature towards the end of drying to make the timing of the end of drying less critical.

Despatch

As soon as the flowers have cooled they should be packed into bags, each containing no more than 65 lb (c. 30 kg) of flowers. Some growers try to pack in more than this amount but this is a bad practice because some of the flowers are broken up and there is a risk of losing the smallest particles during transport. Bags must be sent to the factories as soon as possible, making sure that they are well protected from rain. The pyrethrins content falls at the rate of 3–6% per month during storage so the flowers must be despatched as soon as possible.

Yields

In 1960 the average annual yield in Kenya was about 400 lb of dried flowers per acre (c. 440 kg/ha), but this fell when pyrethrum growing changed hands from experienced large scale farmers to inexperienced smallholders. By the end of the 1960s the average had improved to 250 lb per acre (c. 280 kg/ha). There are several farmers who, with good management, produce 800–900 lb per acre (c. 900–1 000 kg/ha). The best regular field scale production on record is 1 200 lb per acre (c. 1 350 kg/ha).

Pests

Pyrethrum thrips. *Thrips nigropilosus*
These thrips were first seen in Kenya in 1957, since when they have spread to all pyrethrum growing areas of East Africa. They are minute insects, only about $\frac{1}{10}$ in (c. 2–3 mm) long, which normally live on pyrethrum in small numbers and only damage the crop when the population increases in dry weather. At this time the plants are unable to replace damaged leaves.

Thrips damage is characterised by dirty silvery patches on the leaves where insects have scraped back the epidermis to feed on the internal tissues. Minute specks of black frass can usually be seen on these patches. There is no need to spray pyrethrum until there is an average thrip population of one insect on every two leaves sampled. Dimethoate should be used because thrips can become resistant to DDT and dieldrin which were formerly recommended.

Onion thrips (flower thrips). *Thrips tabaci*
These also are a normal constituent of the insect population on pyrethrum and may increase during dry periods; there may be as many as 20 thrips per flower. Onion thrips only inhabit the flowers and although they may cause a browning of the disc florets and petals they cause no reduction in yield or pyrethrins content. They are not, therefore, regarded as serious pests.

Red spider mite. *Tetranychus ludeni*
This is a minute red or yellow mite which only attains any significance after DDT or dieldrin sprays against thrips. Dimethoate sprays control red spider mite effectively and there should be a marked reduction in the number of outbreaks of this pest if this insecticide is used for thrips control.

Root knot nematodes. *Meloidogyne spp.*
These minute organisms are very often found in pyrethrum roots and can cause considerable damage, especially in dry weather. A rotation that includes grassland, bush fallow or cereals should prevent an excessive build-up of nematodes. There is no way in which they can be eliminated except by soil fumigation.

Diseases

Bud disease
This disease is caused by the fungus *Ramularia bellunensis* which enters young buds through the bracts and the receptacle and causes them to die and to turn brown or purple-grey. The stems wither as far as 1 in (c. 2·5 cm) below the buds. Mature buds may also be affected, causing a characteristic browning of the florets on one side of the flower when the bud opens. The period from infection until symptoms are seen is 20 days or more, so brown pin-head buds up to $\frac{1}{5}$ in (c. 5 mm) in diameter cannot have been infected by *Ramularia*; they are more likely to be due to a physiological disorder called false bud disease. True bud disease usually occurs during damp, cloudy weather and there is little that can be done to prevent or cure it. Annual cutting back during the dry season tends to prevent severe outbreaks. Most of the high-content clones now available have been selected for resistance to bud disease so it should be less of a problem in the future.

Root rot
There are at least four different *Sclerotinia* species of fungi which cause root rot; they infect pyrethrum through the soil. A sudden wilting and

browning of the leaves occurs and plants are very easily uprooted. Attacks are especially common in the presence of a high eelworm population, after water-logging or on rich hut site soils. Proper rotations and healthy planting material should prevent severe outbreaks of this disease.

Insecticidal properties of pyrethrum

Pyrethrum is a contact insecticide. It has several unusual properties which have led to its continued demand in the face of competition from synthetic insecticides:

1 Pyrethrum has an outstanding safety record when used on or near humans. This is in marked contrast to many synthetics which are coming under increasing legislative pressure in many countries as a result of their toxicity to mammals. Because it can be used on or near food, pyrethrum is often used in the household.

2 Pyrethrum has a rapid 'knock-down' effect: another advantage for household use.

3 There have been very few cases of insects breeding significant resistance to pyrethrum, as often happens with the chlorinated hydrocarbons, e.g. DDT and dieldrin.

4 Pyrethrum has a repellent effect on insects: a further advantage in the home.

The main disadvantage of pyrethrum is its expense. An apparent disadvantage, i.e. its instability in sunlight, is now recognised to be an advantage because no residues accumulate in flora or fauna.
Most pyrethrum is used in aerosols for household use. Other products are mosquito coils, which produce an insect repellent smoke, grain storage powders and dips for dried fish to kill blow flies.

By-products

The only by-product of the pyrethrum industry is pyrethrum marc, i.e. the ground flowers from which the pyrethrins have been extracted. It can be used as an animal feed; with a protein content of 13% it has approximately the same composition as wheat bran.

Bibliography

1 **Brown, A. F.** (1965). *A pyrethrum improvement programme*. Pyrethrum Post, *8*(1):8.

2 **Brewer, J. G.** (1968). *Flowering and seed setting in pyrethrum; a review*. Pyrethrum Post, *9*(4):18.

3 **Bullock, J. A.** (1961). *The pests of pyrethrum in Kenya*. Pyrethrum Post, *6*(2):22.

4 **Bullock, J. A.** (1963). *Thysanoptera associated with pyrethrum, and the control of* Thrips tabaci. Trop. Agriculture, Trin., *40*:329.

5 **Bullock, J. A.** (1963). *The occurrence, sampling and control of* Tetranychus ludeni *on pyrethrum*. E. Afr. agric. for. J., *28*:252.

6 **Bullock, J. A.** (1964). *A note on the soil fauna of a pyrethrum field*. E. Afr. agric. for. J., *30*:8.

7 **Bullock, J. A.** (1965). *Preliminary observations and experiments on the control of* Thrips nigropilosus. Trop. Agriculture, Trin., *42*:75.

8 **Glynne Jones, G. D.** (1968). *The pyrethrins content of pyrethrum clones and hybrids*. Pyrethrum Post, *9*(3):28.

9 **Head, S. W.** (1963). *An examination of the effect of picking methods on the pyrethrins content of dry pyrethrum flowers*. Pyrethrum Post, *7*(2):3.

10 **Head, S. W.** (1966). *A study of the insecticidal constituents in* Chrysanthemum cinerariaefolium. *(1) Their development in the flower head. (2) Their distribution in the plant*. Pyrethrum Post, *8*(4):32.

11 **Head, S. W.** (1967). *A study of the insecticidal constituents of* Chrysanthemum cinerariaefolium. *(3) Their composition in different pyrethrum clones*. Pyrethrum Post, *9*(2):3.

12 **Kroll, U.** (1961). *The influence of fertilisation on the production of pyrethrins in the pyrethrum flower*. Pyrethrum Post, *6*(2):19.

13 **Kroll, U.** (1962). *The improvement of yields through the application of fertilisers*. Pyrethrum Post, *6*(3):32.

14 **Kroll, U.** (1963). *The effect of fertilisers, manures, irrigation and ridging on the yield of pyrethrum*. E. Afr. agric. for. J., *28*:139.

15 **McKinlay, K. S.** (1959). *Toxicities of various*

acaricides to mites of the Tetranychus *complex.*
E. Afr. agric. J., *25*:28.

16 **Munro, R. J.** (1961). *An experiment on the
drying of pyrethrum flowers.* Pyrethrum Post,
6(2):25.

17 **Nattrass, R. M.** (1950). *Pyrethrum wilt in
Kenya caused by* Sclerotinia minor. E. Afr.
agric. J., *16*:53.

18 **Parlevleit, J. E.** *et al.* (1968). *Influence of
spacing and fertiliser on flower yield and
pyrethrins content of pyrethrum.* Pyrethrum Post,
9(4):28.

19 **Parlevleit, J. E.** (1970). *The effect of rainfall
and altitude on the yield of pyrethrins from
pyrethrum flowers in Kenya.* Pyrethrum Post,
10(3):20.

20 **Pinkerton, A.** (1970). *Visual symptoms of
some mineral deficiencies on pyrethrum.* Exp.
Agric., *6*:19.

21 **Robinson, R. A.** (1963). *Diseases of pyrethrum
in Kenya.* E. Afr. agric. for. J., *28*:164.

22 **Shoemaker, R. L. P. W.** and **Ledger, M. A.**
(1967). *The 'Black rot' condition in pyrethrum in
Tanzania—nematological investigations.* E. Afr.
agric. for. J., *33*:35.

23 **Weiss, E. A.** (1966). *Phosphate—lime trials on
pyrethrum.* Pyrethrum Post, *8*(3):19.

26
Rice

Oryza sativa

Introduction

Rice is a relatively unimportant cereal crop in East Africa; its acreage is small when compared to that of maize, finger millet or sorghum. East Africa's rice is not only purchased by the Asian and Arab communities, but also to a growing extent by Africans, especially on the coastal strip.

Tanzania

An estimated 150 000 acres (c. 60 000 ha) of rice are grown each year in Tanzania. The most important rice growing area is in western Tanzania, to the south of Lake Victoria. Other important areas are the northern shores of Lake Nyasa, the Pangani, Ruvu and Rufiji basins, and Zanzibar. Dry-land rice (sometimes called hill rice) is grown in the wetter parts of the Southern Highlands and on the Kilombero escarpment; the crop is rain fed, as opposed to the bulk of East Africa's crop which receives its water from rivers, lakes or seepage areas, and erosion is usually a serious problem because the slopes are steep and the annual rainfall is often as high as 100 in (c. 2 500 mm). Dry land rice is not important in East Africa and little is known about it; after the introductory section it will not be discussed further in this chapter.

Kenya

Most of Kenya's rice is grown at the Mwea Irrigation Settlement on the plains to the south of Mt. Kenya, and the Ahero Pilot Scheme on the Kano Plains near Kisumu. The Mwea Settlement was initiated in 1953, and by 1969 its rice acreage had expanded to 9 500 acres (c. 3 800 ha). In 1969 1 800 acres (c. 730 ha) were grown at the Ahero Pilot Scheme which was designed to give guidance for further expansion of rice on the Kano Plains. The Bunyala Irrigation Scheme (500 acres, i.e. about 200 hectares, in 1969) is providing information on the production of irrigated rice in the Yala Swamp on the shores of Lake Victoria. In contrast to these carefully supervised irrigation schemes, there is a certain amount of haphazard rice cultivation on the banks of the Tana River, on the shores of Lake

10 cm —

— 1 in

Fig. 102: A rice head.

Victoria and in swamps near Mumias in Western Province. Many rice fields on the lake shore have been abandoned during the 1960s owing to extensive flooding.

Uganda

Rice is of little importance in Uganda. Some is grown on swamp fringes in Bukedi District and the south of Teso District. A little hill rice is grown near Bundibugyo on the western slopes of the Ruwenzori mountains.

Size of fields

Each field must be carefully levelled in order to ensure that each part of it can be flooded to approximately the same depth. Owing to the difficulty of doing this by hand on a large plot, most of East Africa's rice fields are only about $\frac{1}{10}$ acre (c. 0.04 ha). On the Kenya irrigation schemes, where large earth moving machines are used for levelling, the field size is 1 acre (c. 0.4 ha); each farmer has four fields. Whatever their size, each field must be surrounded by a bank (usually called a 'bund') which retains the water; only on seasonal swamps can rice be grown without bunds.

Terminology

The word 'paddy' is often used for the harvested grains before they have been separated from their husks by milling.

Plant characteristics

The height of the plant and the number of tillers are both largely varietal characteristics; most East African varieties grow to a height of about 4 ft (c. 1·2 m) and are intermediate in the number of tillers produced (some varieties produce only a few tillers per plant whilst others form dense clumps with dozens of tillers on each plant). The inflorescence is an open panicle (see Fig. 102) and is 98–99% self pollinated. Using the varieties that are common in East Africa, direct sown rice matures in 3–4 months; transplanted rice takes one or two weeks longer from sowing to harvesting although it takes a shorter time in the field.

Ecology

Water requirements

Rice is an unusual crop in that it can grow not only in water-logged conditions but in standing water. It does not need standing water, however; it is only grown in these conditions to suppress weeds, very few of which tolerate flooding. In the Kenya irrigation schemes water is supplied from rivers; in western Tanzania it is usually supplied by ditches which run from the seepage areas on the gentle slopes of the valleys; on the Tana, Pangani, Ruvu and Rufiji rivers and their tributaries, rice is planted on the banks as the annual floods subside and the water is retained, to a certain extent, by bunds or pits; on lake shores the level of the fields is very near the water table; in seasonal swamps, e.g. in Zanzibar and near Mumias, rice is grown without any levelling or bunding so there is no control on the depth of the water.

The water requirements of paddy are very variable and are affected by several factors, e.g. the permeability of the soil, the length of the growing season, the amount of rainfall in the growing season and the efficiency of water management. At the Mwea Irrigation Settlement water is usually used at the rate of 3–3½ ft per acre (c. 91–107 cm/ha).

Heavy rain at flowering discourages seed setting and is suspected of being one of the main factors causing annual variation in yields at Mwea.

Altitude and temperature

Rice prefers high temperatures and grows best below 4 000 ft (c. 1 200 m); some is grown in western Tanzania and at Mumias, however, between 4 000 and 5 000 ft (c. 1 200–1 500 m). The Mwea Irrigation Settlement, at 3 800 ft (c. 1 150 m) is near the upper limit and long rains crops, which mature in the cold months in the middle of the year, yield very poorly; only one crop is therefore grown each year and this is sown in the period from July to September. At Ahero there is never such a noticeable seasonal drop in temperature so two crops can be grown each year.

Photoperiod

Many varieties of rice are sensitive to the photoperiod, flowering only at times of the year when the day length is decreasing. All the commonly grown East African varieties, however, are non-sensitive and can therefore be grown at any time of the year.

Soil requirements

Sandy soils can support good crops of rice provided that there is a permanent high water table, e.g. near lake shores. Without such a water table, however, a heavy soil is needed to retain the irrigation water.

Varieties

Sindano is the variety which is grown throughout the Kenya irrigation schemes; it yields well, gives a good quality grain and matures quickly. Faya S.L. usually outyields Sindano in experiments but it takes one month longer to mature and has not, therefore, been grown widely; 500 acres (c. 200 ha) were grown at Mwea for the first time in 1969. Basmati strains which are preferred by the Asian community, yield poorly; as Kenya has approached self-sufficiency in rice it has steadily increased the restrictions on imports of Basmati from India. The main effort in Kenya is now being directed to finding better performing varieties than Sindano; a large number of the modern International Rice Research Institute varieties are being tested. In Tanzania the most popular varieties are Afaa (sometimes called Faya or Sifaya) and Kahogo. Afaa varieties differ from place to place and are often given a prefix or suffix according to their origin, e.g. 'Mwanza Afaa', 'Afaa Kilombero', etc.

Direct sowing

In the Mumias swamps and throughout Tanzania (with the exception of the western part) rice is sown directly in the field. The land is cultivated by hand or plough, either during the dry season or shortly before the dry season commences. The seed is usually broadcast, sometimes dibbled, at or before the onset of the rains; about 100 lb of seed are needed per acre (c. 110 kg/ha). Pre-germinated seed, which must be applied to a puddled seedbed, survives a certain degree of flooding and is often used in other parts of the world; in East Africa, however, ungerminated seed is almost always used. It must be applied to a dry seedbed. Early flooding can be disastrous when ungerminated seed is used because it prevents the seeds from getting any oxygen and they are therefore unable to germinate. After germination, which is usually stimulated by the onset of the rains, flooding can be equally harmful; the seedlings are so small that unless the level of the water is very carefully controlled they become submerged and die.

It is obvious, from the above, that germination and early growth are only successful when the depth of the water can be accurately regulated. Unfortunately this is seldom the case in East Africa; the water supply is often inadequately controlled and the fields are almost always poorly levelled so that the seedlings are submerged in the low spots whilst weeds grow prolifically in the higher parts that are not covered with water.

Direct sown rice gives good yields in parts of the world where the fields are well levelled, where the water supply is carefully controlled and where herbicides are used to control the weeds. In East Africa, however, it yields an average of only 1 000–1 500 lb per acre (c. 1 100–1 700 kg/ha). Its only advantage is that it requires considerably less labour than transplanted rice; this is an important factor where a large acreage is grown by one family or where leisure is valued more than yield per acre.

Transplanted rice

All rice in the Kenya irrigation schemes and on the shores of Lake Victoria, and most of it in western Tanzania, is transplanted. The seedlings are raised in small nurseries; $\frac{1}{20}-\frac{1}{30}$ of an acre of nursery is enough for one acre in the field. The nurseries are so small that levelling and water regulation present few problems. The young seedlings are not therefore subjected to the same hazards as the seedlings of direct sown rice. They are transplanted when they are at least 6 in (c. 15 cm) tall and the fields can be flooded immediately to a depth of about 4 in (c. 10 cm), thus covering the soil and suppressing weed growth even if there are minor irregularities in the height of the field. An additional advantage of

transplanted rice is that it remains in the field for a shorter period than direct sown rice; where rice depends on rainfall to supply seepage water, in western Tanzania, for example, transplanted rice is in the nursery in a permanently damp spot for several weeks before the rains begin and suffers less than direct sown crops in a year when the rains end earlier than usual.

Transplanted rice is a highly labour intensive crop; recent labour estimates in Tanzania vary from 550 to 900 man hours per acre (c. 1 350–2 200 man hours per hectare) for land preparation, nursery work, transplanting, weeding, harvesting and threshing. A study at Mwea gave a similar figure of 220 man days per acre.

Nurseries*
Bags of seed are pre-germinated by being soaked in water for 24 hours and then left in the open, covered with straw or grass, for a further 48 hours. 40 lb (c. 18 kg) of seed should be enough for 1 acre in the field if the field spacing is 4 in × 4 in (c. 10 cm × 10 cm). Some rice seed remains dormant for a time after harvesting, the actual period of dormancy differing from one variety to another; most varieties grown in East Africa can be sown safely about 1 month after harvesting. Seed should not be stored for much longer than 6 months otherwise the germination percentage begins to fall. The water is drained from the nursery to leave a thin film immediately before sowing and nitrogen is applied at the rate of 70–100 lb of ammonium sulphate per acre (c. 80–110 kg/ha). The seed is broadcast on to the mud. Local conditions must be considered when deciding on the irrigation procedure; at Mwea, for example, the water may be too cold to allow flooding to any depth at night; alternatively, shallow water may heat up so much during the day, in cloudless weather, that the seedlings are killed.

The main aim in the nursery must be to grow healthy seedlings as quickly as possible; they are ready for transplanting when they are 6–8 in high (c. 15–20 cm) and this stage should be reached after only 3–4 weeks in the nursery. In western Tanzania, seedlings are often left in the nursery for 6–7 weeks, by which time they may be 1 ft (c. 0·3 m) high; the advantage of this system is that the period in the field is reduced in order that it may coincide with the rains. In the lake shore rice growing areas of Kenya, the seedlings are usually

*Both nursery and field techniques described for transplanted rice are those practiced on the Kenya irrigation schemes unless otherwise specified.

left too long in the nursery; they are seldom less than 18 in tall (c. 45 cm) and are sometimes in flower when removed from the nursery; rice never yields well if transplanted as late as this.

Land preparation

For efficient water control the surface of each field must be made as level as possible and bunds, inlets and outlets must be in a good state of repair. On the Kenya irrigation schemes the fields are prepared for transplanting by tractor drawn rotary cultivators; these are mounted and P.T.O.-driven. The fields are flooded four days before rotavating and the tractors operate in about 4 in (c. 10 cm) of water; the tractors have wide tyres but are propelled to a certain extent by the action of the cultivators. If the water is shallower than 4 in, the wheels become fouled with mud; if the water is deeper than this, the cultivators fail to incorporate the weeds into the mud. In western Tanzania and the Kenya lake shore areas the land is prepared by using 'jembes'.

Transplanting

The water is usually drained out of the fields immediately before transplanting to leave a wet mud or, at most, a thin film of water (see Fig. 103). If water is in short supply, however, this is a wasteful procedure when the fields have already been flooded, e.g. for rotavation; in such circumstances the seedlings can be transplanted through the water without any previous draining. Fertilisers (see following section) are broadcast immediately before transplanting. A common fault where supervision is lacking is to use too wide a spacing; this is usually

Fig. 103: Transplanting rice.

prompted either by shortage of planting material or by enthusiasm to finish planting as soon as possible. Varieties differ in their spacing requirements mainly according to their tillering capacity; Sindano and most other East African varieties give satisfactory yields only at around 400 000 plants per acre (c. 1 000 000 plants per hectare), i.e. at a spacing of 4 in × 4 in (10 cm × 10 cm), but Faya S.L. yields well at half this population. A secondary disadvantage of too wide a spacing is that it encourages a prolonged period of tiller formation; tillers therefore ripen over a prolonged period and this causes harvesting problems.

Fertilisers

The usual practice on the Kenya irrigation schemes has been to apply 50 lb of P_2O_5 and 13 lb of nitrogen per acre before transplanting (c. 56 kg P_2O_5/ha and 15 kg N/ha) and to broadcast another 13 lb of nitrogen three or four weeks later. The nitrogen rates continue to give a yield increase of about 10% but they are low by world standards; the reason is that Sindano does not respond as much to heavy dressings of nitrogen as do many other varieties. At Ahero blast has recently become a serious problem (see section on diseases); blast thrives on plants which are well fed with nitrogen so applications at Ahero have ceased. Phosphate applications, which have continued for many years in succession at Mwea, have given declining responses; in 1969 a one year break following two years of application was introduced. Fertilisers are hardly ever used in East African rice outside the Kenya irrigation schemes.

Water control

Immediately after transplanting the fields can be flooded to a depth of several inches. The main aim must be to have the fields level enough and the water deep enough to cover the entire surface of the field; only thus can weeds be suppressed. The water level must not be so high that it comes up to or over the top of the seedlings; this encourages them to float upwards and uproot themselves. A rule of thumb which is often quoted is that the water should be maintained at $\frac{1}{3}$ of the height of the crop; this is satisfactory in the early stages of growth if the fields are level enough to allow such shallow flooding, but it would prove wasteful of water in the later stages of growth when the crop is 3–4 ft high (c. 0·9–1·2 m); in the Kenya irrigation schemes a depth of 6 in (c. 15 cm) is seldom exceeded. The fields are finally drained for the last three weeks before harvesting; a good indication of the time for draining is when the heads begin to bend downwards. Harvesting in a wet soil is an unpleasant task and usually causes the incorporation of small lumps of clay with the crop; these lumps can pass through the mills into the final product.

The water in rice fields must not be allowed to become stagnant; it must either flow very slowly through the fields or must be changed every 2–3 weeks.

Weeding

Weed control does not involve much work if the fields have been properly flooded. Weeds are removed by pulling (the spacing of the crop is so close that 'jembes' cannot be used) and this is usually done only once on the Kenya irrigation schemes. Herbicides are neither used nor needed in East Africa.

Red rice is an undesirable variety of rice which causes considerable problems unless seed selection is very thorough. It is very similar to ordinary rice in all characteristics except that the seed coat is red and can only be removed by setting the mills so close that much of the good rice is broken. Red rice also shatters more readily. The last problem is particularly severe because the seeds may remain dormant in the soil for many years, germinating long after they have fallen to the ground.

Harvesting

If rice has been grown at the correct spacing, tillering should not be excessive and almost all heads should ripen at approximately the same time; this enables the crop to be harvested with a sickle, all the stems being cut at the base (see Fig. 104). If ripening is very uneven, as often occurs in Tanzania owing to too wide a spacing and the sowing of mixed varieties, the oldest heads tend to shatter unless they are harvested individually before the others. Heavy crops often lodge but this is no problem if the fields have been well drained.

Yields

Yields of direct sown rice have been mentioned in a previous section. At the Mwea Irrigation Settlement one crop is taken each year; the average yield* is very constant from one year to the next and is seldom less than 4 500 or more than 5 100 lb of

*At Mwea and Ahero yields exclude the rice which is retained by the farmer for household use: probably an extra 500–600 lb per acre.

Fig. 104: Harvesting rice.

paddy per acre (c. 5 000–5 700 kg/ha). Good farmers regularly produce as much as 7 000 lb and as many as 8 300 lb per acre have been recorded (c. 7 800 and 9 300 kg/ha). At the Ahero Pilot Scheme two crops are taken each year; in 1969 (the first full year of production) the first crop averaged 4 800 lb per acre and the second crop 2 700 lb per acre, i.e. an annual total of 7 500 lb per acre (equivalent figures in kg/ha, 1st crop—5 400; 2nd crop—3 000; total—8 400).

Threshing is done by hand. This is an easy task if the entire plant has been harvested because bundles can be held at the stem end and the heads can be beaten; bits of stalk and leaf are then separated from the seed by winnowing.

Milling and parboiling

The lemma and palea, which adhere closely to the grain, must be removed by milling. During milling the endosperm is polished and the embryo, which contains much more protein, minerals, vitamins and fat than the endosperm, is removed. A parboiling plant has recently been installed at Mwea; parboiling involves soaking the paddy in hot water and then steaming it before milling. The main advantage is that a large proportion of the water-soluble vitamins of the B complex are transferred from the outer layer of the endosperm to the inner part of the endosperm; they are therefore largely retained even after milling which removes the outer layer. A benefit of parboiling is that milling is made more efficient because it hardens the grains and reduces milling losses through breakage.

The milling percentage of rice is 65–70%. The ideal moisture content for storage and milling is 14%.

Pests

Rice is seldom seriously damaged by pests in East Africa. Regular precautions are taken on the Kenya irrigation schemes, however, against the rice hispid (*Trichispa sericea*) by spraying BHC in the nurseries; field applications are sometimes needed. The larvae of the rice hispid beetle feed inside the leaves, making mines which reduce the photosynthetic area. Birds, which feed on the grains when these are in the milky stage, can do considerable damage. Stem borers, e.g. *Chilo suppressalis*, sometimes damage rice but no precautions are taken against them.

Diseases

Blast
This is caused by the fungus *Piricularia oryzae*. It thrives in hot, humid conditions, especially when rice has received heavy dressings of nitrogen. In the lower altitude, humid areas, e.g. Uganda and Ahero, it does a great deal of damage but at Mwea, which is cooler and drier, it is seldom seen. Brown spots with grey centres develop on the leaves, and the stem is often infected just below the inflorescence, causing the head to fall over and preventing the flow of nutrients to the developing grains. The long term solution to the blast problem is the introduction of resistant varieties, although these are often found to have certain undesirable characteristics. In the short term, withholding nitrogen fertilizers sometimes reduces the incidence of the disease. Fungicides are effective in other parts of the world and are currently being tried at Ahero.

Yellow mottling virus
This virus was first noticed in 1968 in Kenya in the rice on the shores of Lake Victoria. In 1969 several insect vectors were identified. It has occurred in patches at Ahero, where insecticides are being used to control the vectors. Rigid precautions are taken to prevent the transfer of seed from western to eastern Kenya as this might bring the virus to the Mwea Irrigation Scheme.

Bibliography

1 **Allnutt, R. B.** (1942). *Rice growing in dry areas.* E. Afr. agric. J., *8*:103.

2 **Doggett, H.** (1965). *A history of the work of the Mwabagole Rice Station, Lake Province, Tanzania.* E. Afr. agric. for. J., *31*:16.

3 **Giglioli, E. G.** (1965). *Mechanical cultivation of rice on the Mwea Irrigation Settlement.* E. Afr. agric. for. J., *30*:177.

4 **MacArthur, J. D.** (1968). *Labour costs and utilisation in rice production on the Mwea Tebere Irrigation Scheme.* E. Afr. agric. for. J., *33*:325.

5 **Suttie, J. M.** (1963). *Rice fertiliser and spacing trials at Mwea Tebere.* E. Afr. agric. for. J., *28*:129.

6 **Wilson, F. B.** and **Tidbury, G. E.** (1944). *Native paddy cultivation and yields in Zanzibar.* E. Afr. agric. J., *9*:231.

27
Simsim

Sesamum indicum

Introduction

Simsim, also known as sesame, is an oilseed crop. Its seeds have an oil content of 45–55%. It is widely grown in the less mountainous areas of southern Tanzania, especially in Mtwara, Ruvuma and Songea Districts. It is also important in the northern parts of Uganda, namely Lango, Acholi and West Nile Districts. In Kenya it is of minor importance in Western and Nyanza Provinces and at the coast. Most simsim is consumed locally but there is an active oil expressing trade in southern Tanzania.

Plant characteristics

Simsim is an annual which grows to between three and six feet (c. 0·9–1·8m) high. It has a taproot and a dense surface mat of feeding roots. The stem is branched or unbranched, depending on the variety. The lower leaves are opposite, broad and palmately lobed whilst the upper leaves are alternatively arranged, narrow and lanceolate. There are 1–3 flowers in each axil, the number depending largely on the variety, and these are white, pink or purple. The fruits are erect capsules about 1 in (c. 2·5 cm) long. Each capsule contains a large number of very small seeds of which there are about 9 000 to the ounce (c. 320 to the gram). The seeds are either white or black. All the local varieties and the successful introductions are dehiscent, i.e. the capsules split from the top downwards for about two thirds of their length when mature; this process is encouraged when the plants are moved about by the wind and unless precautions are taken much seed is shed. Although simsim is predominantly self-pollinated there is up to 5% cross-pollination. In most varieties the period from sowing to maturity is 3–4½ months.

Ecology

Rainfall, and moisture requirements
Simsim is moderately drought resistant; this is

Fig. 105: A simsim plant. Note the pods in the axils of the leaves.

especially true of branching varieties because these have a more prolific root system. 15–20 in (c. 400–500 mm) of rain are needed during the growing season. Simsim is said to be capable of growing in areas which receive an average rainfall of only 25 in per annum (c. 625 mm), but in East Africa it is not grown in areas which receive less than 30 in (c. 750 mm). Moist conditions are needed during the early stages of growth but heavy rain immediately after sowing may do considerable damage by washing away many of the seeds or by capping the soil surface, thus preventing emergence of many of the small seedlings.

Altitude and temperature
Simsim only grows well in a warm climate and in East Africa it is only grown from sea level up to 5 000 ft (c. 1 500 m).

Photoperiod
Most varieties of simsim are photoperiod sensitive. Venezuela 51, an introduced variety, is one of the few commercial varieties known to be day-neutral.

Soil requirements
Simsim does best on fertile soils. It is very intolerant of waterlogging.

Varieties

The local varieties of simsim are branched and drought resistant but have a low yielding capacity and are susceptible to most diseases. As part of a breeding programme carried out in southern Tanzania, many introductions were made in order to find higher yielding varieties. The most successful of these were from Venezuela, the outstanding one being Morada. Morada originated from the Congo but had been further selected in Venezuela; it has purple stems and leaves, is moderately drought resistant (although not to such an extent as the local varieties), has a tendency to bear two or three capsules at each node (compared to one in the local varieties) and is less susceptible to aphids, *Cercospora*, *Pseudomonas* and *Alternaria*. In trials Morada has consistently given double the yields of of the local varieties. It is now the established variety in the Morogoro and coastal regions of Tanzania.

In the Tanzanian breeding programme Morada × Local back-crosses have given good results. They combine the high yielding ability of Morada with the drought resistance of Local and are capable of yielding 34% more than Morada.

Rotations

Simsim is usually sown early in the arable break because it requires fertile soils. It is often sown as an opening crop although grasses must be thoroughly eradicated because it is intolerant of weed competition.

Field operations

Seedbed preparation
A rough seedbed is usually preferred for simsim despite the small size of the seeds. The reason for this is that a fine seedbed is more likely to form a cap if heavy rain falls, thus hindering emergence of the seedlings.

Time of sowing
Simsim must be sown as early in the rains as possible; eight out of ten experiments in Tanzania showed that late sowing led to severe yield reductions. In Uganda it is sometimes sown in the first rains, usually as a pure stand, or may be sown amongst pigeon peas after the finger millet has been harvested from a mixture of pigeon peas and millet. Alternatively it may be inter-sown with maize or sorghum. At the Kenya coast it is often sown amongst maize in June shortly before the cereal crop is harvested.

Sowing
Simsim seed is usually broadcast at the rate of 5–8 lb per acre (c. 5·5–9·0 kg/ha). It is often mixed with soil before sowing in order to achieve an even spread. Thinning is often neglected; if it is done the remaining plants should be about 9 in (c. 23 cm) apart.

Spacing
Sowing in rows is rare; if it is done the optimum population is 70 000–80 000 plants per acre (c. 170 000–200 000 plants per hectare).

Weed control
Simsim grows slowly at first and is very intolerant of competition from weeds when it is young. Efficent weed control is therefore important.

Fertilisers
Almost all fertiliser trials on simsim in East Africa have failed to show significant responses. There are therefore no fertiliser recommendations.

Harvesting

At harvesting simsim presents two problems: the capsules split and shed their seeds when they are mature and they ripen unevenly from the bottom upwards. If harvesting is delayed until the top capsule is ripe, the lower ones split and the wind blows the plants about so much that most of the seed is shed. It is therefore essential to harvest the plants as soon as the lowest capsules ripen; at this stage flowering has usually stopped at the top of the plant and most of the leaves have been shed. The plants are uprooted and the parts of the stems below the lowest capsule are removed; this is best done by holding the stems against a log and chopping them with a 'panga'; this avoids shaking the pods and encouraging them to split or to shed their seeds. The plants are then stooked (see Fig. 106) or tied to a fence which has been previously constructed in the field or homestead. The fences are often as high as 10 ft (c. 3 m) and are usually built parallel to the prevailing wind to prevent them being blown over. The bundles of simsim plants are tied on to each side facing upwards; in this position the seed is not lost because the plants cannot be shaken.

After three or four weeks all the capsules should have matured. The seeds are removed by beating the plants on a mat.

Yields

Yields are usually between 200 and 300 lb of seed per acre (c. 220–330 kg/ha). Good husbandry should produce 400–500 lb per acre (c. 450–550 kg/ha); Morada has yielded 850 lb per acre (c. 950 kg/ha) on a field scale at Nachingwea in Tanzania.

Pests

Simsim webworm. *Antigastria catalaunalis*
The larvae of this pest spin a silken web around the terminal leaves and eat the foliage and the pods. They can be controlled with DDT sprays.

Gall midge. *Asphondylia sesami*
These minute insects lay their eggs in the ovaries of the flowers. When the larvae hatch they devour the inside of the ovaries and in place of capsules round barren galls are produced. Dimethoate sprays are effective but are probably uneconomic.

Flea beetle. *Aphthona bimaculata*
This is the most important pest in southern Tanzania. It eats the foliage during the early stages of growth and can be controlled by using dieldrin as a seed dressing or in the seedbed.

Diseases

Diseases include bacterial leaf spot, caused by *Pseudomonas sesami*, a leaf spot, caused by the fungus *Cercospora sesami*, and *Alternaria spp*. They seldom do serious damage.

Fig. 106: Stooked simsim.

Utilisation

Some seed and oil are exported to Europe for use in edible and pharmaceutical products but most simsim is consumed internally, fetching a high price in local markets. It is usually prepared for eating by pounding until it becomes an oily paste. This can be eaten as it is or can be used for frying vegetables or meat; in both cases it is usually eaten with 'ugali'. Alternatively simsim seeds can be eaten raw, can be eaten fried or can be included in various types of confectionery.

Bibliography

1 **Hill, A. G.** (1947). *Oil plants in East Africa.* E. Afr. agric. J., *12*:140.

2 **Robertson, F. A. D.** (1969). *Insecticide trials to control the pest complex attacking sesame in Eastern Tanzania.* E. Afr. agric. for J., *35*:105.

3 **Tribe, A. J.** (1967). *Sesame.* Field Crop Abstr., *20*:189.

28
Sisal

Mostly *Agave sisalana* but interspecific hybrids are also important (see section on varieties)

Introduction

Sisal leaves provide the most important of the world's hard fibres, i.e. fibres which are so coarse that they can only be made into twines, ropes, sacks or matting. At the time of writing about three quarters of the world's sisal is used for agricultural twine, mostly baler twine.

Tanzania
Sisal was Tanzania's most valuable export for many years; in the 1960s, however, it was overtaken by cotton and coffee owing to a fall in world sisal prices. Most of the country's sisal comes from estates in the Tanga–Korogwe area and, to a lesser extent, from the Kilosa–Morogoro area. Much smaller quantities come from Arusha, Moshi, Lindi and Mtwara. Tanzania's sisal exports reached their peak in 1963 when 219 000 tons were exported, representing a total value of almost £22½ million. In 1967, there were nearly 700 000 acres (c. 280 000 ha) under sisal in Tanzania.

Kenya
The area of sisal estates in Kenya has never exceeded 250 000 acres (c. 100 000 ha). Peak production occurred in 1963 when 70 000 tons were produced and when the exports reached a record value of £7½ million. Most of Kenya's sisal comes from estates between Nairobi and Fort Hall, from the coast and from the Rift Valley.

Uganda
Uganda produces hardly any sisal.

Estates
Almost all East Africa's sisal is grown on estates; most of these have at least 3 000 acres (c. 1 200 ha) of growing sisal. Sisal fibres can only be removed efficiently from the leaves by large decorticating machines; each decorticator costs several thousand pounds and must be kept running for most of the year in order to achieve an annual output of at least 1 200–1 500 tons of fibre. This amount of fibre is produced from 40 000–50 000 tons of leaf and on

Fig. 107: Weeding sisal.

average this can be supplied from about 3 000 acres of growing sisal. The large acreage involved, the importance of organising large labour gangs and transport facilities in order to maintain a regular supply of leaf, and the expense of transporting leaf over long distances make sisal a crop which, in East Africa, is only suited to estate production. An additional factor is that heavy machinery is usually involved in preparing the land for new cycles.

Smallholders
When prices are high smallholders produce fibre from the hedgerow sisal which forms the boundaries of many smallholdings in the lower altitude areas. Leaves are cut into longitudinal strips and are decorticated by pulling them through two pieces of a 'panga' blade set into a stake. This is an exceedingly laborious procedure and it may take as many as 1 600 man hours to produce one ton of fibre. In 1962 hedgerow sisal in Tanzania provided a peak contribution of 12 000 tons of fibre; in Kenya 11 000 tons were produced in 1963. Since then prices have fallen and at the end of the 1960s hedgerow sisal contributed virtually nothing towards East African production.

Other types of smallholder production have

173

usually been unsuccessful. Poor coordination of leaf supply and high transportation costs led to the failure of almost all schemes where smallholders delivered leaf to estates. Small scale decorticators are available but smallholders have been reluctant to provide the large amount of labour that these machines require.

Plant characteristics

Roots

Sisal roots seldom penetrate deeper than 2 ft (c. 0·6 m) below the soil surface; most are found in the top foot (c. 0·3 m) of the soil horizon. The root system is adventitious, each bearer root arising from a leaf scar at the base of the bole. The bearer roots become highly lignified with age and take a considerable time to rot after the completion of a cycle.

Suckers

These grow from rhizomes which are produced from the base of the bole; they may be up to 1 in (c. 2·5 cm) thick compared to $\frac{1}{8}$ in (c. 3 mm) for bearers. Many emerge near the parent plant but some emerge as far as 6 ft (c. 1·8 m) away. When a rhizome emerges it forms a miniature sisal plant, called a sucker. Sucker production begins about a year after planting and is most prolific during the first half of the cycle; as many as twenty suckers may grow from one plant during this period.

The bole

The bole is a hard, woody structure which in a healthy plant reaches and maintains a width of about 8 in (c. 20 cm) at two years after planting and reaches a height of about 4 ft (c. 1·2 m) at maturity. It appears to be thicker than 8 in owing to the presence of the basal ends of the cut leaves. The growing point is situated at the top of the bole. Healthy plants have wide boles; plants with narrow boles produce short leaves at long intervals of time.

Leaves

During its 7–12 year life in the field, a sisal plant should produce between 200 and 250 leaves. Leaves are unrolled from the central vertical spike at the rate of 2–3 per month. As they become older they adopt a more horizontal position. If they are allowed to remain on the plant too long, they wither.

Sisal leaves have been known to reach a length of 6 ft (c. 1·8 m) but 4 ft (c. 1·2 m) is more common in mature plants. In the early stages of growth they are much shorter and several must be

Fig. 108: Poled sisal.

discarded. The shortest leaves which can be decorticated are 2 ft (c. 0·6 m). The fibre from these leaves is finer than that from later cuts and has a special market, especially if machine dried. It usually commands a premium even though it is short. Each mature leaf contains about 1 100 creamy white fibres which run the whole length of the blade. Although very numerous, these fibres form only 3% of the leaf; the remainder consists of cuticle, vascular tissue and fleshy mesophyll. There is a highly cutinised and extremely sharp spine at the end of each leaf.

The pole
This is the name given to the inflorescence. The end of a sisal cycle occurs and the plants are destroyed when 70–80% of the plants have produced poles; after this very little fibre is produced, because a plant can produce no more leaves after poling. The period from planting to poling varies from 7–12 years for *A. sisalana*, depending largely on soil and climatic factors. 10–11 years, however, is the most common cycle length. Poles grow very rapidly and may reach a height of 20 ft (c. 6 m); many horizontal branches, with flowers at the end of each, are produced at the top of the poles (see Fig. 108). Seed setting is rare owing to the formation of an abcission layer below each ovary immediately after the flower has withered.

Bulbils
Bulbils are miniature sisal plants which are borne on the inflorescence; one pole may produce as many as 3 000. When mature, measuring 3–5 in long (c. 8–13 cm), they fall to the ground. With their rudimentary roots, they make excellent planting material. They are easier to collect than suckers and give more uniform growth.

Ecology

Rainfall, and water requirements
Sisal is markedly drought resistant; some is grown commercially in areas with an annual average rainfall of only 25 in (c. 625 mm). In such areas yields are about 60–70% lower than can be expected where the rainfall is nearer the optimum, i.e. an annual average of about 50 in (c. 1 250 mm) well distributed throughout the year. In the drier areas weed control and leaf transport present fewer problems than in wetter areas. Where the rainfall is over 60 in per annum (c. 1 500 mm), little sisal is grown because weed growth is prolific and the soil is often too muddy for efficient leaf transport.

Altitude and temperature
Sisal is grown commercially from sea level up to 6 000 ft (c. 1 800 m). Most of Tanzania's estates are below 3 000 ft (c. 900 m) whilst in Kenya most are above this level.

Soil requirements
Sisal is pale and stunted and yields are low when it is grown in badly drained conditions. On heavy clay soils cambered beds must be formed or ditches must be dug to lead off as much water as possible. During the mid-1960s an estimated 8% of Kenya's sisal was grown on cambered beds. In Tanzania, however, deep ditches are preferred and these are normally spaced about 50 yards apart (c. 45 m) where sisal is grown on heavy soils. Satisfactory yields are only obtained when the nutrient status of the soil is reasonably high. When sisal is grown repeatedly on the same land yields can only be maintained by supplying nutrients in the form of fertilisers or manure (e.g. sisal waste).

Varieties

All Kenya's sisal and most of Tanzania's is *A. sisalana*. It is impossible to breed new varieties because the few seeds produced all give seedlings with spiny leaf margins. Some other *Agave* species, however, set seed more readily and produce seedlings with smooth leaf margins. Several inter-specific hybrids were produced in a breeding programme which started at Amani in Tanzania in 1939. The most promising of these is Hybrid 11648, a cross between *A. amaniensis* and *A. angustifolia*, back-crossed to *A. amaniensis*. When grown at altitudes below 2 000 ft (c. 600 m), 11648 produces 500–600 leaves per plant (compared to 200–250 for *A. sisalana*) and yields as much as 20 tons of fibre per acre per cycle (c. 50 t/ha) under intensive management (compared to 7–8 tons, i.e. about 17·5–20 t/ha). The length of the cycle is considerably longer, however; at sea level it is about 15 years whilst between 1 000 ft and 2 000 ft (c. 300–600 m) it is 10–11 years and yields are correspondingly lower. At high altitudes leaves are very short and the life cycle is greatly reduced; poling has occurred as soon as three years after planting at Thika in Kenya. This means that 11648 is only suitable for the lower altitudes. Even in these areas, however, 11648 presents problems: it is more susceptible than *A. sisalana* to zebra disease, Korogwe leaf spot and sisal weevil and to reduce damage by these to a minimum, standards of husbandry and field inspection must be high and wet soils must be avoided. At the time of writing an estimated 20 000 acres (c. 8 000 hectares) of 11648 have been planted in Tanzania.

Nurseries

Nurseries are essential; direct planting in the field would involve very high maintenance costs and very poor uniformity. Before discussing nursery techniques it is necessary to explain the importance of planting large uniform material from the nursery into the field.

If plants in the field are not uniform, some are ready for their first cut before others; this leads to selective cutting, which is very laborious, or to some leaf on the larger plants being lost whilst waiting for the smaller ones to reach a reasonable size for cutting. An additional disadvantage is that the leaves are of uneven length; this creates decortication problems and reduces the quality of the fibre. Lastly, unevenness causes irregular poling which makes it difficult to decide when to cut out an old field. One of the first and most essential steps in achieving uniformity in the field is to produce uniform planting material from the nurseries.

The second important consideration is the length of time between planting in the field and taking the first cut. It is important to reduce this unproductive period to a minimum, and one of the most effective ways of doing this is to use large planting material from the nursery. If such material is used, and if standards of husbandry are reasonably high, the first cut can be taken as soon as 18 months to two years after planting. With small or medium sized material it is often three or four years before the first cut.

Suckers and bulbils

Sisal is propagated vegetatively by means of suckers or bulbils. The most important disadvantage of suckers is that however carefully they are graded into different sizes, both before and after the nursery, they almost always show a considerable variation of growth in the field. Bulbils, on the other hand, are more uniform. An additional factor in favour of bulbils is that they are easier to collect; they can be picked up from the ground or the branches of mature poles can be shaken in sacks in order to dislodge the bulbils. Suckers must be collected from a much wider area. Their removal from the ground involves more labour and more damage to the young plants and transport costs are higher. Suckers are therefore seldom used as planting material. An exception occurs in the case of Hybrid 11648; up to the time of writing suckers have been mainly used because there is little 11648 that has poled out; this situation will

Fig. 109: A sisal nursery.

change in the near future and bulbils will then probably be used. It is possible to propagate sisal by means of pieces of rhizome and this method is occasionally used for 11648. Each piece must have two scale leaves and must be planted with one scale leaf uppermost. The spacing should be 4 in × 4 in (c. 10 cm × 10 cm) and the depth should be no more than $\frac{1}{2}$ in (c. 1 cm).

By mutation, *A. sisalana* occasionally produces plants with spiny leaf margins. These present problems during cutting, transporting and decorticating so they must not be used as a source of planting material.

Management

Twenty or forty tons of sisal waste per acre (c. 50 or 100 t/ha) speed up growth in the nursery and produce healthy plants; the heavier application is the more effective. The waste should be dug into the soil after cultivation and eradication of perennial weeds. A fairly fine tilth is needed. The bulbils are planted quite shallowly (about $\frac{1}{2}$ in deep, i.e. c. 1·3 cm) after they have been graded into at least three different sizes. Bulbils shorter than 4 in (c. 10 cm) are discarded. The usual spacing is 20 in × 10 in (c. 50 cm × 25 cm). Nitrogen top dressings need only be used (at about 30 lb per acre, i.e. c. 34 kg/ha) if the nursery plants appear pale. Phosphate need only be used if sisal waste is omitted and potassium only if soil analyses show a deficiency. Nurseries should be prepared in time for planting at the onset of the main rains.

Weed control

When weeds are controlled by hoeing in the

nursery, growth is often poor because of the damage done to the roots. Two alternatives have given excellent weed control at the same time as encouraging rapid and healthy growth. The first is the use of soil applied residual herbicides; when applied in the nursery, after planting, bromacil may provide weed-free conditions for as long as 14 months; other herbicides used in sisal nurseries are atraton, diuron and simazine. The second alternative is black 150-gauge polythene mulch; this is applied over the entire surface of the nursery and bulbils are planted through holes punched in it. Polythene mulch conserves moisture as well as suppressing weeds and has therefore proved most beneficial in dry areas. At Mlingano in Tanzania it has given a 30% increase in the size of nursery plants.

Applications of TCA have proved very effective in eradicating couch grass, which is a serious problem in Kenya. There must be enough rain to wash the herbicide into the soil but not so much that it is leached out. TCA is applied to bare soil about three weeks after the last ploughing or harrowing and about three weeks before planting.

Primary nurseries

These are sometimes called smother nurseries. They may be used before transplanting into the main nursery as described above. Bulbils are spaced very close together in a square formation; the distance between plants is only 2–3 in (c. 5·0–7·5 cm). Sisal waste and herbicides are essential. The main advantages of smother nurseries are that weak plants can be eliminated at an early stage when the bulbils are transplanted from the primary nursery to the main nursery; this leads to a high degree of uniformity. The second advantage is that irrigation, which is necessary in dry areas in order to get the bulbils to take root, can be restricted to a small plot instead of being applied to the entire nursery area. Bulbils stay in the primary nursery for only about three months before being transferred.

Duration of nurseries

For rapid growth in the field, plants should be taken from the main nursery when they are 20–28 in long (c. 50–70 cm) and weigh 5–10 lb (2–4 kg). With good standards of husbandry in the nursery, material of this size and weight should be produced in about 18 months.

Land preparation

Whether new land or land previously under sisal is to be prepared for planting, heavy machinery and a considerable amount of labour are needed. The expense involved, coupled with the expenses of raising nurseries, transplanting and early field care, make field establishment much the most costly operation in growing sisal.

When clearing an old field of sisal, the poles are first removed. Being light and strong these are popular for building. The problem then arises of destroying the old boles, each of which weighs 100–200 lb (c. 45–90 kg) and stands as much as 4 ft (c. 1·2 m) high. The original policy was to pile all the plant debris together, using bulldozers, and to burn it. This became unpopular owing to the tendency for bulldozer operators to pile up much of the topsoil. With more skilled operators, bulldozing has recently been re-introduced in Kenya. The most common machines for destroying old boles are rolling choppers, either set parallel or at an angle to each other, and Rome ploughs, which are very heavy disc harrows. These machines are very heavy and must be drawn by large crawler tractors. An alternative, for small areas, is hand clearing. After the boles have been broken up they are either incorporated into the soil or burned after a lapse of two or three weeks.

Transplanting

Sisal can be planted at any time of the year, although it is best to plant as great an area as possible before the long rains in order to give it a good start in its first year. It can survive for as long as three months in completely dry weather immediately after planting. This means that a new block of sisal can be planted systematically throughout the dry season.

A team of 25 men (ten lifting from the nursery, ten planting and five transporting) should be able to plant about 5 acres (c. 2 ha) in a day, provided that the rows have been marked out previously. A recent recommendation for large planting material is to tie the leaves together around the central spike. This involves a little more labour but plants are then much more convenient to transport and transplant and are more easily weeded for the first time in the field. The strings are cut after the first weeding.

The sisal plant is remarkably resistant to damage, in fact experiments have shown that all the roots and horizontal leaves can be removed without causing more than a temporary setback in the field. The normal procedure, however, is to remove the withered leaves (sometimes called sand leaves) and most of the roots, making sure that the bole is unharmed. As the plants are lifted and trimmed they must be graded into at least three sizes; each grade must be planted separately to encourage uniformity. Nursery plants suffer no damage if they are piled up and left for three weeks before planting. A short delay may encourage healthy root development.

Planting holes are dug with a 'jembe'. They need only be about 3 in (c. 7·5 cm) deep. Where sisal weevil is a problem, i.e. in most of Tanzania but only in the lower sisal areas in Kenya, the sides of the planting holes should be dusted with aldrin dust. During planting the soil must be compressed around the bole of each plant after it has been placed in its hole.

Spacing

The following factors affect the crop:

1. A spacing of closer than $2\frac{1}{2}$ ft within the row (c. 0·75 m) causes undesirable competition however far apart the rows are.

2. Provided that the spacing within the row is no closer than $2\frac{1}{2}$ ft, the higher the plant density the higher the yield of fibre per acre, up to a limit of about 2 400 plants per acre (c. 5 900 plants per hectare) on fertile land.

3. Rows must be wide enough to allow workers to move freely during cutting and leaf transport; they must also allow room for tractors if these are to be used.

4. Sisal leaves protrude about 3 ft (c. 1 m) into each inter-row so a single row spacing of 9 ft × $2\frac{1}{2}$ ft (c. 2·7 m × 0·75 m), giving a moderate plant population of 1 936 plants per acre (4 783 plants per hectare), would be impracticable because it would only give a 3 ft gap for manoeuvring in the inter-row. Double row spacings are therefore much more popular; each pair of rows is separated from the next pair by a wide gap of $11\frac{1}{2}$–13 ft (c. 3·5 m– 4·0 m) which allows easy access for workers and tractors and which provides enough space for a cover crop. The spacing between the two rows of each pair is almost always 3 ft (c. 1 m); this helps to suppress creeping grasses near the plants. The spacing within each row varies from $2\frac{1}{2}$–$3\frac{1}{4}$ ft (c. 0·75–1·0 m). The wider spacings mentioned above (both inter-row and

within the row) give a plant population of about 1 700 plants per acre (c. 4 100 plants per hectare) which is suitable for poor soil. The closer spacings give a population of about 2 400 plants per acre (c. 5 900 plants per hectare) which is suitable for fertile soil.

Intercropping

During the first three years in the field there is enough room between the wide inter-row of double row sisal for crops such as beans, maize, cotton, simsim, or pineapples to be grown. These are a useful source of income during the unproductive part of the cycle. They also suppress weeds in the wide inter-rows, although at this stage weeds must be controlled by other means within the rows of sisal. Fertilisers must be applied to the intercrops to ensure that the soil in the inter-row is not exhausted when the sisal roots grow into it.

Fertilisers and manures

Sisal yields are largely maintained by planting on new land, or on land which has been allowed to rest under bush or scrub for many years. Land is rested when yields become so low that it would be unprofitable to plant a new cycle. On poor soil, only one or two cycles may be taken; on good soil several cycles may be taken, although yields inevitably decline unless nutrients are returned to the soil in the form of fertilisers or manures. It is strongly recommended that soil samples should be taken at the end of each cycle in order to give a guide to the nature and degree of nutrient deficiencies.

Nitrogen
Nitrogen applications are uncommon and are only necessary on soils which have a low nitrogen status; these may need 45–90 lb of nitrogen per acre (c. 50–100 kg/ha), depending on the nitrogen content of the soil. Nitrogen fertilisers must be applied to bare soil near the sisal plants.

Phosphate
Sisal seldom gives significant responses to phosphatic fertilisers. On the few soils that are deficient, superphosphate at a rate of 55–110 lb P_2O_5 per acre (c. 62–124 kg/ha) should be incorporated into the planting holes.

Potassium
Potassium deficiency causes banding disease. The

symptoms of this disease are yellow pin-head dots on the underside of the leaf at the neck (the narrowest part of the leaf, near its base); these dots gradually coalesce to form a weak band of necrotic tissue. Banding disease is common in the Korogwe area in Tanzania. It can be prevented by applying muriate of potash at 225–450 lb per acre per annum (c. 250–500 kg/ha). Muriate of potash applications should be split into five dressings and placed on the soil along the lines of sisal.

Lime
Lime deficiencies are common in the sisal growing area of Tanzania and are often corrected by applications of 16–32 cwt of ground limestone per acre (c. 2–4 t/ha). These must be broadcast and incorporated into the soil before planting.

Sisal waste
Well rotted sisal waste is occasionally applied along the rows of young sisal where it acts both as a mulch, conserving soil moisture, and as a manure. Applications of sisal waste are uncommon, even though experiments have shown that they give excellent responses on almost all soils. The reason for this is the expense involved in transporting and applying such a bulky material.

Weed control

In the last few years of the sisal cycle the soil is shaded, there is a competitive root system and the plants are tall, so competition from weeds is not very important. During this stage weeds and scrub are usually allowed to grow unchecked. In the early stages of growth, however, sisal is very susceptible to weed competition; there are indications that it is particularly susceptible to competition for light. Poor weed control in young sisal inevitably leads to a delayed first cut and risks stunting, from which the crop may never recover.

Broad-leaved weeds, which grow prolifically in the wetter areas, can be controlled quite easily in sisal. Buffalo bean (*Mucuna pruriens*), called 'upupu' in Kiswahili, sometimes causes trouble; it has pods covered with minute spines and these cause intense irritation when they come into contact with the skin. The greatest problem, as in all perennial crops, is presented by the perennial spreading grasses and sedges, couch grass (*Digitaria scalarum*), nut grass (*Cyperus spp.*), lalang (*Imperata cylindrica*) and Guinea grass (*Panicum maximum*). These are best controlled by thorough cultivation before planting combined with

spot applications of herbicides during the cycle. An alternative is the use of TCA, mentioned previously in the section on nurseries.

Hand cultivation
This method of weed control is widely used along the rows of sisal during the first few years of the cycle. Before the first cut there may be as many as ten weedings each year. At this stage the soil should be cultivated about 1½ ft (c. 0·5 m) on each side of the rows. In the middle of the cycle, when the crop suppresses weeds more effectively, cultivation along the rows is usually limited to one each year after cutting, when there is plenty of room between the plants for 'jembes' to be manipulated. All hand cultivations must be shallow to prevent damage to the shallow roots of the crop.

Tractor cultivation
The maintenance of weed-free conditions throughout the inter-rows, whether by mechanical cultivations or by herbicides, encourages soil erosion and must therefore be avoided. An additional disadvantage of tractor-drawn implements is that they usually do considerable damage to sisal roots.

Slashing
Tractor drawn rotary slashers are usually used although slashing can be done by hand on small areas. Slashing is particularly effective against broad leaved weeds, especially if it is done before they set seed. It is less effective against grasses because these creep even more vigorously if slashed repeatedly. Slashing does not encourage erosion, neither does it disturb the roots of the crop. However it does cause a certain amount of competition for water and nutrients.

Herbicides
Herbicides can be used along the rows instead of hand cultivation. Their use is likely to become more common as labour costs rise. Bromacil has proved to be the most effective soil applied residual herbicide; atraton, diuron and simazine are alternatives. An important advantage of these, when compared to hand cultivation, is that the roots of the young crop are not disturbed. The use of TCA and dalapon for perennial grasses has been mentioned above.

Cover crops
Sisal is the only important crop in East Africa for which cover crops are recommended. In Tanzania

many of the progressive estates sow tropical kudzu (*Pueraria phaseoloides*), sometimes mixing it with the faster establishing *Centrosema pubescens*. In Kenya, however, cover crops are seldom sown, despite the fact that experiments have shown several of them, especially *Dolichos biflorus,* to be effective. Cover crops are sown in the wide inter-rows. A fine seedbed is prepared and they are then sown by mechanical planters or by broadcasting. Most of them tend to climb amongst the sisal leaves and they must be checked from time to time either by hand or by using herbicides.

The advantages of cover crops are as follows:

1 They smother weeds, including the perennial grasses.

2 They compete very little with sisal as their roots are deep and occupy a lower part of the soil horizon than the roots of sisal.

3 They prevent erosion and improve the structure of the soil by dropping a mulch of dead leaves.

4 All the recommended cover crops are legumes; they may therefore fix nitrogen from the atmosphere.

5 They improve the drainage of the soil by the action of their deep roots.

The only problems involved with cover crops are their establishment and their control. Seed

Fig. 110: Sisal cutting.

scarification, inoculation with a suitable strain of bacteria, heavy fertiliser applications and careful weeding may be necessary in order to get satisfactory establishment. Seed is often expensive and is sometimes in short supply.

Cutting

Harvesting the leaf can only be done by hand and this absorbs a great deal of labour. One labourer can cut about one ton of leaf per day. Most decorticators have a capacity of about 5 tons of fibre (i.e. 165 tons of leaf) per day; this means that over 150 labourers are needed to maintain an adequate supply of leaf to the factory.

Sisal is cut with small knives with straight blades (see Fig. 110). Each leaf is cut near its base after the spine has been removed. Bundles of 30 leaves are made. Labourers' daily tasks are based on the number of bundles cut; the usual daily task is 90 or 100 bundles, i.e. 2 700–3 000 leaves.

Leaves should be decorticated within 24 hours of cutting. If there is much more delay they begin to deteriorate, the flesh adhering to the fibres and thus causing decortication problems. In hot sunny weather when the humidity is high, leaf scorch may occur on the cut leaves and in these conditions it is even more important to transport the leaves to the decorticator as quickly as possible. The symptoms of leaf scorch are brown lesions on the surface of the blades; the lesions often surround islands of healthy tissues. When there is a risk of leaf scorch it is advisable to take certain pre-cautions: the leaves should be bundled as soon as 30 are cut (the normal practice is to bundle at the end of the day); each bundle should be stood on end in the row; when bundles are stacked they should be stacked loosely to allow air circulation and should be covered with grass; if possible cutting and bundling should be done in the evening and early morning.

Trolleys running on movable railway lines are usually used for transporting the leaf out of the fields to the factory. Lorries are becoming increasingly popular, especially where the topography prevents the use of a rail system.

There are three main considerations which affect cutting policy: the time of the first cut, the severity of cutting and the frequency of cutting.

Time of first cut
If the first cut is too early the leaves are too small and the plant suffers a severe setback from which

it may take a long time to recover. If the first cut is too late some of the lower harvestable leaves may wither. The best guide is that the first cut should be taken when leaves of 2 ft (c. 60 cm) in length begin to touch the ground or when they begin to wither. At this stage the plants will have grown about 120 leaves and will stand about $4\frac{1}{2}$ ft (c. 1·4 m) high. The length of time until the first cut depends very largely on the size of the planting material; it may be anything from $1\frac{1}{2}$ to 4 years. The first cut invariably yields much less fibre than succeeding cuts; the fibre is shorter and more leaves are needed to produce a ton of fibre (as many as 100 000 compared with about 50 000 in later cuts).

Severity of cutting
If too many leaves are removed from the plant, photosynthesis is reduced and recovery is consequently slow. 25 leaves should be left after the first cut, but at later cuts only 20 need remain. With Hybrid 11648 70 leaves should remain after the first cut, and 60 leaves after later cuts. It is obviously impractical to count the leaves left on each plant; the general practice, therefore, is to cut all leaves below those which point upwards at an angle of about 45°. If this is done, a satisfactory number of leaves remain. Over-cutting is a serious offence on sisal estates and in Tanzania guilty labourers are liable to summary dismissal for wilful damage to property.

Frequency of cutting
It makes little difference to the plant how often cuts are taken, provided that the correct number of leaves are left at each cut. Very frequent cuts would obviously be wasteful of labour; excessively long intervals, on the other hand, would lead to withering of the lower leaves. A cutting interval of one year is the most convenient and is usually adopted. 5–10 cuts at yearly intervals can be taken during the cycle depending on the length of the cycle and the time of the first cut.

Yields

Yields are usually expressed in tons of fibre per cycle. Average yields of productive sisal are $4\frac{1}{2}$–5 tons per acre (c. 11–12 t/ha). Intensive management should produce 7–8 tons per acre (c. 17·5–20 t/ha) provided that the soil is suitable and the rainfall is good. Some estates in Tanzania have recorded blocks yielding 10 tons per acre (c. 26

t/ha). Yields of Hybrid 11648 have been discussed above in the section on varieties.

Decortication and factory work

Decortication is the process which removes the fleshy tissue (forming about 97% of the leaf by weight) from the fibres. The parenchymatous tissue is attached to the fibres and can only be removed by a harsh beating action (or, in the case of hand decortication, by repeated scraping).

Raspadors

A raspador has a beating drum and a concave. Each leaf is introduced by hand into the gap between the two of these, at right angles to the axle of the drum. The butt end (the base of the leaf) is first fed into the raspador, the other end being held by the operator. The semi-decorticated leaf is then removed and the tip end is fed into the machine whilst the operator holds onto the fibres of the butt end. This process is very laborious and one operator can only deal with about 2 000 leaves each day, i.e. about 65–75 lb (c. 30–35 kg) of fibre. Several small scale raspadors have been developed for smallholders but they were never popular, partly owing to the labour involved, partly to the difficulty of finding a good source of the clean water which is essential for washing the fibre. Raspadors are very rarely used on estates.

Decorticators

These machines (sometimes called automatic decorticators) have a far larger output than raspadors and are almost invariably used on estates. The leaves are passed through the decorticator by a conveyor belt system of ropes which grip the ends of the leaves. The leaves are fed into the machine at right angles to the line of flow and hand labour is only needed to ensure an even feed onto the conveyor belt (see Fig. 111). Each decorticator has two drums; one decorticates the butt end; the other decorticates the tip. Much water is needed at this stage both to wash the fibre and to carry away the waste.

Drying

After leaving the decorticator the fibre is taken to the drying ground where it is spread over wires (see Fig. 112). Lines of three parallel wires are usually used, with the central one slightly raised to prevent the 'kinking' which would result if the fibres were laid over a single wire. For best quality the fibre must be dried to a moisture content of

Fig. 111: Feeding sisal leaves into the decorticator.

Fig. 112: Sisal drying.

about 10% as quickly as possible. In good weather and with the fibre spread thinly on the lines, this may be achieved in only four hours. If the fibre is on the drying lines for more than 24 hours it tends to become slightly yellow, rather than creamy-white; if it remains for as long as three days a great deal becomes severely discoloured. If the moisture content of the fibre is not reduced immediately after decortication, e.g. if it remains piled on the ends of the drying lines for more than 2–3 hours, it may be weakened; this may even make it unsaleable. Several estates have installed driers. The main advantages of these are that they can produce top quality fibre in all weather conditions and that they use less labour.

Brushing

Before baling, sisal must be brushed to remove pieces of flesh which adhere after decortication. Brushing also frees the individual fibres from each other and removes the short fibres, which are called tow. Tow is a saleable product although its price is much lower than that of line fibre. Brushing machines contain revolving metal beaters; hanks of fibre are fed into them by hand, one end first and then the other, in much the same way that leaves are fed into a raspador.

Baling

Sisal fibre is finally baled under great pressure; it is important to achieve a very high density (as high as 60 cubic ft/ton, i.e. about 1·7 cubic m/t) because freight charges are on a volume basis rather than on a weight basis. Bales measure approximately 4 ft × 2 ft × 2 ft and each weighs about $\frac{1}{4}$ ton.

Grades

Sisal is graded according to its length and colour. A simplified description of the six grades of line fibre is given below:

Grade 1   —3 ft (c. 0·9 m) or more, creamy white colour.
Grade A   —3 ft or more, yellowish or spotted.
Grade 2   —2½ ft–3 ft (c. 0·75–0·90 m), creamy white colour.
Grade 3L —3 ft or more, minor defects in cleaning and colour allowed.
Grade 3   —2 ft–3 ft (c. 0·6–0·9 m), minor defects in cleaning and colour allowed.
Grade UG —2 ft or more, defective cleaning and colour allowed.

There are two grades of brushed tow, whilst very short fibres are recovered from the water after decortication and are designated flume tow. Brushed tow is used as padding in mattresses and upholstery whilst flume tow is sometimes used for the manufacture of sacks and cloth.

Pests

A. sisalana is remarkably free of important pests and diseases, except for the sisal weevil. Hybrid 11648, however, is much more susceptible to both pests and diseases.

Sisal weevil. *Scyphophorus interstitialis*

This pest only occurs at altitudes below 4 500 ft (c. 1 400 m) so it does little damage in most parts of Kenya. At the Kenya coast and in almost all parts of Tanzania, however, it can do a great deal of damage in the early part of the cycle. Mature plants are seldom harmed. Most of the damage is done by the larvae, which bore into the boles of young sisal (either in the nursery or in the field) where they make so many tunnels that the bole sometimes looks like a honeycomb. Damaged plants often die and must be replaced; this causes unevenness in the field. Adult weevils sometimes do a little damage by feeding at the base of the leaves.

Sisal weevil is easily and cheaply controlled by spreading aldrin dust around nursery plants and around planting holes. This prevents eggs being laid in the rotting tissue of the old leaves. Hybrid 11648 is more susceptible than *A. sisalana* to sisal weevil.

Scales

Several different species of scale insects may be found on sisal. Attacks are usually localised and occur where plants become dusty owing to a nearby road. Dust discourages some of the parasites of scales. Malathion sprays are effective but are seldom needed.

Diseases

Zebra disease

This disease is caused by the fungus *Phytophthora nicotianae*. It causes striped lesions on the leaves and rotting of the bole and the spike. It often damages Hybrid 11648 but is uncommon on *A. sisalana*. It is usually found in areas with poorly drained soil. No effective fungicides have been found. The only precaution is to prevent any part of the plants coming into contact with surface drainage water.

Bole rot

This disease is caused by the fungus *Aspergillus niger*. It attacks the boles, causing a yellowing of the leaves which is followed by death of the plant. *Aspergillus niger* is only a weak parasite on sisal and causes serious damage only when plants are weakened by nutritional deficiencies, cutting in wet conditions, etc.

Korogwe leaf spot

This disease has only been noticed in the Korogwe area since 1955. No causal organism has been isolated, nor is any control known. The symptoms are round, corky lesions which appear on the lower leaves. Hybrid 11648 is much more severely affected than *A. sisalana*.

By-products

Sisal waste, i.e. the flesh which constitutes 97% of the weight of the leaf, can be fed to cattle or can be ensiled. The crude protein content is only about 6% so it can only be used as a maintenance ration. Composted sisal waste makes excellent organic manure. Waxes, sodium pectates, alcohol, methane and hecogenin (containing cortisone) can be extracted from sisal waste. Hecogenin is produced on two sisal estates (one in Kenya and one in Tanzania), but in general all by-products of sisal can be obtained more economically from other sources.

Bibliography

1 **Anon.** *A Handbook for Sisal Planters— compiled by the staff of the Sisal Research Station, Mlingano.* Published by the Tanganyika Sisal Growers' Association.

2 **Burgwin, W. A.** (1965). *The importance of nurseries in sisal cultivation.* Kenya Sisal Bd. Bull., *54*:32.

3 **Constantinesco, I.** (1964). *Mechanical application of fertiliser to sisal.* E. Afr. agric. for. J., *29*:330.

4 **Diekmahns, E. C.** (1957). *Boron deficiency in sisal.* E. Afr. agric. J., *22*:197.

5 **Diekmahns, E. C.** (1960). *Cultivations in sisal fields.* Kenya Sisal Bd. Bull., *34*:19.

6 **Eckstein, A.** (1962). *Sisal selection and breeding.* Kenya Sisal Bd. Bull., *39*:22.

7 **Grundy, G. M. F.** (1959 and –60) *Leguminous cover crops (with reference to sisal in Tanganyika).* Kenya Sisal Bd. Bull., *30*:25, *31*:27, and *32*:31.

8 **Hopkinson, D.** (1968). *Experiments on mulching sisal with black polythene in Tanganyika.* Expl. Agric., *4*:143.

9 **Hopkinson, D.** (1969). *Leguminous cover crops for maintaining soil fertility in sisal in Tanzania, I—Effects on growth and yield.* Expl. Agric., *5*:283.

10 **Hopkinson, D.** and **Materu, M. E. A.** (1970). *The control of sisal weevil in Tanzania, III—Trials with insecticides in field sisal.* E. Afr. agric. for. J., *35*:273.

11 **Hopkinson, D.** and **Materu, M. E. A.** (1970). *The control of sisal weevil in Tanzania, IV—Field trials with insecticides in bulbil nurseries.* E. Afr. agric. for. J., *35*:278.

12 **Hopkinson, D.** and **Materu, M. E. A.** (1970). *The control of sisal weevil in Tanzania, V—Ways of reducing weevil attack.* E. Afr. agric. for. J., *35*:286.

13 **Lerche, K.** (1960–61). *Spacing and cutting.* Kenya Sisal Bd. Bull., *34*:38 and *35*:28.

14 **Lock, G. W.** (1957). *A manurial trial relating to banding disease of sisal.* Emp. J. exp. Agric., *25*:219.

15 **Lock, G. W.** (1969). *Sisal: Thirty years' sisal research in Tanzania.* Longmans, Green and Co. Ltd., London.

16 **Materu, M. E. A.** and **Hopkinson, D.** (1969). *The control of sisal weevil in Tanzania, I—Laboratory experiments with contact insecticides.* E. Afr. agric. for. J., *35*:79.

17 **Materu, M. E. A.** and **Webley, D. J.** (1969). *The control of sisal weevil in Tanzania, II—Laboratory experiments with systemic insecticides.* E. Afr. agric. for. J., *35*:88.

18 **Osborne, J. F.** (1967). *The prospects for, and limitations of, long fibre Agave hybrids.* Kenya Sisal Bd. Bull., *62*:16.

19 **Pinkerton, A.** and **Bock, K. R.** (1969). *Parallel streak of sisal in Kenya*. Expl. Agric., *5*:9.

20 **Richardson, F. E.** (1965). *Cover crop recommendations*. Kenya Sisal Bd. Bull., *53*:13.

21 **Richardson, F. E.** (1965). *Bromacil: a herbicide suitable for sisal nurseries*. Kenya Sisal Bd. Bull., *54*:41.

22 **Richardson, F. E.** (1966). *Plant populations and field spacing*. Kenya Sisal Bd. Bull., *57*:16.

23 **Richardson, F. E.** (1967). *Chemical control of couch grass by heavy application of sodium trichloroacetate (Na. TCA)*. Kenya Sisal Bd. Bull., *59*:29.

24 **Richardson, F. E.** (1968). *Hybrid 11648 in Kenya*. Kenya Sisal Bd. Bull., *63*:7.

25 **Sanders, M. B.** (1962–63). *Herbicides for sisal*. Kenya Sisal Bd. Bull., *42*:14 and *43*:14.

26 **Strange, R.** (1963). *Some factors affecting inter-row cropping in young sisal*. Kenya Sisal Bd. Bull., *44*:20.

27 **Terry, P. J.** (1969). *Herbicides in sisal*. Kenya Sisal Bd. Bull., *70*:26.

28 **Wallace, M. M.** and **Diekmahns, E. C.** (1952). *Bole rot of sisal*. E. Afr. agric. J., *18*:24.

29 **Wienk, J. F.** (1968) Phytophthora nicotianae: *a cause of zebra disease in* Agave *hybrid No. 11648 and other agaves*. E. Afr. agric. for. J., *33*:261.

30 **Wienk, J. F.** (1968). *Observations on the spread and the control of zebra disease in* Agave Hybrid No. 11648. Kenya Sisal Bd. Bull., *64*:30.

29
Sorghum

Sorghum vulgare

Introduction

Sorghum is an important cereal crop. It is drought resistant and can regularly outyield maize in many of the drier parts of East Africa; it is more resistant to waterlogging than the other important cereals apart from rice; it yields reasonably well on infertile soils and it can be ratooned. It has several disadvantages, however; current varieties have a lower yielding potential than maize on free draining soils in areas of good or moderate rainfall; sorghum is often heavily attacked by birds and there are fewer children available, now that most of them go to school, to scare them away; finally, the harvesting, threshing and cleaning of sorghum are more labour consuming than these operations on maize.

Tanzania
An estimated one million acres (c. 400 000 ha) of sorghum are grown annually in Tanzania. It is the staple food in the dry central part of the country.

Uganda
It was estimated that 565 000 acres (c. 230 000 ha) of sorghum were grown in Uganda in 1968. The main concentrations are in Kigezi and Karamoja Districts; it is also important in Eastern and Northern Regions, especially in Lango and Teso Districts. In the banana growing areas around Lake Victoria a small amount is grown solely as an ingredient for banana beer.

Kenya
Sorghum is important in Kenya in the western part of the country and in Turkana. Kenya's acreage of sorghum has been estimated as being 500 000 (c. 200 000 ha).

Plant characteristics

Sorghum has a very efficient, well-branched root system and this may be an important cause of its drought resistance; in a given volume of soil it produces about twice as many roots as maize. The

Fig.113: Sorghum.

endodermis contains considerable amounts of silica, an element which is almost absent in the endodermis of maize and which may prevent sorghum roots from collapsing in dry soil. Sorghum tillers freely. The stem height depends on the variety and may be as much as 15 ft (c. 4·5 m). The heads vary greatly according to the variety; the panicles may be open or compact (see Figs. 114 and 115); they may be erect or 'goosenecked', i.e. hanging over (see Fig. 116); glumes may be large or small; awns may be present or absent; the seed coat may be persistent or it may be easily removed by pounding; it may be white or almost any shade of brown or red and the endosperm may be flinty or mealy. These characteristics are important because they affect either bird resistance, weevil resistance

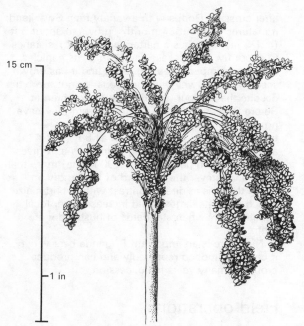

15 cm

1 in

Fig. 114: Sorghum; a variety with an open panicle.

or taste. 5% is an average figure for cross-pollination but it may be as high as 25% in varieties with open panicles.

Sorghum matures in 2½ to 8 months at 3 500 ft (c. 1 100 m) above sea level; the period depends on the variety. At higher altitudes it matures more slowly.

Ecology

Rainfall, and water requirements

Sorghum is very drought resistant; it can withstand droughts almost as well as bulrush millet. The growth and the structure of the roots, discussed above, may be partly responsible for this but other characteristics are also important. Sorghum can reduce its transpiration during periods of water shortage by rolling its leaves and possibly by stomatal closure; in this condition it can remain dormant when other crops would be killed and when the rains start again it recovers rapidly.

Sorghum needs a rainfall of at least 12–15 in

15 cm

1 in

Fig. 115: Sorghum; a variety (Serena) with a fairly compact panicle.

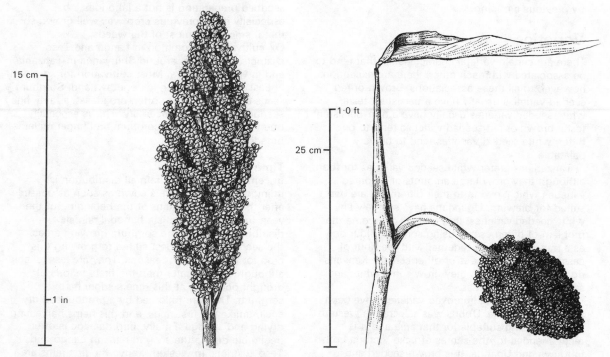

1·0 ft

25 cm

Fig. 116: Sorghum; a "goose necked" variety with red grains.

(c. 300–380 mm) during its growing period; however, an ideally distributed rainfall of only 7 in (c. 175 mm) has given a reasonable yield of 1 000 lb per acre (c. 1 100 kg/ha). It grows well in central Tanzania where the average annual rainfall in parts is only 25 in (c. 650 mm).

Sorghum is one of the few crops that withstands short periods of waterlogging; it is therefore popular on heavy clay soils, e.g. on the Kano Plains in Kenya.

Altitude and temperature

Most sorghum varieties prefer warm conditions but high altitude varieties are grown as high as 8 000 ft (c. 2 400 m) in the Kigezi Highlands in Uganda. With the exception of that grown in Kigezi District, most East African sorghum is grown between the altitudes of 3 000 and 5 000 ft (c. 900–1 500 m); if the varieties grown in this zone are taken to higher levels they give poor yields and are heavily attacked by diseases and shoot fly.

Soil requirements

Sorghum grows best on reasonably fertile soils but is second only to bulrush millet in its ability to give satisfactory yields on soils that have been exhausted by previous cropping.

Varieties

There are certain varietal characteristics that tend to be associated with each other; there are exceptions, however, to all these associations. Brown or red seeded varieties usually have a persistent testa; white seeded varieties usually have a non-persistent testa; brown or red seeded varieties tend to be bitter; white seeded varieties tend to be very palatable.

Tanzanians prefer white seeded varieties for food, although they grow large amounts of coloured varieties which, owing to their bitterness, are best suited for brewing. Ugandans have scarcely any white seeded varieties; they usually overcome the problem of bitterness by using sorghum flour only as a minor ingredient together with the flour of cassava, sweet potatoes or other cereals. Kenyans do the same although they grow some white and some coloured varieties.

The most notable improved varieties have been Dobbs and Serena. Dobbs was selected in western Kenya as being suitable for that area and it is recommended for the shores of Lake Victoria both in Kenya and Uganda. It is brown seeded and matures in about four months. Serena was selected

after crossing Dobbs with a variety from Swaziland; it matures in about 3½ months, grows to about 5 ft (c. 1·5 m) high, has a certain amount of resistance to shoot fly, leaf diseases and *Striga*, its panicles are fairly compact (see Fig. 000) and it has brown seeds. It yields well and the seeds are not too bitter despite their colour. Hybrids have been bred at Serere Research Station in Uganda but they have not yet been released or named. The most promising hybrid grows as high as 8 ft (c. 2·4 m) and is slightly later maturing than Serena. It gives 500 to 1 000 lb per acre (c. 1 100 kg/ha) more than Serena whatever the standard of husbandry; in doing this it is in direct contrast with hybrid maize which gives its largest yield increases over local varieties only when standards of husbandry are high.

Namatare is an important Buganda beer variety; it can be ratooned repeatedly and can produce crops for many years in succession.

Field operations

Seedbed preparation

Where sorghum is grown in the second rains seedbed preparation is not a laborious job, especially if the previous crop was well grown so that it smothered most of the weeds. Ox-cultivation is common in Lango and Teso Districts in Uganda, around Shinyanga in Tanzania and in western Kenya. Most cultivation for sorghum in East Africa, however, is done by hand. Sorghum seeds are small and are often broadcast; a fairly fine seedbed is therefore necessary. This is especially true when a mixture of sorghum and finger millet is broadcast.

Time of sowing

In central Tanzania the rainfall distribution is unimodal and sorghum is sown as soon as possible after the start of the rains at the beginning of the year. In Uganda there is a bimodal rainfall distribution in all of the sorghum growing areas; the second rains are best suited to growing this crop for the following reasons. They are less reliable although often greater than the first; periods of drought occurring at this time seldom harm sorghum. They are followed by a pronounced dry spell, unlike the first rains, and this helps harvesting, drying and storage. Finally, bird damage is often negligible during the second rains in Lango and Teso Districts. In western Kenya the first rains are more popular for sowing sorghum and it is very common for a second crop to be established in the second rains by ratooning.

Sorghum responds well to sowing as early as possible during the rains; in Tanzania eleven out of twelve experiments, in various parts of the country, showed that yields decline markedly, sometimes dramatically, with delayed sowing.

Spacing

Most sorghum in East Africa is sown at random and a large proportion is intercropped. In Tanzania and Kenya the most common practice is to interplant it with a combination of other cereals, pulses, simsim or sweet potatoes. In Uganda about half of the crop is grown as a pure stand and half is interplanted with finger millet; Ugandans seldom grow cereals and pulses together as Kenyans and Tanzanians do.

The recommended spacing is 2 ft × 6 in (c. 0·6 m × 15 cm) although in moist conditions a closer spacing gives the highest yields. The seed rate depends mostly on the quality of the seed; with a high germination percentage as little as 3 lb of seed per acre (c. 3·3 kg/ha) may be enough but for average quality seed 7 lb per acre (c. 7·7 kg/ha) should be sown.

Sowing

In Uganda almost all sorghum is broadcast with the exception of the small proportion that is sown in rows. In Kenya and, to a lesser extent, in Tanzania broadcasting is common, especially when a pure stand or a mixture of sorghum and finger millet is being sown. Dibbling, however, is more common and is the usual practice when sorghum is included with pulses.

Fertilisers

On moist soils sorghum responds well to farmyard manure, nitrogen and phosphate. In Uganda 100 lb per acre (c. 110 kg/ha) of single superphosphate are recommended as a seedbed application and 100 lb per acre of calcium ammonium nitrate as a top dressing about three weeks after sowing. Smallholders rarely adopt these recommendations.

Weed control

The main problem is *Striga* which is also known as witchweed. There are two species, *S. hermonthica* and *S. asiatica*; the first is more common and has purple flowers whilst the second occurs less frequently and has bright red flowers. Both are parasitic weeds whose roots penetrate those of sorghum, reducing yields considerably. Each *Striga* plant can produce a very large number of seeds;

these are stimulated to germinate by the root exudates of cereal crops and to a smaller extent by those of some other crops; sorghum exudates give greater stimulation than those of any other crop. *Striga* seeds can remain dormant and yet viable for as long as ten years.

There is a degree of *Striga* resistance in Dobbs and Serena; farmers cannot rely on this alone, however, and where *Striga* is a problem they should use other methods of control. Rotations which include crops other than sorghum, finger millet or maize help to prevent the build up of an excessive *Striga* population. Rotations cannot give complete control of this weed because of the long period that the seeds can remain dormant in the soil. Hand pulling before the plants have a chance to set seed is the best method of reducing *Striga* populations; if there is a heavy infestation this may be a laborious job because pulling must be done many times. An alternative, which is unlikely to appeal to smallholders, is to sow a trap crop; this involves sowing sorghum at a close spacing to encourage as much *Striga* germination as possible and then to plough in the sorghum and *Striga* before the weeds set seed.

Harvesting

Sorghum is harvested by breaking off the heads by hand; these are usually sun dried before being stored.

Yields

Sorghum yields are usually between 500 and 1 500 lb of dried grain per acre (c. 550–1 700 kg/ha). Farmers growing Serena or hybrids and practising good husbandry should get 3 000–4 000 lb per acre (c. 3 400–4 500 kg/ha). A yield of 6 000 lb per acre (c. 6 700 kg/ha) on a field scale would be considered extremely good.

Pests

Birds

These are one of the main causes of crop loss in sorghum; bird susceptible varieties sometimes give no yield at all. The most devastating species is the Sudan Dioch, *Quelea quela aethiopica*. Quelea birds breed in vast colonies in the drier parts of East Africa, notably near Dodoma in Tanzania, where they build their nests in the scrub. From their breeding centres they spread through Tanzania and Kenya, following well known migration routes. At first each flock may consist of several million birds but it gradually divides and subdivides so smaller flocks occur further away from the breeding centres. Weavers, starlings, bishop birds and many others also attack sorghum but they are not as devastating as the Sudan Dioch.

There are several kinds of resistance to birds. The first, and the most effective, is the persistent, bitter tasting seed coat found in most varieties with coloured grains; these varieties are third in the birds' order of food preference and are only eaten in the absence of either non-bitter varieties or a good supply of grass seed. Goose-necked varieties are slightly resistant; birds may perch on top of them and eat the upper grains but they only hang upside down and eat the lower grains in the absence of a more convenient food supply. Varieties with large glumes or prominent awns are very slightly resistant; birds attack them heavily in the absence of more popular food.

A Quelea control unit kills large numbers of birds annually by using explosives, flame-throwers or poison sprays in breeding colonies or roosting sites.

Sorghum shoot fly. *Antherigona varia*
This pest attacks sorghum in the early stages of crop growth, often when the plants are only one or two inches high. The adult fly lays eggs on the underside of young leaves and the developing larvae enter the funnel and move down to feed on the meristem. The central shoots become yellow and then die; affected plants often compensate for this damage by producing several tillers but these may also be attacked. There is resistance to shoot fly in some varieties; this depends on the ability of the plants to produce new and vigorous tillers which rapidly grow beyond the stage at which they can be attacked. Early planting is important; late planted crops are very frequently devastated by this pest. Six endosulfan sprays at intervals of three days give effective but expensive control; these must be applied in the very early stages of growth.

Stem borers
Three species of stem borer attack sorghum. *Busseola fusca*, the stem borer most common in maize, is relatively easy to control because distinct larval populations appear at intervals of about five weeks and these feed in the funnels before moving down to the developing tissue. A timely application

of insecticide therefore kills them. *Chilo zonellus* (*partellus*), the most important stem borer of sorghum, is more difficult to control because there are no distinct population peaks and young plants may be attacked at any time. *Sesamia calamistis* is almost impossible to control, for instead of entering the plant by way of the funnels where they can pick up insecticide, the larvae bore straight into the centre of the stem after hatching behind the lower leaf sheaths. Fortunately, heavy outbreaks occur only occasionally.

Sorghum midge. *Contarina sorghicola*
The adult midge lays eggs behind the glumes at the times when anthers are protruding from the heads. The seeds are hollowed out and only the seed coats develop properly. The sorghum midge seldom causes serious crop loss in East Africa although it is a major pest in the U.S.A., India and some other parts of Africa.

Diseases

Several diseases can cause serious crop loss. The most important are leaf blight (*Helminthosporium turcicum*), anthracnose (*Colletotrichum graminicola*), sooty stripe (*Ramulispora sorghi*), downy mildew (*Sclerospora sorghi*), head smut (*Sphacelotheca reiliana*), loose smut (*Sphacelotheca cruenta*) and covered smut (*Sphacelotheca sorghi*). The first four are leaf diseases; the rest are diseases of the inflorescence. The first five seldom cause much damage when improved varieties are grown owing to effective varietal resistance. The last two are best controlled by seed dressings which should be a routine precaution.

Utilisation

When sorghum is cooked as a staple food the dry grains are usually ground, either alone or with other cereals, cassava chips or dried sweet potatoes. The flour is used for making 'ugali' or 'uji'. The inhabitants of the Kigezi Highlands have an unusual way of preparing a cereal crop; they mix the sorghum grains with wood ash, soak the mixture overnight, cover it for about five days to encourage germination, pound it and winnow it to remove the small roots and then dry it. The flour from the ground grains is used for 'uji' or brewing; it cannot be used for 'ugali'. They do this because their sorghum varieties are brown seeded and bitter; these would produce an unpalatable food without the sugars which are formed in the grains during germination. There are many variations in brewing practice; when sorghum alone is used it is most commonly soaked, germinated and coarsely ground to produce malt. Meanwhile freshly ground ungerminated grains have been made into a thick paste and have been fermented under water for about five days. The fermented paste is then fried and is again submerged; this time malt is included and the period of fermentation is shorter, usually two or three days, after which the beer is ready for drinking. Brewing is often done with sorghum as a minor ingredient (see Chapters 1 and 16).

Bibliography

1 **Akehurst, B. C.** and **Sreedharan, A.** (1965). *Time of planting—a brief review of experimental work in Tanganyika, 1956–62.* E. Afr. agric. for. J., *30*:189.

2 **Disney, H. J. de S.** and **Haylock, J. W.** (1956). *The distribution and breeding behaviour of the Sudan Dioch in Tanganyika.* E. Afr. agric. J., *21*:141.

3 **Doggett, H.** (1953). *The sorghums and sorghum improvement in Tanganyika.* E. Afr. agric. J., *18*:155.

4 **Doggett, H.** (1957). *The breeding of sorghum in East Africa, I—Weevil resistance in sorghum grains.* Emp. J. exp. Agric., *25*:1.

5 **Doggett, H.** (1957). *Bird resistance in sorghum and the* Quelea *problem.* Field Crop Abstr., *10*:153.

6 **Doggett, H.** (1958). *The breeding of sorghum in East Africa, II—The breeding of weevil resistant varieties.* Emp. J. exp. Agric., *26*:37.

7 **Doggett, H.** (1964). *A note on the incidence of American bollworm in sorghum.* E. Afr. agric. for. J., *29*:348.

8 **Doggett, H.** (1965). Striga hermonthica *on sorghum in East Africa.* J. Agric. Sci., *65*:183.

9 **Doggett, H.** (1969). *Yields of hybrid sorghums.* Expl. Agric., *5*:1.

10 **Doggett, H.** (1970). *Sorghum.* Longman.

11 **Doggett, H.** and **Jowett, D.** (1966). *Yields of maize, sorghum varieties and sorghum hybrids in the East African lowlands.* J. Agric. Sci., *67*:31.

12 **Dowker, B. D.** (1963). *Sorghum and millet in Machakos District.* E. Afr. agric. for. J., *29*:52.

13 **Evans, A. C.** (1960). *Studies of intercropping, I—Maize or sorghum with groundnuts.* E. Afr. agric. for. J., *26*:1.

14 **Glover, J.** (1948). *Water demands by maize and sorghum.* E. Afr. agric. J., *13*:171.

15 **Glover, J.** (1959). *The apparent behaviour of maize and sorghum stomata during and after drought.* J. Agric. Sci., *53*:412.

16 **Ingram, W. R.** (1960). *Experiments on the control of stalk borers on sorghum in Uganda.* E. Afr. agric. J., *25*:184.

17 **Jowett, D.** (1965). *The grain structure of sorghum related to water uptake and germination.* E. Afr. agric. for. J., *31*:25.

18 **Swaine, G.** and **Wyatt, C. A.** (1954). *Observations on the sorghum shoot fly.* E. Afr. agric. J., *20*:45.

19 **Wheatley, P. E.** (1961). *The insect pests of agriculture in the Coast Province of Kenya, V—Maize and sorghum.* E. Afr. agric. for. J., *27*:105.

20 **Williams, J. G.** (1954). *The* Quelea *threat to East Africa's grain crops.* E. Afr. agric. J., *19*:133.

30
Sugar cane

Saccharum spp.

Introduction

Sugar cane is one of the few plants that stores its carbohydrate reserve in the form of sucrose. It is grown throughout the tropics and provides more than half of the world's sugar; the remainder is supplied by sugar beet which is a root crop grown in temperate areas.

Milling, jaggery, chewing and brewing
The most economically important product made from sugar cane in East Africa is white sugar. This is manufactured in mills; the milling process is described briefly at the end of this chapter. Jaggery manufacture is more widespread in East Africa than the manufacture of white sugar which is restricted to the vicinity of the dozen or so mills. Jaggery factories are small and simple. The cane is crushed to extract the juice which is then boiled until it is thick enough to set in moulds when cooled. If good quality cane is used and if the juice is well clarified, good quality, or 'superfine', jaggery can be produced. This is a sweetening material used by the Asian community for cooking and sweetmeats, although its use is declining. Poor quality cane is usually used for jaggery and clarification is given little attention; the result is a bitter product, sometimes called 'black jaggery', which is widely used for the illicit distillation of Nubian gin. Cane is grown in small plots or as isolated plants throughout the lower altitude, wetter areas of East Africa for chewing and brewing. The remainder of this chapter deals solely with growing cane for the white sugar industry.

Kenya
Of the three East African countries, Kenya has the highest demand for sugar. It has been unable to meet this demand and has had to import sugar each year, mainly from Uganda. By the mid-1970s, however, expanded production should meet the country's needs if current development plans are implemented. The main producing area is the northern part of the Kano Plains and the adjacent foothills and is based on three mills at Miwani,

Chemelil and Muhoroni. About 65 000 acres (c. 25 000 ha) of cane were grown in this area in 1969. At Ramisi, on the coast, there is an estate of 11 500 acres (c. 4 500 ha) which accepts outgrowers' cane from about 2 000 acres (c. 800 ha). A nucleus estate of 8 000 acres (c. 3 200 ha) is planned at Mumias. There are already approximately 5 000 acres (c. 2 000 ha) of smallholder cane in this area but in the absence of a mill it is sold to jaggery factories. The smallholder acreage is planned to rise to 18 000 acres (c. 7 300 ha) shortly after the mill opens.

Uganda
In Uganda there are two large estates, each with approximately 20 000 acres (c. 8 000 ha) of cane. One is at Lugazi; the other is at Kakira. There is a smaller estate of about 7 000 acres (c. 2 800 ha) at Sango Bay and a further scheme is planned for Kinyala, near Masindi.

Tanzania
The two main estates are at Arusha-Chini (8 500 acres of cane, i.e. about 3 500 ha) and at Kilombero (7 500 acres of cane, i.e. about 3 000 ha). There are smaller estates at Kagera, Mtibwa and Karangai.

Estates and smallholders
At each place mentioned above there is (or will be) a mill with a nucleus estate. Each estate, with the exception of Arusha-Chini, accepts cane from outgrowers. Most outgrowers are smallholders but some, especially in the Kano Plains area in Kenya, have 50–100 acres (c. 20–40 ha) of cane. (Arusha-Chini is an irrigated estate in a very dry area where rainfed smallholder cane could not be grown).

Plant characteristics

For commercial purposes in East Africa sugar cane is propagated solely by short pieces of stem which are called 'setts'. Each sett usually has three nodes. When a sett is buried the buds on the nodes produce stems, provided that the soil is adequately

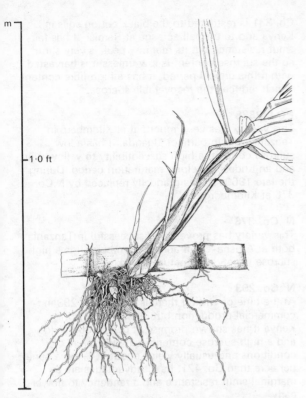

m

—1·0 ft

Fig. 117: A two-node sugar cane sett. For clarity, the shoots from the right-hand node have been removed.

moist and warm. Roots first grow from the root primordiae which are situated in a band around each node. Later, however, these roots rot, as does the sett, and the developing plant gets its water and nutrients from roots which grow from newly produced nodes. These nodes also produce new shoots so one sett gives a cluster of shoots which is usually called a 'stool' (see Fig. 117).

Each stem may be 1–2 in (c. 2·5–5·0 cm) thick and 10–15 ft (c. 3·0–4·5 m) high when mature; both the thickness and the height depend to a large extent on the variety, but also on the growing conditions. The inflorescence is a loose, white, feathery panicle, usually known as the 'arrow'. Arrowing can be forecast with some accuracy because the leaves have successively shorter blades and longer sheaths as arrow emergence approaches. Arrowing depends on the interaction of a number of factors, not all of which are fully explained; variety, altitude and photo-periodism are amongst the most important. The seed, which is usually called 'fuzz', is difficult to collect because it is easily blown away. Viability is of short duration under natural conditions so seed is only used for breeding.

Most of the popular varieties are self-stripping canes, i.e. their leaf sheaths are non-persistent and form a considerable amount of trash on the ground. Stripping canes are easier to harvest than non-stripping canes whose leaf sheaths are persistent and have to be removed by the harvest gang.

Life cycle

The first crop after planting is called the 'plant crop'. Its growing period to maturity varies with the ecology, being as short as 13–14 months at the coast and as long as 21–22 months in parts of Nyanza and Western Provinces in Kenya. After the plant crop has been harvested, the old stools regenerate rapidly, producing a 'ratoon crop'. The first ratoon crop usually yields only $\frac{2}{3}$–$\frac{3}{4}$ as much cane as the plant crop; yields almost always decrease steadily with successive ratoons. Nevertheless, ratoon crops are more profitable than the plant crop, provided that there are not so many in succession that they bring the yield to a very low level. The reasons for their profitability are that they do not have to bear the cost of establishment as in a plant crop and they mature in a shorter time (12 months at the coast, rising to 18 months in the Kano Plains area). In most parts of East Africa only two ratoon crops are taken; at Kilombero, however, three or four are usually taken.

Ecology

Rainfall and water requirements

Sugar cane is fairly drought resistant but it needs a steady supply of soil moisture throughout the year if it is to give maximum yields. The optimum rainfall depends on many factors, especially the soil type; 60 in (c. 1 500 mm), evenly distributed, is often quoted as the minimum annual average. In Kenya, other than in the Mumias area and in the foothills of the Nandi escarpment, rainfall is normally deficient and severe damage by drought is incurred in one or two years out of five. Excessive rain can be a problem, especially if drainage is poor; cane may suffer from waterlogging and excessive weed growth, and cane transport may become difficult.

Altitude and temperature

High yields have been obtained in experimental

plots at Kakamega in Kenya at 5 200 ft above sea level (c. 1 600 m) and the highest yielding commercial cane in the Kano Plains area is situated slightly above 5 000 ft (c. 1 500 m). 5 000 ft is usually regarded as being near the upper limit for commercial cane growing. Arrowing is less common at the higher altitudes than it is at sea level.

Soil requirements

Sugar cane tolerates a wide variety of soil conditions but only yields well under conditions of free drainage. On heavy soils, such as the black cotton soils of the Kano Plains in Kenya, cambered beds, ditches or furrows must be formed in order to lead off surplus water. Sugar cane can give good yields on sandy soils provided that nutrient deficiencies are rectified by the applications of fertilisers or manures and provided that there is an adequate water supply. In Africa, however, nematodes have sometimes prevented healthy growth of cane on sandy soils.

Varieties

The number of each variety usually has a prefix of one or more letters which denote its origin. The prefixes of the most commonly grown varieties in East Africa are explained below:

Co = Coimbatore (India)
N:Co= Natal (South Africa) from Coimbatore
    seed
POJ = Proefstation Oost Java (Indonesia)
B = Barbados (West Indies)
Q = Queensland (Australia)

Co. 421

This has been the most important variety in Kenya, Uganda and at Arusha-Chini in Tanzania since the smut epidemic starting in the late 1950s precluded several other popular varieties. It has fair smut resistance and yields well in a variety of conditions although there are several varieties with a higher yielding potential which will probably replace it during the 1970s. Its leaf sheaths are, to a certain extent, persistent; they encircle the stem loosely, however, and can be removed during cutting with very little effort or loss of time.

Co. 331

This variety has the virtue of giving moderate yields on badly drained or infertile soils; in these conditions most other varieties would yield poorly.

Co. 331 is restricted to the black cotton soils in Kenya and to a small acreage at Ramisi. It has fair smut resistance but its maturity peak is very short so the sucrose content is low unless it is harvested within this critical period. It has a high fibre content which reduces the recoverable sucrose.

POJ. 2878

This variety was very important at Kilombero in Tanzania and in parts of Uganda. It has a low sucrose content, a high susceptibility to yellow wilt and an undesirably long maturation period. During the late 1960s it was gradually replaced by N:Co. 376 at Kilombero.

N :Co. 376

This variety has proved very successful in Tanzania both at Arusha-Chini and at Kilombero. It is a high sucrose variety and matures quickly.

N :Co. 293

At the time of writing there is little N:Co. 293 in commercial production but in experiments in Kenya it has shown promise. It gives good yields and a high sucrose content in a wide range of conditions and usually yields about 10% more cane per acre than Co. 421. Its disadvantages are a marginal smut resistance and a tendency to flower early.

B. 41227

This variety is very resistant to smut and yields well when there is a good supply of soil moisture, e.g. under irrigation in Uganda or in high rainfall areas such as Western Province in Kenya.

Other varieties which have shown promise in experiments and which will probably be grown commercially in the 1970s include Co. 440, Co. 467, Co. 680, Co. 746, C0. 1001, B. 47419 and Q. 47.

Land preparation

Preparing the land for the plant crop is both time consuming and expensive. After the last ratoon has been harvested, the soil has usually remained virtually undisturbed for about five years; during this time it becomes compacted and the amount of dead roots increases steadily. These conditions must be changed by deep and thorough cultivation in order to ensure healthy development of the following plant crop.

Fig. 118: Cultivating for sugar cane.

The machinery used and the number of operations depend largely on the soil conditions. On the heavy clay soils of the Kano Plains crawler tractors (see Fig. 118) are needed to pull heavy cultivators and subsoilers which penetrate up to 2 ft (c. 0·6 m) deep; on more manageable soils, e.g. at Mumias in Kenya, conventional wheeled tractors and conventional ploughs and harrows can be used successfully. Most cane in East Africa is grown on the flat; after preparing a reasonable tilth shallow furrows are drawn and the cane is planted in these. At Arusha-Chini in Tanzania large ridges, 5 ft (c. 1·5 m) from centre to centre, are formed in order to allow furrow irrigation. Large ridges have fallen out of favour in Kenya because they dry out too quickly in a dry spell. On the heavy soils in Kenya cambered beds, usually 25 ft (c. 7·5 m) wide, have proved successful. An alternative to these, which is also used in poorly drained areas in Uganda, is to dig deep ditches at intervals of 20–25 ft (c. 6·0–7·5 m).

Green manures and gypsum applications are sometimes used to improve the structure of the soil.

Sunn hemp (*Crotalaria juncea*) is used as a green manure at Kakira in Uganda, at Ramisi in Kenya and on some of the heavier soils in the Kano Plains area. 50 lb of seed are broadcast per acre (c. 55 kg/ha) and the fields are then harrowed. The crop is ploughed in about six weeks later. Enough time should be allowed for decomposition between ploughing in and planting the cane. Gypsum, applied at 1½–2 tons per acre (c. 3·8–5·0 t/ha) is beneficial on montmorillonitic clays but is less popular than green manure owing to its expense. It is more lastingly beneficial than green manure, however; experiments on black cotton soils have shown that it can increase yields by as much as 10 tons per acre (c. 25 t/ha) and at the time of writing, 13 years after the start of the experiments, there is no noticeable decline in this effect.

The 'turn round time', i.e. the time taken from harvesting the last ratoon crop to planting the plant crop of the next cycle, need be no longer than 3 months if the weather allows cultivation at the desired times. In practice it is usually longer. The reason for this is often that growers are reluctant to plant at the onset of the rains when weed growth may be excessive.

Planting

Nurseries for planting material should be established with heat treated cane in order to reduce the risk of ratoon stunting disease (see section on diseases for details of heat treatment). Heat treatment is not universal, in fact it has not yet become commercially accepted in East Africa in spite of the evidence in its favour. One acre of nursery should provide enough setts for ten to fifteen acres of plant crop. $1\frac{1}{2}$–$2\frac{1}{2}$ tons of planting material are needed for one acre (c. 3·8–6·3 t/ha).

As mentioned previously, setts with three nodes are usually used. Longer setts have proved unsatisfactory in East Africa because few of the buds germinate and because, as the upper surface dries faster, the setts bend, lifting some of the developing shoots out of the ground. Setts can be taken from the lower part of the cane, provided that the buds are undamaged, but the green part at the top should not be used because it tends to rot when buried. The age of the cane used for setts depends on the growth period; in areas with a short growth period 8 month old cane can be used but where the growth period is long the cane should be 14 months old. Planting material is usually dipped into organo-mercurial fungicide; this ensures even germination owing to protection from soil fungi.

In East Africa setts are always buried horizontally 2–3 in (c. 5·0–7·5 cm) deep. Planting deeper than this has proved unsatisfactory, probably because the setts do not receive enough warmth for good germination. Setts are usually placed end-to-end in a single line in the planting furrow (see Fig. 119); some estates, however, insist on two lines of setts or on a single line with the setts overlapping each other. Any method is satisfactory provided that it produces one stool approximately every 2 ft (c. 0·6 m). Infilling must be done if there are gaps of over 3 ft (c. 0·9 m). A large proportion of infills is undesirable because they seldom catch up with their neighbours.

Time of planting is not a very important factor because most cane is grown in areas that receive a good rainfall (or can be irrigated) in most months of the year. In the Kano Plains in Kenya, short rains planting, i.e. in the rains preceding the January/February dry period, has proved more successful than planting at other times of the year. The drought starts when the crop is established but at a time when the moisture

Fig. 119: Covering sugar cane setts.

demand is low, whilst the heavy rains of March/April coincide with a period of greater moisture demand.

Spacing

The spacing between rows varies from 4 ft to 6 ft (c. 1·2–1·8 m). Between these limits there is very little variation in yield. The 6 ft spacing leads to weed problems, especially in the areas where cane grows slowly and takes a long time to form a canopy over the soil. Cane rows spaced 4 ft apart suppress weeds more rapidly but they are inconvenient for most machinery and cane transport vehicles. 5 ft (c. 1·5 m) rows are therefore the most common.

Fertilisers

Nitrogen applications are essential for good yields. 60–70 lb of nitrogen per acre (c. 65–80 kg/ha) is an average application but some estates apply two or three times as much as this. Nitrogenous fertilisers should be applied not only to the plant crop but also to each ratoon crop; responses are usually greater in ratoon crops than in plant crops. Nitrogen must be top-dressed in moist conditions along the lines of cane (not on the trash) shortly after the shoots have started to grow vigorously. In a normal wet season split dressings are unnecessary.

Phosphate and potassium applications seldom give economic responses. They are much less common than nitrogen application but are adopted as an insurance policy by some estates. 60 lb of P_2O_5 per acre (c. 65 kg/ha) gives economic responses in the Mumias area and at the coast in Kenya.

Weed control

Weeding is an important operation in the plant crop, which grows more slowly in the early stages than the ratoon crops. Three to four weedings are usually needed unless herbicides are used. Tractor drawn implements are often used for weeds in the inter-rows whilst weeds within the rows are controlled by hand; alternatively, hand labour may be used throughout.

For ratoon crops, only one or two weedings are usually necessary. Ratoons suppress weeds not only by the quickly growing canopy but also by the trash from the preceding crop. In ratoon crops the soil is sometimes drawn away from the stools during the first weeding in order to encourage early tillering. In later operations the soil may be piled around the stems to prevent further tillering.

Herbicides such as the triazines, MCPA and 2,4–D are used to a very limited extent; their use may become more widespread as labour costs rise.

The parasitic weed *Striga* can cause severe crop losses in Kenya, especially in areas where cereal crops were formerly grown. Long term control is the prevention of seeding. Short term control is not completely effective; it takes the form of hand pulling or spot applications of 2,4-D in high concentration and supplementary application of fertiliser to affected cane.

Irrigation

Irrigation is essential at Arusha-Chini in Tanzania where the average annual rainfall is less than 20 in (c. 500 mm). Most of the cane is watered by furrow irrigation but a small area is grown with overhead irrigation. Overhead irrigation is used at Kilombero in Tanzania, on parts of the Ramisi, Chemelil and Miwani estates in Kenya and on the Kakira estate in Uganda.

Cane quality

Four factors are usually considered when determining cane quality:

1 *Brix*
This is the percentage of soluble solids in the juice. For millable cane the figure should be over 20.

2 *Pol*
This is the percentage of apparent sucrose in the juice. Mill operators want the Pol figure to be as near the Brix figure as possible; a Pol figure of 16 or more is generally acceptable.

3 *Purity*
This is the percentage of apparent sucrose in the soluble solids, i.e. $\dfrac{Pol}{Brix} \times 100$. For efficient operation most mills require a purity of 80% but some insist on a minimum of 84%. A low purity would mean a high proportion of impurities: mostly invert sugars, i.e. glucose and fructose. Invert sugars are undesirable because they remove sucrose with them in the separation process.

4 *Fibre content*
The higher the fibre, the lower the juice content. In addition, fibre retains some sucrose however thoroughly it is milled. The fibre content is relatively constant, and depends largely on the variety, although it is also affected by age, climate and agricultural practices. It is usually between 10% and 18%. Fibre content is not discussed in the following section.

Factors affecting cane quality

1 *The age of the cane*
Young cane has watery juice with a low Brix figure. The cane ripens from the base upwards until at maturity the quality is uniformly high throughout. After maturity the process of inversion begins: the disaccharide sucrose is inverted, with the help of the enzyme invertase, into the monosaccharides glucose and fructose which are usually referred to as invert sugars. This means that both Pol and Purity fall. Inversion begins at the base of the cane and works its way upwards.

2 *The part of the cane*
If the cane is immature the Brix figure is higher at the base than near the top. If the cane is over-mature the position is reversed (see 1). Even when mature, however, the green top is undesirable because it contains comparatively

large quantities of invertase; this ensures that the carbohydrates are kept in a mobile mono-saccharide form for growth. Immediately after cutting the base of the cane, invertase moves down the stem from the tops in order to mobilise carbohydrates for future growth; for this reason removal of the top must not be delayed after cutting the base.

3 *The length of time from cutting to milling*
Even if the top is removed immediately after cutting, the cane contains enough invertase for inversion to proceed gradually. A delay of 24 hours is the maximum that should be allowed, ideally, between cutting and milling; 48 hours is the usually accepted limit before serious loss of quality occurs.

4 *Burning*
Cane is burned intentionally before harvesting at Arusha-Chini in Tanzania and on some of the estates on the Kano Plains in Kenya. The main advantage of burning is that it makes harvesting less laborious because the trash is destroyed and the workers have only the unburned green tops to remove. Other points are that burning is essential when non-stripping varieties are grown and when mechanical loading is used to put the cane into the vehicles that transport it to the mills; without burning, mechanical grabs pick up an excessive amount of trash. The main problem involved is that very rapid inversion starts in the standing cane about 48 hours after burning. No loss is caused if the fires can be controlled and if a limited acreage can therefore be burned. If, however, fires get out of control and a large acreage is burned, growers are faced with the problem of cutting and trans-porting large quantities of cane within 48 hours. Labour and transport facilities seldom allow this, so large amounts of cane are cut late and are sold for a low price based on the reduced quality.

5 *Climate*
Cane quality is at its optimum under conditions of sunshine, low humidity, low night temperatures and low rainfall, i.e. conditions favouring maximum photosynthesis and minimum mobilisation of stored carbohydrate for physiological purposes.

6 *Arrowing*
Cane quality falls after arrowing. In some

varieties, e.g. B. 41227, it falls immediately after arrowing; in others, e.g. N:Co. 293, the drop in quality is delayed and the cane may be harvested safely as late as one month after arrowing. Some N:Co. varieties have a bad reputation for early arrowing; upon inspection, however, it is usualy found that fewer than 5% of the plants have arrowed.

7 *Variety*
The genetic constitution of the cane affects not only its inherent quality but also the speed at which quality declines after maturity. Co. 421, for instance, maintains its quality for several months after maturity whilst Co. 331 deteriorates almost immediately.

8 *Agronomic factors*
Agronomic factors such as fertiliser rates, moisture supply, drainage and depth of rooting can affect cane quality.

9 *Pests and diseases*
Some pests and diseases can affect cane quality, e.g. soil insects can remove many of the feeding roots and can thus cause droughting before the cane is mature.

Harvesting

Most of the recommendations concerning harvesting are guided by the quality considerations mentioned above. Cane must be cut when it is mature, as judged by uniformity of quality at the base and towards the top of the cane. Each cane must be cut as near ground level as possible. The green tops must be removed immediately (see Fig. 120). The dead leaves must be stripped off unless they have been removed by burning. The cane must be trans-ported to the mill as soon as possible and certainly not later than 48 hours after cutting; this rule is especially important when the cane has been burned.

All East African cane is cut by hand; 'pangas' are invariably used. In some countries cane is harvested mechanically but this will not be necessary in East Africa until there is a dramatic rise in the cost of labour.

Lorries, movable rail systems or tractors with trailers are used for transporting the cane from the field to the mill.

Post-harvesting operations

If a ratoon crop is to be taken, the trash must be

Fig. 120: Cutting sugar cane.

irrigation under sub-optimal rainfall, 45 tons of cane per acre (c. 110 t/ha) is considered a reasonable yield for a plant crop. At the Arusha-Chini estate in Tanzania, where irrigation is practised, the average plant crop yield is 75 tons of cane per acre (c. 190 t/ha) and as many as 120 tons per acre (c. 300 t/ha) have been recorded. First ratoon crops should yield 35 tons of cane per acre (c. 90 t/ha) on the Kano Plains and second ratoon crops 20 tons per acre (c. 50 t/ha) if standards of husbandry are reasonable. In areas where conditions are more favourable, ratoon crops should give 50–60 tons of cane per acre (c. 125–150 t/ha).

Owing to the great variability of growing period, a comparison of crop yields between different areas may be misleading. Hence sugar experts usually refer to production as tons of cane per acre per month (or other standard units of time). Yields per acre per month, under reasonable standards of husbandry, vary in East Africa from 2–8 tons of cane in the plant crop and 1–5 tons of cane in the ratoon crops (c. 5–20 t/ha/month in the plant crop and 2·5–13 t/ha/month in the ratoon crops).

A more meaningful criterion than tons of cane per acre is tons of sugar per acre. The sucrose percentage in cane ranges from 9–11% in poor quality varieties to 13–15% in good quality varieties. Yields of 15 tons of sugar per acre (c. 38 t/ha) are therefore possible from a very heavy plant crop but yields of 4–8 tons per acre (c. 10–20 t/ha) are more common.

Pests

Pests are seldom a major problem in sugar cane. Some of those most commonly found are mentioned below.

White scale. *Aulacapsis tegalensis*
This pest, which may be confused by the layman with a comparatively harmless mealybug, forms a mat of scales beneath the leaf sheaths. In this position it is protected from contact insecticides; it is, in any case, extremely resistant to insecticides, even systemics. White scales cause serious damage only at the Kenya coast, where they cause severe wilting during droughts. No economic control measures are currently available except stripping the drying leaf sheaths in order to expose the scales to insect predators and parasites.

White grub
Cochliotis melolonthoides causes damage at Arusha-Chini in Tanzania whilst *Schizonycha spp.*

raked into the inter-rows. This leaves the soil bare along the cane lines for fertiliser application and allows the sun to warm the soil around the stubble and so encourage germination; it also leaves the stools clear for stubble shaving. Unless the stools are trimmed down to ground level, buds may develop above the ground at the base of the old canes. Such development is undesirable because the new shoots would have to rely solely on the old root system; for healthy ratoon growth it is essential that buds develop below ground level and grow their own roots. Some estates subsoil or rip between alternate rows before each ratoon; the trash is piled into alternate inter-rows to allow this. When the cane is burned there is no trash and mechanical cultivation in each inter-row is usually practised.

After the last ratoon crop the trash is usually burned so that the soil surface is left bare for subsequent cultivations. The loss of organic matter due to burning is not very important because trash and tops are far less by weight than the stools; these are chopped up and incorporated and are not affected by fire.

Yields

Yields vary from one area to another. In the Kano Plains area in Kenya, where cane is grown without occur on the Kenya coast. White grubs live in the soil and feed on sugar cane roots. They can be controlled to a limited extent by BHC, dieldrin or aldrin sprayed onto the walls of the planting furrows and onto the setts. Resistance to dieldrin has recently been reported from Arusha-Chini.

Heteronychus andersonii and *H. licas.*
The adult beetles attack the base of the shoots below the ground and the larvae attack the roots. Damage is seldom serious enough to justify chemical control by soil applied insecticides; when it is, the treatment is the same as for white grub.

Termites. *Pseudocanthotermes militaris*
Termites may attack sugar cane setts. They are easily controlled by dipping the setts in an insecticide solution before planting. Termites occasionally attack mature cane during a drought; there is little that can be done about this on a field scale except to improve the moisture status of the soil.

Diseases

There are several diseases, e.g. Fiji disease and gumming disease, which cause extensive damage in other parts of the world but which have not yet appeared in East Africa. Strict quarantine regulations and the careful screening of introductions at Muguga in Kenya are designed to protect the East African sugar industry from such diseases. Quarantine regulations are occasionally contravened; the introduction of smut in the late 1950s was probably due to cane smuggled from the Congo.

Ratoon stunting disease
This disease is endemic in East Africa and is caused by a virus. There are no overt symptoms other than a decrease in vigour and a decline in yields, both of which are especially noticeable in ratoon crops. Red spots can sometimes be seen in the vascular tissue, especially in the nodes.

Ratoon stunting disease can be spread not only by infected planting material but also during cutting; if a 'panga' is used to cut one diseased stool it can spread the virus to a great many stools which are cut subsequently. Sterilising 'pangas' between each cut is impracticable although they are sometimes sterilised at the end of each row when cutting a nursery for planting material.

The amount of virus in the planting material can be reduced, although it is unlikely to be completely eliminated, by heat treatment. Heat treatment involves immersing the setts in water for two hours at 122°F (50°C). The temperature must be precisely controlled because lower temperatures fail to kill the virus whilst higher temperatures kill the setts. Even when the temperature is kept at exactly 122°F there is a certain reduction in the germination percentage; the reduction depends on the variety being treated. Heat treated cane cannot be planted straight into commercial fields for two reasons: its germination is likely to be poor and so much cane would be involved that the treatment would be prohibitively expensive. Heat treated setts are therefore planted in nurseries which are harvested about a year later to provide planting material for the commercial fields. Varietal resistance to ratoon stunting disease is unknown, although its effect is more serious in some varieties than in others.

Smut
This disease, caused by the fungus, *Ustilago scitaminea,* was first officially diagnosed in East Africa in 1956. There was a severe smut epidemic in the early 1960s and several susceptible varieties, e.g. Co. 419, were withdrawn. The main symptom of smut is a long whip-like structure with a fibrous stem, covered in black spores and contained in a silver coloured skin; this emerges at the top of the stem in place of the inflorescence. Accompanying symptoms are stunting and the production of thin, horizontal leaves.

Smut is spread by infected planting material and by spores which infect young shoots. Resistant varieties are now grown throughout East Africa, although whips are occasionally found even on these. The heat treatment recommended for ratoon stunting disease kills the smut fungus and ensures healthy planting material.

Mosaic
Mosaic is a virus disease which causes stunting and a characteristic yellow mottling of the leaves. It is spread by infected planting material and by aphids. Mosaic is prevented by growing resistant varieties and by rejecting planting material which shows any symptoms.

Yellow wilt
At the time of writing this disease is virtually restricted to the Kilombero estate in Tanzania. It is encouraged by conditions of imperfect drainage and may be caused by a virus. POJ. 2878 proved to be susceptible and has been replaced by other varieties.

Leaf spots

Several fungi cause leaf spotting on sugar cane but only two, *Cercospora longipes,* causing brown spot, and *Puccinia erianthi,* causing rust, do any economic damage. Brown spot is the more important and is common on the estates around the shores of Lake Victoria. At a tentative estimate it is causing a constant 5% loss on dry-farmed cane, while the loss on irrigated cane, where the main variety grown Co. 421, is somewhat susceptible, is nearer 10%. Damage from rust is less common because the main commercial varieties are resistant to it. Growing resistant varieties is the only practical method of controlling leaf spot diseases.

Milling

The cane is chopped and squeezed to express the juice. The juice is then led to the clarifier where fine suspended matter and soluble non-sugars are precipitated and separated as a dark coloured mud. The juice is next led to evaporators where it is boiled under reduced pressure until it becomes a thick syrup. In the next container, the vacuum pan, the concentration of sucrose in the syrup is so high and the pressure is so low that sucrose crystals are formed; the vacuum pan produces massecuite which is a dark brown mixture of molasses and sucrose crystals. The massecuite is next stirred in open tanks called crystallisers; in these the crystals are encouraged to grow. Finally the massecuite is led into centrifuges which separate the crystals from the molasses. Even this highly simplified account of the various stages in sugar manufacture gives some idea of the complexity of the process. The machinery is large, heavy and expensive and the construction of a large mill costs several million pounds.

By-products

Bagasse

Bagasse consists of the fibrous residue after the cane has been milled and the juice extracted. It is usually used as fuel for the factory boilers but in some countries it is used in the manufacture of fibreboard or paper.

Filter cake

Filter cake is the product resulting from filtration and washing of mud from the clarifier. It has a high calcium content and has approximately the same concentrations of nitrogen and phosphate as farmyard manure; it is sometimes applied to the cane fields nearest to the mills.

Molasses

Molasses is a heavy, dark brown liquid which consists largely of sucrose (the brown colour is caused by caramelisation of sucrose) and invert sugars. In East Africa the main uses of molasses are in the distilling of industrial alcohol and incorporation in livestock feeds.

Bibliography

1 **Burnett, G. F.** and **Leather, T. H.** (1967). *Applying insecticide to control soil pests in irrigated sugar cane.* E. Afr. agric. for. J., *32*:419.

2 **Clarke, R. T.** (1961). *A manurial trial on sugar.* E. Afr. agric. for. J., *26*:184.

3 **Dutt, A. K.** (1962). *Sulphur deficiency in sugar-cane.* Emp. J. exp. Agric., *30*:257.

4 **Hocking, D.** (1966). Cordyceps barnsii, *a fungal parasite of white grub in sugar cane.* E. Afr. agric. for. J., *32*:75.

5 **Park, P. O.** and **McKone, C. E.** (1966). *The persistence and distribution of soil-applied insecticides in an irrigated soil in Northern Tanzania.* Trop. Agriculture, Trin., *43*:133.

6 **Robinson, R. A.** (1959). *Sugar cane smut.* E. Afr. agric. J., *24*:240.

7 **Sheffield, F. M. L.** (1956). *Ratoon stunting— a degeneration disease of sugar cane.* E. Afric. agric. J., *22*:70.

8 **Sheffield, F. M. L.** (1959). *Sugar cane importation.* E. Afr. agric. J., *25*:16.

9 **Sheffield, F. M. L.** (1962). *Infectious diseases of sugar cane.* E. Afr. agric. for. J., *27*:207.

10 **Waller, J. M.** (1967). *Varietal resistance to sugar cane smut in Kenya.* E. Afr. agric. for. J., *32*:399.

31
Sunflower

Helianthus annuus

Introduction

In East Africa sunflower is only important in Kenya where it is grown by large scale farmers in Trans Nzoia District and by small scale farmers in Western Province, notably in Kakamega District. It is grown to a smaller extent in Coast Province.

Sunflower is a potential oil crop; the oil content of the seeds is 25% to 50% depending on the variety. During the 1960s, however, most of the Kenyan production was exported to the U.S.A. as food for caged birds. Other uses include grinding the dried heads as a cattle feed, ensiling the young plants or ploughing them in as a green manure.

Fig. 121: Giant black sunflowers.

Plant characteristics

Sunflowers are annuals which grow from 2–15 ft (c. 0·6–4·5 m) high, depending on the variety. They have strong tap roots with dense surface mats of feeding roots. The stems are seldom branched and bear large ovate leaves. The flowers are often more than 1 ft (c. 0·3 m) in diameter and have yellow petals. Plants flower three to four months after planting and take a total of $3\frac{1}{2}$–6 months to mature, depending on the variety. There is a high percentage of cross-pollination.

Ecology

Sunflowers are very drought resistant, possibly because of their deep tap roots, and grow well in areas which receive an annual rainfall of 30 in (c. 750 mm) or more. For best yields they need a reasonable rainfall during the three or four weeks that coincide with flowering. It is important that there should be dry weather during ripening otherwise the heads rot. Sunflowers can be grown from sea level up to 8 500 ft (c. 2 600 m). Any soil that will produce a good crop of maize is suitable for sunflower.

Varieties

The varieties grown in Kenya are tall and have an oil content of only about 28%. In Russia varieties have been bred with an oil content as high as 50%. The names of the Kenya varieties refer to the colour of the seeds, i.e. dark stripe, grey stripe and white; the last fetches the highest price because it is best for bird food. High yielding hybrids with a high oil content have been produced in a breeding programme at Kitale in Kenya. These have not been released at the time of writing but have given promising results in trials.

Field operations

Time of sowing
In Trans Nzoia District in Kenya the best time for

sowing is from mid-June to mid-July. This allows enough rain at flowering yet the crop can ripen in dry weather.

Sowing
Sunflower can be sown with a maize planter.

Spacing
Spacing trials carried out at Kitale suggest that the optimum plant population is from 13 000 to 17 500 plants per acre (c. 32 000–42 000 plants per hectare). These populations can be achieved by spacings of 30 × 16 in and 30 in × 12 in (c. 0·75 m × 0·4 m and 0·75 m × 0·3 m).

Fertilisers
150 lb of single superphosphate per acre (c. 170 kg/ha) applied to the seedbed has given economic yield increases on most soils in Rift Valley Province, Kenya.

Weeding
It is necessary to weed sunflower only until the crop is about 3 ft (c. 0·9 m) high; after this stage weeds are suppressed by shading. Lodging is discouraged if soil is drawn up around the stems whilst weeding.

Harvesting

Harvesting presents three main problems: the local varieties of sunflower ripen unevenly so more than one operation is usually needed; the crop is very prone to shattering and bird damage can be severe, especially if the heads face upwards. To prevent shattering and bird damage, and to encourage rapid drying, the plants should be cut when the disc florets turn brown and the backs of the heads turn from green to yellow. They can be stooked, or each head can be impaled, upside down, on its stem: for this diagonal cuts are made near the bases of the stems. Threshing is usually done by beating individual heads with sticks; a maize sheller can also be used. The moisture content of the heads should be reduced to 14% for safe storage.

In other countries short stalked varieties of sunflower are harvested by combines. There is no experience, as yet, of mechanical harvesting in East Africa.

Yields

Average yields in Trans Nzoia District are 800 to 1 000 lb of seed per acre (c. 900–1 100 kg/ha).

In the future improved varieties and good husbandry should give about 2 000 lb per acre (c. 2 200 kg/ha).

Pests

Birds are the main pests. The American bollworm, *Heliothis armigera,* sometimes damages the developing seeds and DDT sprays may be necessary if there is a heavy infestation.

Diseases

Rust
This disease is caused by the fungus *Puccinia helianthi.* It is the most serious disease of sunflowers. Small lesions on the leaves produce red pustules on the lower surfaces. Until new varieties are bred with rust resistance the only precaution that can be taken against rust is to destroy all crop residues at the end of the year.

Leaf spot and white blister
These are caused, respectively, by the fungi *Septoria helianthi* and *Cystopus tragopogonis.* They cause, respectively, brown leaf spots and white leaf blisters. Their incidence should not be too high if crop residues are destroyed and if sunflowers are grown in a rotation with cereals and grasses.

Root and stem rot
These are caused by *Sclerotinia spp.* The main symptom is wilting; the stems and heads contain a black fungus. Healthy seed, crop rotations and destruction of crop residues should prevent severe outbreaks.

Bibliography

1 **Hill, A. G.** (1947). *Oil plants in East Africa.* E. Afr. agric. J., *12*:140.

2 **Childs, A. H. B.** (1948). *Sunflower production in the Iringa District.* E. Afr. agric. J., *14*:77.

3 **Weiss, E. A.** (1966). *Sunflower trials in Western Kenya.* E. Afr. agric. for. J., *31*:405.

32
Sweet potatoes

Ipomea batatus

Introduction

Sweet potatoes and cassava are the most important root crops in East Africa. Sweet potatoes are the more widely distributed of the two; they are grown in all areas below 7 000 ft (c. 2 100 m) provided they do not experience such a long dry period that the vines cannot survive from one growing season to the next.

Where standards of smallholder stock husbandry are high, e.g. in Central Province in Kenya and on the slopes of Mt. Kilimanjaro in Tanzania, sweet potato vines are an important animal feed, especially during the dry season because they are very drought resistant. In these areas many of the varieties produce large amounts of vine and few tubers and are grown specifically for stock feed.

Fig. 122: A field of sweet potatoes.

Plant characteristics

The sweet potato plant is a perennial vine, although it is often treated as an annual crop. Each plant produces many trailing stems which seldom rise more than 18 in (c. 45 cm) above the ground. Where nodes touch the ground or when they are buried during weeding they may produce root tubers although these are usually much smaller than those from the buried end of the planting material. Leaf shape, tuber shape, tuber colour and tuber: vine ratio all differ widely from one variety to another. Most East African sweet potato varieties flower freely and it has been shown recently in Uganda that seedlings often grow in the field.

The first tubers can usually be harvested about 4–5 months after planting. The duration of the crop depends on the variety, the climate and local custom. In Uganda, Tanzania and the western parts of Kenya sweet potatoes are seldom left in the ground for longer than a year; in Central Province in Kenya, where the rainfall distribution is favourable, they usually remain in the ground for two years, sometimes for as many as six.

Ecology

Rainfall, and water requirements
Sweet potatoes are very drought resistant; the vines remain green and healthy during severe droughts although tuber growth is negligible. Sweet potatoes grow well in areas which have an average annual rainfall of 30 in (c. 750 mm) or more. Where there is a long dry period, e.g. in most of Tanzania, they are grown as an annual and are replanted near swamps, rivers or seepage areas during the drought. Where the rainfall is high and well distributed, e.g. in Central Province in Kenya, they are usually a perennial crop.

Altitude and temperature
Sweet potatoes grow well both in the warm and the cool areas of East Africa. They are widely

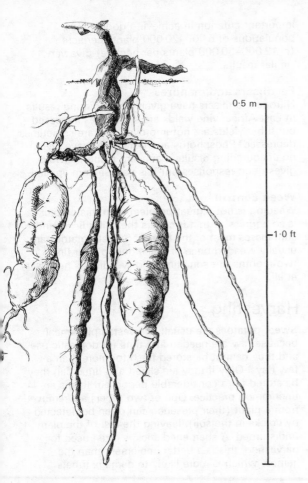

0·5 m

1·0 ft

Fig. 123: The underground structure of a sweet potato plant.

Varieties

There are so many names of varieties, each of which may be applied to a number of similar clones, that they cannot be mentioned individually here. There are great varietal differences, not least in yielding capacity, as has been shown in the collection of over 2 000 clones at Ukiriguru in Tanzania. There is great scope for selecting, breeding and issuing improved varieties.

One of the problems of sweet potato growing in East Africa is that varieties often degenerate over the years; this is well recognised by the growers. The cause of this degeneration is suspected to be virus diseases. Seedlings are thought to be the source of new varieties.

Rotations

When sweet potatoes are planted in a pure stand, as they usually are, they may be grown more or less continuously on the same plot or they may be grown in a crop rotation. Examples of continuous cultivation occur on steep slopes (where sweet potatoes cover the ground and prevent erosion better than most other crops), on small plots near the homestead (conveniently situated for ease of transport), along roadsides (especially true in parts of Kenya where local bye-laws prohibit growing tall crops such as maize directly alongside roads) and in valley bottoms.

When sweet potatoes are grown in a crop rotation they are often grown as the first crop after clearing the land from grass, e.g. in Uganda and the western parts of Kenya; they cover the ground quickly and suppress much of the grass that has not been killed during clearing. In Central Province in Kenya, on the other hand, sweet potatoes are often grown on land which has been exhausted by previous crops of maize and beans.

Field operations

Land preparation
Methods of land preparation differ widely in East Africa. In Uganda and the lakeshore areas of Kenya sweet potatoes are often grown on mounds; the mounds are about 3 ft (c. 0·9 m) apart and about four cuttings are planted on each. In Central Province in Kenya they are almost always grown on the flat. Ridges are the general rule in Tanzania and are sometimes seen in Uganda and western Kenya. Mounds and ridges provide ideal soil

grown from sea level up to 7 000 ft (c. 2 100 m) and some are found as high as 8 000 ft (c. 2 400 m). They therefore replace cassava, which is seldom grown over 5 000 ft (c. 1 500 m), at the higher altitudes.

Soil requirements
No details of the soil needs of sweet potatoes are available in East Africa. They are grown in a wide variety of soils, from swamps to eroded areas. Judging from the good responses which they show to farmyard manure, sweet potatoes need fertile soils for good yields.

conditions for tuber expansion. Where sweet potatoes are grown on the flat the soil is usually sufficiently friable and well drained to make mounds or ridges unnecessary.

Time of planting

Little experimental evidence is available to suggest the best time for planting sweet potatoes. Growers very seldom give them priority over cereals or important cash crops such as cotton. They plant at any convenient time when there is sufficient soil moisture for establishment.

Planting material

Propagation by seed is possible but is only likely to be used by breeders; vegetative propagation is the only method practised by smallholders. Throughout East Africa apical pieces of vine, i.e. the pieces furthest from the base of the plant, are preferred and are taken from mature plants. Local custom and individual preference dictate the length of the planting material; pieces may be anything from 9 in to 3 ft (c. 23 cm–0·9 m) long. When planting material is in short supply it may be taken from the middle or even the basal part of vines. In some parts of East Africa, e.g. Uganda, the lakeshore areas of Kenya and the Southern Highlands of Tanzania, the planting material is commonly wilted for a few days before planting to encourage root initiation; it may be covered with a layer of soil or vegetation or may be kept in the shade with no direct cover. In other parts the cuttings are planted without wilting.

Planting

Cuttings are planted at any angle with $\frac{1}{2}$–$\frac{2}{3}$ of their length, including the cut end, buried in the soil. Long cuttings, however, are usually planted almost flat with the end pointing up and out of the soil at an angle. Occasionally, as is sometimes done in Central Province in Kenya, cuttings are planted with both ends exposed and only the middle part buried.

Spacing

Sweet potatoes are usually planted in pure stands or with perennial crops. They spread so rapidly that they are seldom interplanted with annual crops and when they are, planting is usually delayed until the other crops are well established. When pure stands are planted on the flat a random spacing is adopted. When grown on ridges the ridges are 3–5 ft apart (c. 0·9–1·5 m) and the vines are planted 1–2 ft apart (c. 0·3–0·6 m) along them. Experiments have shown that spacing is not an important criterion in achieving good yields; populations of 5 000–20 000 plants per acre (c. 12 000–50 000 plants per hectare) give very similar results.

Fertilisers and manures

Nitrogen fertilisers have given disappointing results in East Africa; vine yields are sometimes increased but tuber yields are not improved and are sometimes depressed. Phosphorus and potassium have given no encouraging results. Farmyard manure always gives good responses but it is very seldom applied.

Weed control

Weeding is not a great problem with this crop; it is so aggressive that it covers the soil quickly and suppresses most of the weeds. Sweet potatoes are usually weeded once or twice during their first two months but sometimes no weeding is necessary at all.

Harvesting

Sweet potatoes are usually harvested piecemeal because few are needed at a time for domestic use and they cannot be stored fresh for more than a few days. Only if they are sliced and dried can they be stored for a considerable period but this is an uncommon practice. One or two tubers are removed from a plant (their presence can often be detected by cracks in the soil), leaving the rest of the plant undisturbed. A sharpened stick is often used for harvesting; this is a better implement than the 'jembe' which is more likely to damage tubers.

Yields

There are few figures available on average yields but a recent estimate for Tanzania suggests that they are as low as 1 ton of fresh tubers per acre (c. 2·5 t/ha). 5–7 tons per acre (c. 12·5–17·5 t/ha) should be obtained on fertile soils and with good husbandry. 15 tons per acre (c. 38 t/ha) would be considered an exceptionally good yield, although over 20 tons per acre (c. 50 t/ha) have been obtained from experimental plots.

Pests

Sweet potato weevils

These are undoubtedly the most damaging pests of sweet potatoes in East Africa. There are two species, *Cylas formicarius* and *C. puncticollis;* the former is the more common. The larvae tunnel inside vines and tubers causing discolouration and bitterness in the tubers and dwarfing and yellowing of the vines. Crop rotation and dipping planting material in a DDT solution help to reduce weevil damage although neither eliminates it.

Diseases

Sweet potato virus B causes the only disease that poses a great threat to sweet potatoes. It causes stunting, excessive branching and yellowing of the vines and cork symptoms may develop in the tubers. Virus B is spread by white flies (*Bemisia spp.*). Varieties differ widely in their susceptibility to virus and any sweet potato breeding programme will have to concentrate largely on virus resistance. Another virus, sweet potato virus A, which is transmitted by aphids, is of very little significance except in western Tanzania.

Utilisation

The most common method of preparing sweet potato tubers is to boil them unpeeled; they are then peeled immediately before eating. An alternative, which is popular in Central and Eastern Provinces in Kenya, is to roast them unpeeled in the ashes of a fire. When cooked in either of these ways, sweet potatoes are eaten alone or with tea, milk or 'uji'; other types of food are very seldom included. A less common method of cooking is to peel and slice the tubers before boiling or frying with other root crops, cereals, legumes or vegetables. In a few areas, e.g. Nyanza in Kenya and western Tanzania, sweet potatoes are peeled, sliced, dried, and then ground into a flour with sorghum or maize for making 'ugali' or 'uji'. Tubers are occasionally eaten raw but this is not an important method of consumption.

Bibliography

1 **Aldrich, D. T. A.** (1963). *The sweet potato crop in Uganda.* E. Afr. agric. for. J., *29*:42.

2 **Ingram, W. R.** (1967). *Chemical control of sweet potato weevil in Uganda.* E. Afr. agric. for. J., *33*:163.

3 **MacDonald, A. S.** (1963). *Sweet-potatoes, with particular reference to the tropics.* Field Crop Abstr., *16*:219.

4 **MacDonald, A. S.** (1965). *Variation in open pollinated sweet potato seedlings in Buganda, East Africa.* E. Afr. agric. for. J., *31*:183.

5 **Wheatley, P. E.** (1961). *Insect pests of agriculture in the Coast Province of Kenya, III—Sweet potato.* E. Afr. agric. for. J., *26*:228.

33
Tea

Camellia sinensis

Introduction

Tea was first planted in East Africa in 1900 at Entebbe in Uganda. In 1903 small acreages were planted in Limuru (Kenya) and Amani (Tanzania). The growth of the tea industry was at first slow and by 1925 there were only 325 acres (c. 130 ha) in East Africa. In that year, however, a rapid expansion of estates began when international tea companies bought large acreages of land, particularly in the Kericho area in Kenya. This expansion was checked by the depression of the 1930s and by the second world war, but was renewed in the post-war period. In 1945 Kenya had only 16 000 acres (c. 6 500 ha) of tea, but by the end of 1969 it had approximately 90 000 (c. 35 000 ha). Between these years Uganda's tea acreage grew from 3 500 to 37 000 (c. 1 400 to 15 000 ha) and Tanzania's grew from a very low figure to 28 000 (c. 11 000 ha).

Kenya

In most years during the 1950s and 1960s tea has been Kenya's second most valuable export crop after coffee. The main tea area is Kericho where there is a strip of estates about twenty five miles (c. 40 km) long. There are also estates near Sotik, Nandi Hills, Endebess, Subukia, Limuru, Makuyu and Meru. The smallholder acreage, which totalled 36 000 (c. 14 500 ha) in 1969 and which is planned to reach 70 000 (c. 28 000 ha) in 1973, is distributed between Kisii, Kericho, Mandi, Kakamega, Kiambu, Murang'a, Nyeri, Kirinyaga, Embu and Meru Districts, with a small acreage in the Cherengani foothills. Smallholder tea, which has been more successful in East Africa than in any other part of the world, was initiated in Kenya in 1950 when a few acres were planted in Mathira Division in Nyeri District. The development of Kenya's smallholder tea has been closely supervised; the result is that yields and quality, especially the latter, have been excellent. Kenya's production of made tea, from both estates and smallholdings, was approximately 80 million lb (c. 36 million kg) in 1969.

Uganda

Uganda's main tea growing area is in the west, centred around Fort Portal, although quite widely dispersed. Tea is also grown near the northern shores of Lake Victoria and near Mityana, Uganda's smallholder tea totalled 9 000 acres (c. 3 600 ha) in 1969; the total tea acreage was 36 500 (c. 14 500 ha). In that year Uganda produced approximately 39 million lb (c. 18 million kg) of made tea.

Tanzania

Much of Tanzania's tea is produced near Mufindi on the eastern facing edge of the Southern Highlands. Tea is also grown near Njombe and Tukuyu in the Southern Highlands, in the Usambara Mts and at Bukoba on the western shores of Lake Victoria. In all these areas tea is grown both on estates and smallholdings; the smallholder acreage was approximately 4 800 (c. 1 950 ha) in 1969 from a total tea acreage of 27 500 (c. 11 000 ha). Tanzania produced only 19 300 000 lb (c. 8 800 000 kg) of made tea in 1969; her production is lower than Kenya's, partly because of the cool, dry weather which restricts growth in the southern part of the country for some of the year.

Plant characteristics

Roots

It has been stated by some authorities that tea is a shallow rooted crop. This is not true of East African tea for each plant usually produces several deep roots, some of which have been traced as deep as 20 ft (c. 6 m) below the soil surface (see Fig. 124). Most of the roots, however, are found in the upper three feet (c. 0.9 m). The roots of any crop adapt themselves to the soil conditions; in parts of Assam, in India, for instance, tea grows

Fig. 124: The root system of a tea bush. For clarity, the roots have been painted white. Strings have been placed at 1 ft (c. 0·3 m) intervals.

amongst rice fields with its roots confined to the 1 ft (c. 0·3 m) layer of soil which lies above the water table.

Stems
If left unpruned, the stems of Assam varieties grow upwards to form a single trunk; the height of a mature tree is usually about 30 ft (c. 9 m) (see Fig. 126). China varieties, however, produce many main stems even when unpruned; mature bushes seldom exceed 15 ft (c. 4·5 m) in height. When pruned, both Assam and China varieties branch profusely and form a dense bush; a flat top or 'plucking table' is encouraged (see Fig. 133) and pruned regularly to prevent it growing out of reach of the pluckers.

Flowers
China varieties flower prolifically; flowers can also

often be seen in fields of Assam varieties derived from seedlings, as there is invariably a proportion of plants which are nearer the China type. Assam varieties only produce large quantities of flowers on mature trees which have been left unpruned. The flowers are white and sweet smelling; each produces a capsule which usually contains three seeds. The seeds are brown, about ½ in (c. 1·3 cm) in diameter, spherical or, if two seeds only have developed, hemispherical; they have tough seed coats.

Leaves
The leaves of Assam varieties are wide, positioned horizontally or pointing slightly downwards and pale green; they have a glossy upper surface and are bullate; i.e. there is a series of bulges on their upper surface. The leaves of China varieties, on the other hand, are narrower, shorter and darker green; they point upwards and have a dull, flat surface.

Ecology

Rainfall, and water requirements
Although tea tolerates dry spells, it only gives continuous flush growth when there is adequate soil moisture throughout the year. In long dry spells, in the absence of irrigation, flush growth ceases, the bushes wilt and eventually defoliate. In southern Tanzania there is often a six month dry period; the first part of this occurs from June to August when the weather is so cool and misty that the bushes suffer little because evapotranspiration is low; from September to November, however, the weather is hot and sunny and at this time the crop suffers considerably. In Kenya the dry season occurs at the beginning of the year and the crop soon suffers in the hot bright weather and the low humidity.

Mature tea requires a minimum of 4 in (c. 100 mm) of rainfall or irrigation water per month; in some areas, however, this figure may be as high as 6 in (c. 150 mm). The minimum average annual rainfall is sometimes quoted as being 55 in (c. 1 400 mm) but tea is grown successfully at Limuru in Kenya with an annual average of only 50 in (c. 1 250 mm); in this area evapotranspiration is restricted for several months in the middle of the year by mists and low cloud. Most tea in East Africa is grown in areas that receive a well distributed annual rainfall which averages 60–70 in (c. 1 500–1 750 mm).

Hail
Much damage is done by hail at Kericho and Nandi

Hills in Kenya. Before hail suppression measures were taken annual losses from the Kericho estates were estimated as being two million pounds (c. 900 000 kg) of made tea. At the time of writing hail storms in the Kericho and Nandi Hills area are located by radar; aeroplanes equipped with silver iodide flares are directed to them and fly around inside them. This results in a reduction in the damage done by hail because much of it falls in a soft form which does little harm to the foliage.

Altitude and temperature

Tea needs an acid soil and high rainfall; both of these tend to occur at high altitudes in East Africa. In Kenya tea is grown between 5 000 ft and 7 400 ft (c. 1 500–2 200 m). Above 7 400 ft Assam varieties are liable to frost damage; below 5 000 ft the rainfall is usually inadequate. In Uganda and Tanzania, however, tea is grown successfully at 4 000 ft (c. 1 200 m) near the shores of Lake Victoria because there is sufficient rain.

Growth is faster and yields tend to be higher in warmer areas and at warmer times of the year, e.g. as the dry weather begins in December or January in Kenya. Flavour, however, tends to be inferior in the warmer areas; it increases during periods of slow growth, e.g. at high altitudes or during cold seasons.

Shade

Planting shade trees was a practice which proved to increase tea yields in India and Ceylon before the introduction of nitrogenous fertilisers. It was adopted as a standard practice by the early tea planters in East Africa (see Fig. 125), many of whom had gained their experience in these countries. In the absence of nitrogenous fertilisers, shade of any kind, whether planted or artificial, can still increase tea yields, probably by limiting photosynthesis so that it matches the limited nutrient supply. With moderate applications, e.g. 40–80 lb per acre per annum (c. 45–90 kg/ha), shade has little effect on yield. At heavier rates, however, e.g. 120 lb (c. 135 kg/ha) or more, shade reduces yields by 10–25% and quality is also reduced. Shade is no longer recommended because heavy nitrogen applications are now standard practice. No shade trees are established in newly planted tea, and estates in all parts of East Africa are reducing their shade (mostly *Grevillea robusta*) in their mature tea; some are doing this quickly whilst others are doing it gradually or partially.

Fig. 125: Shaded tea. The plucking table is not yet fully formed because the bushes are still young as can be seen by their height in relation to the pluckers.

Wind

Wind can do much damage on eastern facing slopes, e.g. at Mufindi in Tanzania and in most of Central Province in Kenya. In these areas wind breaks are essential to prevent the high evapotranspiration and water stress which can occur in unprotected tea. The recommended plants for windbreaks are *Hakea saligna* and tea itself. Tea is comparatively slow growing but if one row is left unpruned it will eventually grow up into a satisfactory barrier. Some pine species must be avoided because their needles can drop onto the the crop and taint the leaves. *Eucalyptus* species compete so strongly for water that no tea should be grown near them; their leaves can also taint tea.

Windbreaks must be planted in straight lines, otherwise the wind is funnelled towards certain points, causing great turbulence on the other side. Gaps in windbreaks cause a similar funnelling effect. Windbreaks must be planted at right angles to the direction of the prevailing wind. The protection afforded by a windbreak extends approximately as far as ten times its height. Thus *Hakea saligna*, which grows to a height of 20–25 ft (c. 6–7·5 m), should be planted in rows 200–250 ft (c. 60–75 m) apart.

Soil requirements

A deep, well drained soil with a good water retaining capacity is essential because tea is a deep rooted crop and it requires an uninterrupted supply of water. The minimum soil depth is usually quoted as being 6 ft (c. 1·8 m).

Most tea is grown on soils which have a pH of between 4·0 and 6·0. Above pH soils tend to have too much calcium; comparatively small quantities of this element can suppress the uptake of potassium, thus preventing healthy growth. An additional disadvantage of a high pH is that it discourages early root growth when stumps are used as planting material. Tea grows successfully at Makuyu, in Kenya, with a soil pH of 7·0 provided that it is established by using sleeved plants rather than stumps. The soil in this area is unusual because it has a low calcium/potassium ratio even though the pH is high.

Some tea in East Africa is grown on soils with a pH as low as 3·6. The danger in these conditions is that phosphate becomes locked up in unavailable forms; this is particularly true in laterised soils but is less so if there is a reasonable amount of organic matter in the soil.

Varieties

Assam and China varieties
Assam and China teas, whose structural characteristics have been described above, have been naturally cross-pollinated for so long (since the middle of the 19th century) that there is now virtually no pure Assam or China seedling tea grown. One can only refer in general to Assam variety and China variety; the latter are usually called China hybrids. In any seedling population of either type it is possible to find a small proportion of plants which conform more to the other type. China hybrid is grown widely in Darjeeling in India where it produces a distinctively flavoured and highly valued tea. In East Africa, however, the ecological conditions prevent the full development of this flavour. Owing to its poor quality and to the problem of plucking the small leaves, the small proportion of China hybrid in East Africa is gradually being replaced by Assam tea.

Clones
During the 1960s the planting of vegetatively propagated, clonal material has increased to such an extent that at the time of writing seedling material is virtually never used except in Tanzania. In tea plants raised from seedlings there is great genetic variation between plants, as regards both yields and quality. On a field scale, maximum yields of best quality tea are therefore prevented if seedlings are used. Clones, on the other hand, are selected from the rare plants that combine both yielding ability and quality. Because vegetative propagation is used, each plant in a clone is genetically identical to the mother plant which was originally selected. A field of clonal material is therefore capable of giving higher yields of better quality tea than a field derived from seedlings.

Details of individual clones are not discussed in this chapter because they would soon be made irrelevant by the introduction of new clones. Details of clone selection are also omitted because this is a complicated and specialised procedure. Briefly, however, selecting clones involves the screening of many thousands of plants; in a seedling population only one plant in 100 000, as an approximate average, is suitable for selection as the mother plant for founding a new clone.

Seedlings and stumps

Owing to the decline of seedling planting material, this method of propagation is not discussed in detail in this chapter.

Seed is produced in 'baries', i.e. in special plots where the tea plants are allowed to grow unchecked (see Fig. 126). The spacing is usually 20 ft × 20 ft (c. 6 m × 6 m). After being collected from the ground, the seed should be subjected to a flotation test; all seed still floating after 24 hours should be discarded whilst all that sinks should be graded over a $\frac{1}{2}$ in (c. 1·3 cm) mesh. Tea seed loses its viability very quickly so it must be despatched as quickly as possible.

Germination is uneven, largely owing to the toughness of the seed coat. Pregermination is therefore necessary in order to ensure uniform growth in the nursery. The seed may be covered with wet sacks or laid on polythene sheets in the sun and sprayed with water several times a day or laid on beds of rotting vegetation which provide heat to stimulate germination. The seeds are inspected daily and those which have cracked seed coats or which have a little root showing are planted about 1 in (c. 2·5 cm) deep in the nursery at a 5 in (c. 13 cm) triangular spacing. Before planting the beds must be cultivated to a depth of at least $2\frac{1}{2}$ ft (c. 0·75 m) to encourage healthy root development. Overhead shade may be erected but in cloudy areas, e.g. around Mt. Kenya or in the Usambara mountains in Tanzania, nothing more is needed than bracken fronds around the seedlings during the early stages of growth. After about two years in the nursery the seedlings should be ready for transplanting; their stems should be $\frac{3}{8}-\frac{1}{2}$ in thick at ground level (c. 1·0–1·3 cm thick). They are lifted from the ground and the part which was above the ground is cut back to leave a 4 in (c. 10 cm) stump. To ensure that only the best material is planted, at least 25% of the seedlings should be discarded as being too short or thin or having too short or bent a root.

Seedlings can be raised in sleeves using the techniques used for cuttings (see below).

Vegetative propagation

Cuttings
Mother trees are allowed to grow for about six months after pruning, thus providing long stems for cuttings. Single leaf internode cuttings are usually used (see Fig. 127) although cuttings with two or three leaves grow more quickly under polythene and may become more popular in the future whenever there is an ample supply of material for cuttings. The top two or three internodes on each stem must be discarded because they are too short; so must any part towards the base of the stem which has flaky bark as opposed to smooth reddish-brown bark because cuttings from this part do not strike readily. Razor-sharp knives must be used for making the cuttings; the top cut must be made as near to the axillary bud as possible and sloping away from it; the lower cut must also be sloping and must allow for $1-1\frac{1}{2}$ in (c. 2·5–3·8 cm) of stem below the leaf. The leaf area must not be reduced nor must cuttings with damaged leaves be used. Application of rooting hormones to the base of the cuttings has not proved to be of any benefit. Cuttings must be kept shaded and wet, preferably floating in water, for the entire time from cutting to planting.

Rooting medium
Clones vary in their requirements for a rooting medium. Most, however, only take root when there is virtually no humus in the soil, when the pH is below 5·5 and when there is little clay or sand present. A friable, acid subsoil is therefore ideal. For each cubic yard of rooting medium, 1 lb (c. 0·45 kg) of single superphosphate and $\frac{1}{2}$ lb (c. 0·23 kg) of potassium sulphate should be incorporated.

Sleeves
Cuttings are always planted in polythene sleeves. The main advantage of these is that they allow a minimum of soil disturbance when the cuttings are transferred from the nursery to the field. Sleeves are

Fig. 126: A tea barie, i.e. a plot of trees left unpruned for seed production.

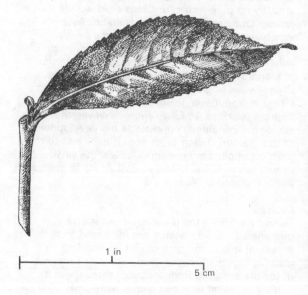

1 in

5 cm

Fig. 127: A cutting ready for planting.

Fig. 128: Planting tea cuttings in the nursery. Note the polythene sleeves and the walls of split bamboo. A polythene tent is positioned over the cuttings immediately after planting is completed.

Fig. 129: A sleeved cutting which has recently been planted in the field.

usually 10 in (c. 25 cm) long and 3 in (c. 7·5 cm) in diameter; they must be stapled or sealed only at one point at the bottom, thus allowing free drainage. The sleeves are sometimes filled to the top with rooting medium. The cuttings may grow best, however, if the sleeves are ⅔ filled with fertile topsoil and if about 3 in (c. 7·5 cm) of rooting medium is placed on top.

Planting
One cutting is planted in each sleeve. In most soils the stems can be pushed in without making any hole for them; only when gritty subsoils are used should holes be made before planting. The stems must sometimes be pushed in at an angle to ensure that the tips of the leaves do not touch the soil. They must not be pushed in so far that the petioles touch the soil.

Environment
Cuttings only grow well if they are shaded, if they are protected from the wind and if the humidity is high. Polythene covers are usually used and these are stretched over hoops above the cuttings; their edges are buried in the soil so that very little moisture escapes and the humidity is constantly high. The cuttings need only be watered once every three months provided that there are no holes in the polythene and that its sides are buried deeply. Shade is usually provided either by an overhead framework which supports a roof of dead vegetation or by a lath frame which closely surrounds the polythene tent (see Fig. 128). Hardening off can be done when the roots are about 4 in (c. 10 cm) long and shoots have developed; both the polythene cover and the shade are gradually removed (polythene first). Mist propagation, an alternative method of controlling the environment for cuttings, has been adopted by at least one company.

Duration in the nursery
Cuttings are ready for transplanting into the field when their roots reach the bottom of the sleeves and when their tops are about 8 in (c. 20 cm) high (see Fig. 129); this should take only 6–10 months in the nursery.

Mortality
If nurseries are well managed, no more than 10% of the cuttings should die.

Land preparation

During land preparation special care must be taken

to eradicate couch grass; during later stages this weed can seriously reduce yields and can be difficult to remove without damaging the crop. Precautions against *Armillaria* root rot are essential and these are discussed in the section on diseases.

On old hut sites, cattle 'bomas' or charcoal burning sites, the soil pH is too high for healthy tea growth. On such areas planting holes of double the normal dimensions should be dug and sulphur should be incorporated into the soil used for refilling the hole; the amount of sulphur needed depends on the soil pH. In cases of extreme alkalinity, however, applications of sulphur may need to be so great that they become uneconomic. If hut sites, etc. cannot be located before planting they are shown up later by patches of unhealthy tea bushes; in such places repeated applications of aluminium sulphate have sometimes proved successful in increasing the acidity of the soil.

The usual dimensions of planting holes for sleeved plants are 6 in (c. 15 cm) deeper and 6 in wider than the sleeves; for stumps the depth must be at least 2 ft (c. 0·6 m). A standard practice, proved to encourage early growth, is to include ½–1 oz (c. 14–28 g) of double superphosphate with the soil used for refilling each hole. A recent recommendation is to include some potassium sulphate and, in the case of grassland soils, di-ammonium phosphate instead of double superphosphate. Holes are dug as soon as possible before planting and, unlike coffee planting holes, are only refilled at planting time.

Planting

The staple is removed from the bottom of the sleeve and the sleeved plant is held in the middle of the planting hole; the top of the sleeve must be level with the surface of the soil in the field. The polythene is cut and removed, making sure that as little soil as possible is disturbed, and the hole is gradually filled with topsoil which must be compressed gently at regular intervals. Plants are sometimes shaded with bracken fronds, pieces of banana leaf or other suitable material, but this is unnecessary if the plants have been properly hardened off in the nursery. Planting as early as possible in the main rains gives the best results.

Spacing

Almost all tea is contour planted, although there is little justification for this if there is a broadcast cover crop or a mulch. The most common spacings are 5 ft × 2½ ft (c. 1·5 m × 0·75 m) and 4 ft × 3 ft (c. 1·2 m × 0·9 m). These give populations of approximately 3 500 plants per acre (c. 8 700 plants per ha), although this is low by modern standards in Assam where tea is now planted at up to 5 000 plants per acre (c. 1 230 plants per hectare).

Nurse Crops

Nurse crops are often grown between the rows of young tea; they may be established before or shortly after planting. A densely sown band of oats 6 in (c. 15 cm) wide prevents erosion and wind damage and provides mulching material. Broadcasting the nurse crop gives even better control of soil erosion. Other plants, e.g. lupins, are sometimes used; these are less convenient than oats but are satisfactory if they are prevented from growing too high and shading the tea.

Mulching

Mature tea is never mulched because the bushes themselves and their prunings give adequate protection to the soil. Young tea is often mulched, however; oats can be used if this has been sown as a nurse crop, or mulching material can be brought in from outside. Mulch should be applied along the inter-rows or in a ring around each plant; in neither case must it be allowed to touch the stems of the tea plants. Mulch reduces crop growth during wet weather unless nitrogen is applied because as it decomposes it locks up the natural supply of nitrogen in the soil.

Frame Formation (Bringing young tea into bearing)

During the first few years of the tea plant's life in the field it must be encouraged to grow laterally in order to provide a permanent, wide frame for the plucking table. The frame must be wide enough to allow a continuous plucking table with no gaps between bushes. There are two methods of frame formation: formative pruning and pegging.

Formative pruning
This is the traditional method; it relies on repeated cutting back to restrict upward growth and to encourage lateral growth.

When sleeved cuttings are used, each usually produces one dominant stem. This should be cut

Fig. 130: A plant ready for pegging. Note the number of stems which have been produced by pruning the main stem. The point of pruning can be clearly seen although it is a little lower than the 6 in (c. 15 cm) above ground level that is generally recommended. The plants shown in figs. 130 to 132 have been somewhat damaged by hail.

back to 6 in (c. 15 cm) above ground level when it reaches a height of 12–14 in (c. 30–35 cm). This encourages the growth of lateral shoots. When most of them are about pencil thickness (i.e. about $\frac{3}{8}$ in or 1 cm thick) at a height of 11 in (c. 28 cm) above the ground, they should all be cut back at that height. This procedure should be repeated when the next shoots reach a suitable thickness at a height of 16 in (c 41 cm) above ground level. These pruning operations occur at approximately 12 month intervals. Some clones are very free branching and may need only one pruning at an intermediate level.

The main disadvantage of formative pruning is that it provides such severe and repeated checks to the growth of the crop that plucking cannot commence until 2½–3 years after planting.

Pegging
This method of frame formation is becoming increasingly popular throughout East Africa. The first step is the same as in formative pruning, i.e. the stem of each plant is cut back to 6 in (c. 15 cm) above ground level when it has reached a height of 12–14 in (c. 30–35 cm). This stimulates the growth of several lateral branches on each plant and these are pegged down when they are 20–26 in long (c. 51–66 cm) (see Figs. 130 and 131). One peg is usually used for each branch but other methods have been tried, e.g. pegging down two rows of poles parallel to a line of plants, with one row of poles on each side of the line, using these to press the branches down. Great care must be taken during pegging to ensure that the branches radiate outwards from the stem and point slightly upwards. If they are horizontal or point slightly downwards most buds remain dormant; if they point upwards too steeply a low spreading frame is not achieved. If, on the other hand, a branch is correctly positioned, buds break along its length producing shoots which eventually form the foundation of the plucking table. After pegging, the terminal leaf and bud of each branch is removed.

The main advantage of pegging is that a wide, healthy frame is produced quickly because the plants are not continually subjected to the severe check of pruning. Pegged bushes therefore come into bearing more quickly than pruned bushes and a small crop can be taken in the second year after planting. Pegging produces an extra 1 000–2 000 lb of made tea per acre (c. 1 100–2 200 kg/ha) in the first 3–5 years of the crop's life when compared to formative pruning. If the plants are only pegged once (it has recently been found that this is all that is necessary) the labour involved is equivalent to that involved in two formative prunings.

Fig. 131: A plant which has recently been pegged. Note that the tips of the branches have been removed.

Plucking table formation

As soon as the frame has been formed the new shoots are allowed to grow up for about three months, after which they are checked by the procedure known as 'tipping' or 'tipping in'. Tipping in involves the removal of three leaves and a bud from each shoot when this amount of growth has appeared above the desired height of the plucking table (see Fig. 132). When frames are formed by pruning the initial plucking table should be 24 in (c. 60 cm) above ground level; when they are formed by pegging the frame is lower so the initial plucking table should be 20 in (c. 50 cm) above ground level. These heights allow a sufficient depth of maintenance foliage, i.e. the foliage below the plucking table. If there is less than 8–10 in (c. 20–25 cm) of maintenance foliage, photosynthesis is restricted and there are not enough carbohydrates to maintain an adequate reserve in the root system. If there is more than 8–10 in of maintenance foliage the plucking table becomes unnecessarily high.

Tipping in is done by hand rather than by using knives and great care must be taken to get the plucking table parallel with the ground. Wooden frames or Y-forked sticks are usually used to give an accurate guide to the tipping height. Pruned plants need to be tipped in at least three times at 2–3 week intervals whilst pegged plants need to be tipped in about five times; after this plucking can commence.

Plucking

Plucking is the harvesting of the tips of the shoots as they appear above the plucking table. A light straight stick, often a piece of bamboo, is usually laid across the top of the bushes in front of the plucker to give him a guide to the height of the plucking table (see Fig. 133). The leaf is transferred to a basket on the plucker's back. Leaf must not be compressed into the basket nor must it be collected in one hand whilst plucking with the other; both of these practices crush the leaf and rupture the surface, thus initiating fermentation. When the baskets are emptied and weighed, the leaf must be kept in the shade.

Frequency of plucking
The length of time between one plucking operation and the next, i.e. the plucking round, depends on a number of factors, e.g. whether the plucking is fine

Fig. 132: A pegged plant which has received its first "tipping-in". The production of upright stems from the pegged branches is clearly seen on the left.

Fig. 133: Plucking tea. Note the bamboo wand.

or coarse, hard or light (see below) or whether the bushes are in flush growth or not. If hard fine plucking is practised, as is the general rule, the plucking round should be 5–7 days in the flush season but only 10 days or so at slacker times of the year.

Fine and coarse plucking
Good quality tea can only be produced when fine plucking is practised, i.e. when two leaves and a bud are removed (see Fig. 134). At the time of writing almost all East Africa's tea is fine plucked because growers are conscious of the need to

maintain quality. Coarse plucking is the removal of more than two leaves and a bud. Its main advantage is that plucking can be done less frequently because it takes longer for three leaves and a bud to be produced above the plucking table. Coarse plucking was almost completely abandoned in East Africa during the late 1960s for reasons of quality.

Hard and light plucking

Hard plucking involves breaking off the tips of the shoots at exactly the height of the plucking table. If hard plucking is practised continuously there are two undesirable results: the maintenance foliage becomes gradually less productive and the plucking table becomes denuded of growing points, with an

Fig. 134: Two leaves and a bud.

Fig. 135: A recently pruned, mature plant..

accumulation of twigs, sometimes called 'cross feet'. To add some new leaf to the maintenance foliage and to create a healthier plucking table, the plucking table must be occasionally raised by light plucking. An example of light plucking is the removal of two leaves and a bud after three leaves and a bud have been produced above the plucking table. Light plucking must not occur too often, otherwise the plucking table would soon be out of reach; it is usually done only two or three times each year.

Breaking back

If tea is plucked too seldom, shoots with three or more leaves and a bud grow above the plucking table. There are then three alternatives for plucking; coarse plucking down to the plucking table, which could lead to rejection of the leaf; fine plucking which would maintain quality but which would raise the plucking table (i.e. light plucking) or breaking back, i.e. fine plucking followed by breaking off and discarding the part of the shoot which is still above the plucking table. Breaking back is obviously a wasteful practice; it must be avoided by short plucking rounds.

Banjhi shoots

After a shoot has produced several leaves, usually four or five, it becomes dormant or 'banjhi'. In the banjhi state, the bud is very short (compared to a bud in flush growth which is at least half the length of the top leaf; (see Fig. 134) and the top leaf gradually becomes hard. The tips of shoots usually appear above the plucking table before they go into the banjhi stage. When banjhi shoots appear above the plucking table they can be plucked and sent to the factory if the top leaf is still soft; only one leaf and a bud should be plucked. If banjhi shoots are overlooked when they are in the soft stage, they harden and they should then be plucked and discarded. Sometimes, when soil or climatic conditions or both are unfavourable, whole bushes go into the banjhi stage; at such times they should not be plucked until flush growth resumes.

Labour

Skilled workers can pluck $\frac{1}{5}$ acre (c. 0·08 ha) in a day.

Pruning

Reasons

At regular intervals, usually every three years but

sometimes every four, tea bushes must be pruned; all the maintenance foliage is removed and only the frames remain (see Fig. 135). This is to prevent an excessive decline in yields. During the pruning cycle, i.e. the time from one pruning operation to the next, yields usually decline after the second or third year. (In the first year yields are low because the bushes are recovering from pruning). The reason for the decline in yields is not fully understood. Without pruning, however, yields would fall so far that production would be uneconomic. An additional reason for pruning is to keep the plucking table within reach of the pluckers. During a pruning cycle the plucking table rises gradually because of the need for light plucking from time to time; after three or four years, pruning is necessary to bring it back to a manageable height.

Pruning height
If bushes were brought into bearing by formative pruning, they should be pruned three or four years later at 20 in (c. 50 cm) above ground level. Pegged plants should be pruned 3–5 years after planting at 16 in (c. 40 cm) and, after the next pruning cycle, at 20 in (c. 50 cm). All subsequent prunings, whether for pegged plants or otherwise, should be 2 in (c. 5 cm) above the previous one. If bushes are pruned at the same height every time, large lumps of callus tissue develop where the branches are cut and these prevent healthy regrowth. After several pruning cycles, each of which raises the plucking table by 2 in (c. 5 cm) it becomes necessary to prune right down to about 18 in (c. 45 cm) in order to prevent the plucking table growing out of reach.

Technique
Pruning is done with very sharp knives with blades hooked at the end. Each worker usually carries a stick which is marked at the correct height in order to give an accurate guide to the level of pruning. A common practice in the past was to clean out the frames after pruning, i.e. to remove stems where necessary to allow a gap of two fingers' width between neighbouring stems. Nowadays, however, a straight cut across is recommended with no cleaning out. It is important to leave the prunings on the ground; these contribute nutrients and improve the moisture relationship of the topsoil. When tea is pruned in sunny weather some of the prunings are sometimes laid on top of the bushes to prevent sunscorch. If this is done the bushes should be covered within a few minutes of pruning.

Fig. 136: A mature plant. The method of frame formation was formative pruning. Pegging would have produced a lower frame.

Timing
Taking only the health of the tea bush into consideration, the optimum time for pruning is just before the onset of a dry period. At this time the starch reserves in the roots are high, thus allowing rapid regrowth; in addition the period of least water demand (i.e. during regrowth) coincides with the period of least water supply. However, the onset of the dry weather in Kenya and Uganda in December or January usually coincides with a flush of growth brought on by the warm weather. Growers usually prefer to postpone pruning, therefore, until they have made best use of this flush. Pruning during the dry season means that regrowth is less rapid because starch reserves are lower.

Fertilisers

Nitrogen
Nitrogen, whose main effect in most crops is to increase leaf growth, is the most important element in tea nutrition. Responses vary according to soil type, climate, etc., but in general annual yields of made tea increase by 4 lb per acre for each pound of nitrogen applied per acre. In warm wet areas this response may be as high as 8 lb of made tea per acre, whilst it may be considerably lower than 4 lb if the wrong type of fertiliser is used, if weed growth is excessive or if there is a deficiency of phosphate or potassium in the soil. Nitrogen

applications give a cumulative response in tea, i.e. the responses increase for several years provided that nitrogen is applied regularly. It is important to start the annual application as early as three months after planting. In the first and second years nitrogen should be applied at half and threequarters, respectively, of the rate for mature tea. Experiments have shown that there are linear responses to considerably higher rates than this; most estates apply between 100 and 200 lb per acre annually (c. 110–220 kg/ha).

Compound and mixed fertilisers became increasingly popular during the 1960s with the increasing accent on phosphate and potassium. The ratio of the three elements should be 5:1:1; 15:3:3 was the first NPK fertiliser recommended in tea but more concentrated forms are more common now, e.g. 25:5:5. Calcium ammonium nitrate must never be used; it contains enough calcium to restrict the potassium uptake of the crop.

The time of fertiliser applications is not a critical factor in tea; fertiliser can be applied at any time of the year and there is no benefit in splitting the annual applications, except in Bukoba District in Tanzania where soil and rainfall can cause rapid leaching.

Phosphate
The use of superphosphate in planting holes and during vegetative propagation has been mentioned in previous sections. This, particularly in planting holes, has long been recognised as being beneficial, but this is not true of applications of phosphate to mature tea. Most growers, especially the estates, started to apply phosphate in the form of NPK fertilisers only in the 1960s. The widespread use of herbicides on estates is generally considered to be responsible for the excellent responses of tea to phosphate; instead of the surface roots being periodically cut by hand weeding and thereby being prevented from reaching the phosphate in the top few inches of the soil, they are encouraged to grow upwards (as described below in the section on weed control) and they therefore come into contact with the fertiliser.

The rate of phosphate application is largely governed by the common proportions of 5:1:1 NPK in the compound and mixed fertilisers used for tea. If 120 lb of nitrogen are applied per acre (c. 135 kg/ha), such fertilisers provide 24 lb of P_2O_5 per acre (c. 27 kg/ha).

Potassium
The symptoms of potassium deficiency are dark olive green leaves which become scorched in severe cases without any previous yellowing. These symptoms are followed by defoliation of the maintenance foliage, starting from the bottom, and by the production of thin branches, which may die back, becoming blackened at the tips. These symptoms occur more commonly towards the end of a pruning cycle. They are more commonly seen in Uganda and Tanzania than in Kenya. In the Amani area in Tanzania they became so severe that the bushes were only being plucked once every six months.

If deficiency symptoms are serious, a heavy application of potassium sulphate should be given immediately after pruning. Most growers, however, are now guarding against potassium deficiency by the regular use of compound or mixed fertilisers.

Sulphur
Sulphur deficiencies are common around Tukuyu in southern Tanzania and in parts of Kenya and Uganda (e.g. Kericho and Toro Districts). Most compound and mixed fertilisers contain at least 5% of sulphur in order to correct any deficiencies.

Copper
Copper is an essential constituent of the enzyme which induces fermentation. Deficiencies cause slow fermentation and occur in parts of southern Tanzania and Kericho District in Kenya. They can be corrected by foliar sprays of copper sulphate.

Zinc
Zinc deficiency is generally distributed throughout East Africa. It can be corrected by foliar sprays of zinc sulphate.

Weed control

Young tea
Weeds grow prolifically between the bushes before these have formed a complete soil cover. Herbicides can be used but the most common ones, paraquat and simazine, do not control all weeds so either pulling or hand cultivation is necessary to remove those that are resistant. Weeds need only be controlled immediately below the young bushes. Great care must be taken to avoid damaging the roots of young plants when using 'jembes'. Young tea cannot be sprayed with dalapon; couch grass can therefore only be removed by hand during the early stages of the crop's life, but as mentioned above this weed should be eradicated before planting.

Mature tea

In mature tea weeds are suppressed both by the shade and the roots of the tea bushes and by the mulch of prunings. This is especially true when herbicides are used because the prunings are not disturbed by regular hand weeding and they therefore persist for longer. In addition, the roots of the tea bushes grow to the surface of the soil and even into the mulch itself because they are also undisturbed by hand weeding; weed seedlings therefore meet more competition from the roots of the crop. The use of herbicides, especially paraquat, is standard practice on almost all estates and on some of these the weed population has fallen so much that only one spot application of herbicide or one hand pulling is needed each year. Paraquat is a contact herbicide which desiccates foliage in a matter of hours and which is inactivated immediately upon contact with the soil; it kills most annual weed seedlings with the exception of *Polygonum spp.* (Black bindweed etc.). Simazine and diuron are soil applied persistent herbicides; they can only kill weed seedlings as these germinate so the ground must be cultivated before they are applied and rain must fall shortly after application to wash them into the soil. Dalapon is used for perennial grasses, especially couch grass, but if the land has been properly prepared before planting only spot application should be needed.

Irrigation

Overhead irrigation during the dry season has proved to give economic increases in yields and it is likely that the acreage of irrigated tea in East Africa, which is about 1 000 acres (c. 400 ha) at the time of writing, will increase in the future.

Yields

Uganda's average annual yield of made tea per acre is 1 400 lb (c. 1 550 kg/ha). In Kenya this figure is lower, 1 180 lb per acre (c. 1 300 kg/ha), because most of the country's tea is grown at a higher altitude than Uganda's. Tanzania's average annual yield is only 800 lb per acre (900 kg/ha) partly because of the unfavourable climate in the southern tea growing areas. Yields of 3 000 lb per acre (c. 3 400 kg/ha) have been achieved in Uganda.

Five pounds of green leaf produce, on average, 1 lb of made tea.

Quality

The climate in the higher tea growing areas of East Africa, i.e. in the Kenya highlands, around Fort Portal in Uganda and in southern Tanzania, encourages the production of good quality tea. The prices paid for this never reach those paid for some of the Indian early flushes, but when the average price throughout the year is considered, East African high grown teas are superior to those from the major exporting countries, India, Ceylon, Indonesia and Japan. In the lower altitude areas the quality is not so high.

Details of tea quality criteria are omitted from this chapter because, unlike those of coffee, they are the concern of specialists and need not be understood by field workers. The only way that the field worker can ensure good quality tea is by practising fine plucking.

Pests

Black tea thrips. *Heliothrips haemorrhoidalis*
These insects cause silver patches with black spots on the lower sides of the leaves. Attacks usually occur in dry weather. Thrips can be controlled by fenitrothion sprays.

Red spider mite. *Oligonychus coffeae*
The upper surfaces of the leaves, where these mites feed, darken and then turn brown. Dimethoate or dicofol applications are recommended.

Red crevice mite. *Brevipalpus phoenicis*
There have been some severe outbreaks of this pest near Limuru in Kenya. Corky patches appear on the lower sides of the leaves; affected leaves soon fall. This mite seldom attacks more than a few isolated bushes; it is probably checked by predaceous species of mites. Chlorobenzilate and dinocap applications are recommended.

Yellow tea mite. *Hemitarsonemus latus*
Young leaves become distorted and curl inwards. Corky patches may appear between the main veins on the lower sides of the leaves. The yellow tea mite occasionally damages shaded nurseries. Dicofol applications are recommended.

Weevils, cockchafer larvae, etc. often damage young sleeved plants, especially if there is a heavy layer of mulch. They can be partially controlled by mixing a suitable insecticide into the planting holes.

Nematodes occur in various places, especially near Tukuyu in Tanzania and near Lake Victoria in Uganda.

Diseases

With the exception of *Armillaria* root rot, tea is very seldom attacked by diseases. Several diseases which cause damage in other parts of the world, notably blister blight (caused by the fungus *Exobasidium vexans*), do not occur in East Africa.

Armillaria root rot

This disease is caused by the fungus *Armillaria mellea*. It is confined to the soil and is normally a saprophyte, living on dead stumps and roots of forest trees. In the Kenya highlands it is also found in an epiphytic association (i.e. an association which does not harm the host) with tea roots and shade tree roots. In this case it forms rhizomorphs which are black, root-like structures which look like leather boot laces. These are often in close contact with tea roots but in the epiphytic stage they never penetrate far because healthy roots are able to seal off the portions of the root surface which have been invaded by the fungus. If the root surface of the tea bush is damaged by cultivation or by boring insects, *Armillaria* can enter the roots and become parasitic. Once in this stage, the main root system rots away and the bush eventually dies. The leaves turn yellow and fall and sheets of white mycelium can be found between the bark and the wood.

Armillaria should be prevented by ensuring that there are no stumps or pieces of wood in the field which could serve as a reservoir for the fungus. This can be done by ring barking forest trees and leaving them for 18–24 months before felling. Treatment with 2,4,5-T may be included. Ring barking ensures that the starch reserves of the trees are thoroughly exhausted; this makes the stumps and roots much less capable of supporting fungus growth. In addition, as much root as possible must be removed from the soil during land preparation. If *Armillaria* occurs in established tea, diseased trees must be thoroughly removed so that they do not become a source of infection for the neighbouring bushes.

Stem canker

This disease, caused by the fungus *Phomopsis theae*, occurs near Mufindi in Tanzania and is probably connected with the incidence of drought. No cure is known.

Processing

Details of processing are not given in this chapter because this subject is the concern of specialists and need not be understood by field workers. Briefly, however, in orthodox manufacture the leaf is withered in troughs through which either hot or cold air can be blown in order to control the degree of wither. The leaf is then passed through the rotor vane machine; this breaks up the leaf cells and is the latest development replacing the old practice of rolling. Fermentation follows and in modern factories this process is controlled by a flow of air through the rotorvaned and sifted leaf. Firing, i.e. drying to stop fermentation, is the final stage of manufacture; during firing the moisture content is reduced to about 3%.

The process of C.T.C. manufacture varies only in that Cut, Tear and Crush machines are used in addition to rotorvane machines and, generally, instead of a second pass through the rotorvane. The end result is basically smaller grades of tea.

Bibliography

General
1 *Tea Growers' Handbook* (1969). Published by Tea Boards of Kenya, Tanzania and Uganda.

2 **Hainsworth, E.** (1970). *The East African tea industry*. Production and Productivity. Tea, *10* (4):23.

Plant characteristics
3 **Kerfoot, O.** (1962). *Tea root systems*. World Crops, *14*:140.

Ecology
4 **McCulloch, J. S. G.** *et al.* (1964). *Measurements of shade-tree effects in tea gardens*. Tea, *5* (1):9.

5 **McCulloch, J. S. G.** (1967). *Shade tree effects in tea gardens*. World Crops, *18* (3):26.

6 **Ripley, E. A.** (1967). *Effects of shade and shelter on the microclimate of tea*. E. Afr. agric. for. J., *33*:67.

7 **Smith, A. N.** (1963). *The chemical and physical characteristics of tea soils*. E. Afr. agric. for. J., *28*:123.

Varieties
8 **Green, M. J.** (1966). *Clonal selection in seedling stump nurseries*. Tea, *7* (3):11.

9 **Green, M. J.** (1967). *The release of TRI selected clones*. Tea, *7* (4):31.

Propagation

10 **Green, M. J.** (1965). *Seed management.* Tea, *5* (4):11 and *6* (1):11.

11 **Hartley, R.** (1969). *The application of mist propagation for the vegetative propagation of tea.* Tea, *10* (2):13.

12 **Green, M. J.** (1970). *Clonal selection.* Tea, *11* (1):21.

Frame formation

13 **Brandram-Jones, R.** *et al.* (1965). *Bringing young tea into bearing.* Tea, *6* (3):9.

Weed control

14 **Hainsworth, E.** (1969). *A system of soil management for tea production.* Tea, *10* (3):14.

15 **Willson, K. C.** (1963). *The future of chemical weed control.* Tea, *4* (3):29.

Fertilisers

16 **Chenery, E. M.** and **Schoemaekers, J.** (1959). *Magnesium deficiency in East African tea.* E. Afr. agric. J., *25*:25.

17 **Ellis, R. T.** (1966). *Copper deficiency and fermentation.* Tea, *7* (1):20.

18 **Green, M. J.** (1964). *The effect of phosphate on young tea.* Tea, *5* (1):13.

19 **Smith, A. N.** (1962). *The effect of fertilisers, sulphur and mulch on East African tea soils, I—The effect on the pH reaction of the soil.* E. Afr. agric. for. J., *27*:158.

20 **Smith, A. N.** (1962). *The effect of fertilisers, sulphur and mulch on East African tea soils, II—The effect on the base status and organic matter content.* E. Afr. agric. for. J., *28*:16.

21 **Willson, K. C.** (1965). *The long term effects of fertilisers.* Tea, *6* (2):9.

22 **Willson, K. C.** (1966). *Potassium in East African tea.* Tea, *7* (2):15.

23 **Willson, K. C.** (1967). *NPK fertiliser and other nutrients.* Tea, *8* (2):17.

24 **Willson, K. C.** (1967). *Forms of nitrogen.* Tea, *8* (3):11.

25 **Willson, K. C.** (1970). *Foliar analysis of tea.* Expl. Agric., *6*:263.

26 **Willson, K. C.** and **Choudhury, R.** (1968). *Fertilisers and tea quality.* Tea, *9* (3):17.

Irrigation

27 **Carr, M.** (1969). *Irrigation. Where and when will it pay?* Tea, *10* (1):25.

28 **Winter, E. J.** (1966). *Practical suggestions for the irrigation of tea in East Africa.* Tea, *6* (4):15.

Pests

29 **Benjamin, D. M.** (1969). *Insects and mites on tea in Africa and adjacent islands.* Tea, *9* (4):16.

30 **Crowe, T. J.** (1967). *Black tea thrips.* Tea, *8* (1):11.

31 **Hainsworth, E.** (1970). *Tea nematodes.* Tea, *11* (1):15.

Diseases

32 **Goodchild, N. A.** (1960). Armillaria mellea *in tea plantations.* Tea, *2* (1):43.

34
Tobacco

Nicotiana tabacum

Introduction

Tobacco is an important cash crop in parts of
Tanzania and Uganda; it is relatively unimportant in
Kenya. Almost all East Africa's tobacco is grown by
smallholders, the only exception being that from
Iringa in Tanzania. Flue-cured and fire-cured
tobacco are the only types of commercial
importance in East Africa and are the sole concern
of this chapter after the introductory section.
Aromatic tobacco, nicotine tobacco, cigar tobacco
and methods of preparing tobacco for home
consumption are described briefly below.

Flue-cured tobacco
When a variety of flue-cured tobacco is grown in a
suitable environment and is given adequate care
during harvesting and curing, the result is light
tobacco: a yellow, orange or light brown product
which is used in the manufacture of mild
cigarettes.

In Tanzanian the main growing areas are Urambo,
Tabora and Iringa with subsidiary production near
Mpanda and Lupa Tingatinga. Production has risen
steadily in the 1960s and in 1965 Tanzania made
her first exports outside East Africa. In 1969
production was approximately 17 000 000 lb
(c. 7 700 000 kg) of cured leaf. The area under
flue-cured tobacco was about 20 000 acres
(c. 8 000 ha) in that year.

In Uganda the main growing area is West Nile
District where approximately 5 000 000 lb
(c. 2 300 000 kg) were produced in 1969. 1 200 000
lb (c. 540 000 kg) were produced in Acholi and
Lango Districts, mainly around Gulu, and 200 000 lb
(c. 90 000 kg) were produced in Kigezi District.
Exports outside East Africa started in 1968. The
area under flue-cured tobacco in Uganda was
about 7 500 acres (c. 3 000 ha) in 1969.

In Kenya flue-cured tobacco has been grown
experimentally, with poor results, in many areas.
Unsuitable ecological conditions explain not only
these failures but also, to a large extent, the poor
production and quality in the existing flue-cured
tobacco growing areas. Kenya's annual acreage is

Fig. 137: Young flue-cured tobacco.

only about 1 000 (c. 400 ha), divided between
Kitui, Ena and Sagana. In 1969 300 000 lb
(c. 140 000 kg) of cured leaf were produced.
Kenya relies largely on her East African neighbours
for her supply of flue-cured tobacco.

Fire-cured tobacco
When a variety of fire-cured tobacco is grown in a
suitable environment and is given adequate care
during harvesting and curing, the result is dark
tobacco, a dark brown product which is mostly
used for pipe tobacco and in unsophisticated,
strong flavoured cigarettes. The latter are popular in
East Africa although there is virtually no demand
for them in Europe or the U.S.A. The market for
fire-cured tobacco is smaller than that for flue-
cured tobacco and is growing less rapidly.

In Tanzania, the main growing area is in the
south, around Songea. The annual production from
this area in the late 1960s was usually in the region
of 7 000 000–8 000 000 lb (c. 3 000 000–3 600 000
kg), most of which was exported outside East
Africa. Subsidiary production around Biharamulo
has now ceased, partly because the climate was

unsuitable and partly because of attacks of tobacco weevil.

In Uganda, Bunyoro District is the only fire-cured tobacco area. 1 200 000 lb (c. 540 000 kg) were produced in 1969. Production in West Nile District has ceased, partly because of an unsuitable climate and partly because it was inconvenient to grow both fire- and flue-cured tobacco in the same district. The last fire-cured crop was grown in 1968. There is a little grown, unofficially, in Kigezi District.

In Kenya there are a few growers near Migori in South Nyanza. They started growing fire-cured tobacco in the late 1960s.

Aromatic tobacco

This type of tobacco, also called Turkish tobacco, has, as its name suggests, a distinctive aroma. It must be grown on very infertile soils with little rain; it is cured in the sun after a short period of yellowing in the shade. Considerable quantities were grown near Tabora in Tanzania and as many as 750 000 lb (c. 340 000 kg) were produced in some years. Productions has now ceased, owing to marketing difficulties; the last crop was grown in 1967.

Nicotine tobacco. *Nicotiana rustica*

Contains larger quantities of nicotine than *N. tabacum* and has been grown in Kigezi District in Uganda for nicotine extraction. Nicotine was popular as an insecticide but its demand has dwindled with the advent of the synthetic insecticides. Approximately 2 000 acres (c. 800 ha) were grown in Kigezi District in 1947. Production has now ceased.

Cigar tobacco

Missionaries at Mutolere in Kigezi District grow cigar tobacco for making cheroots. This is the only cigar tobacco in East Africa.

Methods of preparing tobacco for home consumption

In almost all parts of East Africa below 8 000 ft (c. 2 400 m) small plots of tobacco for home consumption can be found. Patches of fertile soil, e.g. ant hills or old cattle bomas, are usually used. The leaves are cured either in the sun or inside the houses and are sold either whole, shredded, in hard cakes or in coils. They may be chewed, smoked in pipes or made into cigarettes. Snuff is popular in some areas, e.g. Central and Eastern Provinces in Kenya. For snuff, tobacco leaves are chopped,

dried, and then ground into a powder, often with the addition of goats' fat. An unusual method of taking tobacco is found in Kisii District in Kenya and Kibondo District in Tanzania. The leaves are placed in a horn together with water; the tip of the horn, which has a small hole in it, is inserted into one nostril; the leaves are squeezed downwards and the liquid is sniffed in.

Plant characteristics

The tobacco plant is an annual which grows from four to eight feet (c. 1·2–2·4 m) high, largely depending on the variety. It has a tap root unless, as often happens, this is broken during transplanting. The roots seldom penetrate deeper than 4 ft (c. 1·2 m). The leaves number between 18 and 30. The leaves are broader and less pointed towards the bottom of the plant. Glandular hairs on the surface of the leaves secrete gums and oils which contribute to the flavour of the finished product. When the top of the plant is removed suckers grow in the remaining axils. If the inflorescence is allowed to develop it produces many pink flowers. The seeds are minute and may number as many as 300 000 per ounce (c. 10 000 per gram).

Life cycle

Tobacco seedlings usually remain in the nursery for about two months. The period from transplanting to final harvesting is 4–4½ months.

Ecology

Rainfall, and water requirements

For both flue-cured and fire-cured tobacco the rainfall must be reliable but it must not be excessiv otherwise there is a high incidence of diseases. 15 in (c. 38 mm), well distributed, in the 3½ months after transplanting is ideal. One of the ecological problems in the tobacco growing areas of Kenya is that the rainfall is unreliable and its time of onset cannot accurately be forecast.

Altitude and temperature

Temperatures over 80°F (c. 27°C) or below 55°F (c. 13°C) reduce the quality of the final product; tobacco is therefore a medium altitude crop. Above 5 000 ft (c. 1 500 m) the leaves become thick and leathery, whilst below 3 000 ft (c. 900 m) they are too light. Fire-cured tobacco has a lower tolerance of cool conditions than flue-cured

tobacco and should not be grown above 4 500 ft
(c. 1 400 m).

Soil requirements

Flue-cured tobacco must be grown on light, well-
drained soils with a reasonable water holding
capacity. Sands and loamy sands are ideal; heavy
soils must be avoided. There must be a reasonable
supply of nitrogen in the early stages of growth but
this must be almost completely taken up or leached
out by the time the crop ripens. If the crop is
supplied with nitrogen during ripening, growth
becomes rank, the leaves yellow poorly, curing is
difficult, and the protein content is undesirably
high at the expense of carbohydrates, thus reducing
the combustibility. An excessive supply of chlorine
leads to poor quality tobacco; the cured leaves are
grey with white midribs, they hold moisture, thus
reducing the combustibility, and have an unpleasant
smell. One of the ecological problems of growing
tobacco in Kenya is that in most places where the
climate is favourable the soils are only suitable in
scattered patches. The ideal soil pH is between
5·5 and 6·5.

Fire-cured tobacco must be grown on medium
to heavy structured soils in order to give the leaf
sufficient body. There should be a steady supply
of nitrogen throughout the season.

Varieties

The main varieties of flue-cured tobacco in East
Africa are White Gold and Kutsaga 51. The former
requires a higher standard of management and has
a lower yielding potential. NC. 95 has been
introduced in parts of Kenya and Tanzania; it is
reputed to be resistant to nematodes.

Heavy Western is the only variety of fire-cured
tobacco grown in East Africa.

Propagation

Tobacco seedlings are raised in nurseries (see Fig.
138). The principles are the same for both types of
tobacco, although fire-cured crops often receive less
attention owing to their lower potential value. The
production of a healthy crop in the field, which
resumes growth quickly after transplanting, is
dependent on care and attention in the nursery.

Site

The site should be near a supply of clean water and
should be sheltered from the wind. If there are no
sheltered positions a windbreak should be

Fig. 138: A tobacco nursery.

constructed around the nursery. Ideally the nursery
should be situated in a different place each year in
order to reduce the risk of nematode damage.
Nursery rotation is often ignored, however.

Preparation

The raised beds are usually 4 ft (c. 1·2 m) wide
and about 25 yards long (c. 23 m); one of these
should provide enough seedlings to plant an acre
(c. 0·4 ha) in the field. After preparing a fine
seedbed the soil is sterilised. On small scale farms
this is done by burning a 2 ft (c. 0·6 m) deep layer
of brushwood and dry grass; the ash is later
removed. On large scale farms methyl bromide or
nematocides are usually used. Compound or mixed
fertilisers are recommended, e.g. 6 lb (c. 2·7 kg)
of a 6:18:6 fertiliser per bed. Fertilisers should be
gently raked into the seedbed immediately before
sowing.

Sowing

The seed, being so small, is best sown by mixing it
with water and pouring it on from a watering can
with a good rose. After this the seedbed may be
gently raked but the seed must not be buried too
deep; very few seeds emerge if they are placed as
deep as $\frac{1}{5}$ in (c. 5 mm).

Maintenance

A chopped grass mulch must be applied
immediately after sowing in order to maintain moist
conditions in the surface soil. The mulch can be
thinned as the seedlings grow through it. Watering
is necessary two or three times a day at first but it
can be reduced to once a day after about two

weeks. Towards the end of the nursery period watering should be done only on alternate days in order to harden the seedlings. Precautions must be taken against pests and diseases (see below). If burning has been done efficiently most of the weed seeds in the surface soil should be killed and weed control should not be an arduous task. Given good care and attention the seedlings should be about 6 in (c. 15 cm) high and ready for transplanting two months after sowing.

Rotations

Two crops of tobacco should only be grown in succession immediately after the land has been cleared from bush. Normally tobacco should be preceded and followed by other crops to prevent the build-up of nematodes. Arable crops should not be grown for more than three or four years before resting the land under grass or a bush fallow, otherwise there is a risk of the soil structure deteriorating.

Fig. 139: A crop of fire-cured tobacco, ready for harvesting.

Field operations

Land preparation
Both flue-cured and fire-cured tobacco should be grown on ridges. Ridges are used throughout East Africa although their quality varies; they are usually poorly made in Kenya.

Time of planting
Late planted tobacco is almost always low yielding and of poor quality, partly for environmental reasons and partly because of the high incidence of diseases. Planting early in the rains or about two weeks before the onset of the rains, if this time can be reliably predicted, is therefore recommended.

Planting
The seedlings must be thoroughly watered before lifting from the nursery. They must be planted in the early morning or in the evening and must be protected from the sun. When tobacco is planted before the onset of the rains, water is poured into the planting hole before planting; as little as $\frac{2}{3}$ pint (c. 0·4 litres) is sometimes used per hole but better results are obtained if 2 or 3 pints (c. 1·1–1·7 litres) are used. The tap root should be kept as straight as possible during transplanting and the seedlings should be buried to such a depth that their buds are only about 1 in (c. 2·5 cm) above the ground.

Spacing
The spacing recommendation for flue-cured tobacco is $3\frac{1}{2}$ ft (c. 1 m) between the rows with 22 or 24 in (c. 54–59 cm) between plants within the row. This allows enough room for healthy leaf growth yet not so much that the leaves become heavy bodied.

The spacing recommendation for fire-cured tobacco, which has larger leaves, is $3\frac{1}{2}$ ft × $3\frac{1}{2}$ ft (c. 1 m × 1 m).

Fertilisers
Nitrogen is the most critical element, especially in flue-cured tobacco which must not have too little nitrogen, otherwise the leaves become small, yellow and brittle, nor too much (see section on ecology). 20–30 lb per acre (c. 22–34 kg/ha) of nitrogen should be applied within a week of planting flue-cured tobacco, whilst as much as 40 lb per acre (c. 45 kg/ha) can safely be applied to fire-cured tobacco.

Phosphate is essential on most soils for both types of tobacco; without it yields are usually low and ripening is poor. There is no danger of excessive applications; 100 and 150 lb per acre (c. 110–170 kg/ha) of P_2O_5 are common recommendations.

Potassium gives good burning properties to flue-cured tobacco. Potassium sulphate is usually used because muriate of potash contains too much chlorine. Potassium is sometimes included in compound or mixed fertilisers such as 5:35:10 and 3:27:15 which are recommended in Tanzania.

Fertilisers should be placed about 4 in (c. 10 cm) deep and about 4 in away from each plant as soon as possible, and no later than a week, after planting.

Topping and de-suckering
Both types of tobacco should be topped, i.e. their inflorescences should be removed, when about 20% of the plants have started to flower. This gives wider leaves with more body. With flue-cured tobacco 16–20 leaves should be left on each plant; with fire-cured tobacco 8–12 leaves should be left. Topping stimulates the growth of suckers in the axils of the leaves; these should be removed at regular intervals and should not be allowed to grow longer than 4 or 5 in (c. 10–13 cm).

Weed control
At least three weeding operations are usually needed between planting and the time the crop is knee-high. During weeding the soil should be heaped around the base of the plants.

Harvesting

Tobacco leaves ripen from the base of the plant upwards. They are removed as they ripen; this process is called priming. A few of the lower leaves may be primed and discarded if they are diseased or too small. The leaves to be cured must be protected from the sun after harvesting and must not be bruised or folded.

In flue-cured tobacco one to three leaves are primed at a time; the first are taken at about the time of topping, i.e. $2–2\frac{1}{2}$ months after planting. From then onwards priming is done at approximately weekly intervals, although the interval between the last primings may be longer because the uppermost leaves take longer to ripen. There are usually six to nine primings in all. Ripeness is indicated by a lightening of the colour of the leaf blade, a whitening of the midrib and an angle of approximately 90° between the stem and the base of the midrib. At this stage senescence is sufficiently advanced for a large proportion of the nitrogenous compounds to have been passed back into the plant; the carbohydrates are less mobile and largely remain in the leaf at this stage.

In fire-cured tobacco more than two leaves are primed at a time and only two or three operations may be needed. They are usually about two weeks apart. Ripeness is indicated by the tips and edges

of the leaf turning downwards and a yellow mottling of the lamina.

Curing

Flue curing
Flue curing is done in carefully constructed brick, stone or mud and wattle barns (see Fig. 140) in which the temperature and humidity can be accurately controlled. It is a relatively quick process, taking four to six days. It allows for the oxidation of chlorophyll, causing the disappearance of the green colour, and for the breakdown of starch into sugars. After this the leaf is killed so quickly by desiccation that there is no chance for the sugars to be hydrolysed nor for browning to occur by the oxidation of phenolic compounds; there is thus a high sugar:nitrogen ratio and a pale colour. The barns should be big enough to hold one day's reaping. In East Africa they are usually between 12 ft (c. 3·5 m) square and 15 ft (c. 4·5 m) square and high enough to allow four to six tiers. They must be airtight, apart from ventilators at the top and the bottom, so that there can be a constant flow of air from the bottom of the barn to the top. A metal flue (see Fig. 141), which is the chimney of a furnace outside the barn, leads across the floor. Wood is used as fuel for the furnaces.

The leaves are strung on sticks either by the single-string method (see Fig. 142) or by the

Fig. 140: A flue curing barn. Note the ventilators at the top of the building.

double-string method. The sticks are 4 ft (c. 1·2 m) long and are placed in tiers 2½–3 ft (c. 7·5–9·0 m) apart.

The first stage of curing, after the barn has been filled from the top downwards, is yellowing. High humidity is essential for this process so water is usually scattered on the floor or wet sacks are placed on the flue pipes. All ventilators should be closed or very slightly open to retain as much water vapour as possible and the temperature should be between 90°F and 100°F (c. 32–38°C). Yellowing normally takes 24–36 hours and is complete when only the veins show any trace of greenness.

The next stages are fixing and drying, when the colour is fixed by killing the leaf and destroying the oxidising enzymes. The two stages overlap. The ventilators are first opened and the temperature is raised. After about 24 hours, when the temperature should have reached about 120°F (c. 49°C) and most of the moisture from the leaf blades should have been dispelled, the ventilators are closed in order to raise the temperature as high as possible to dry out the midrib. The temperature should rise to about 160°F (c. 71°C) but should not exceed 170°F (77°C). Drying is complete when the

Fig. 142: Flue-cured tobacco, tied by the single-string method. Note the slight incidence of barn spot.

midribs can be snapped near their bases. Fixing and drying normally take 3–4 days. It is essential that during these two processes the temperature is never allowed to fall, otherwise condensation occurs and the leaf is damaged.

Fire curing
Fire curing is done in thatched barns (see Fig. 143). It is a much longer process than flue curing, taking 4–7 weeks. The leaves are killed less quickly so the phenolic compounds are oxidised, giving a brown colour, and the sugars are largely hydrolysed, giving a low sugar/nitrogen ratio. The smoke, which is all-important, adds greatly to the brown colour and to the flavour.

The barns usually contain only two tiers. There are pits in the floors for the fires (see Fig. 144).

The leaves are first yellowed, either by hanging them in the barns without a fire for 4–7 days or by piling them in a heap in the shade before placing them in the barn. After yellowing fires are lit and are kept going by day for several weeks. The fires should give much smoke but little heat.

Yields

Average yields of cured leaf in Kenya are as low as

Fig. 141: The inside of the barn in fig. 140, showing the metal flue

Fig. 143: A fire-curing barn.

Fig. 144: Inside a fire-curing barn. Note the size of the logs in the fire pit.

300 lb per acre (c. 340 kg/ha). This is due partly to the poor ecological conditions but must be largely due to poor standards of husbandry: control plots in fertiliser experiments give yields of 800 lb per acre (c. 900 kg/ha). In West Nile District in Uganda average yields have risen to 1 000 lb per acre (c. 1 100 kg/ha) and skilled growers, both in this district and in Iringa in Tanzania, achieve yields of 1 800–2 000 lb per acre (c. 2 000–2 200 kg/ha). Average yields of fire-cured tobacco in Songea District in Tanzania are about 500 lb per acre (c. 560 kg/ha).

It should be stressed that the return per acre is as reliant on the quality of the cured leaf as on the yield per acre.

Pests

Nematodes

Meloidogyne javanica, *M. napta* and *Heterodera marioni* are the most important nematodes. Fumigation of seedbeds and rotations have been discussed above and are the only precautions recommended in East Africa.

White fly. *Bemisia tabaci*

White flies are often seen in the nurseries. They

transmit leaf curl virus and should be controlled in the same way as aphids.

Termites, ants, cutworms, millipedes and crickets

These often do damage in the nurseries. They can be controlled by soil applications of aldrin which can be applied via the watering can during sowing.

Diseases

Frog-eye and barn spot

This fungal disease, caused by *Cercospora nicotinae,* is widespread in Tanzania and Uganda but seldom occurs in Kenya. Frog-eye occurs in the nurseries and in the field; infections in the field may develop as barn spot during curing. The symptoms of frog-eye are spots with pale coloured centres and dark margins on the leaf blades; the symptoms of barn spot are small dark spots (see Fig. 000). The disease is spread by rain-borne spores and is most serious if there is heavy rain. Early planting, avoiding excessive applications of nitrogen and the removal of the lower seedling leaves (if they are infected) are important precautions. Routine copper sprays in the nursery should ensure the planting of clean seedlings.

Brown spot

This fungal disease, caused by *Alternaria tenuis,* occurs in all parts of East Africa. It usually appears on ripening leaves and therefore works its way up the plant from the bottom. The lesions are circular brown spots which are zonate, with concentric rings. Unlike frog-eye, lesions can also be found on the midribs and stems. There is no practical means of control although early planting, priming leaves as soon as infection is seen and the use of balanced fertilisers are important precautions.

Anthracnose

This fungal disease, caused by *Colletotrichum tabacum,* is found in Tanzania and usually occurs in the nursery. The symptoms are water-soaked patches on the lower leaves; the lesions may turn brown or white with a dark margin. Zineb, maneb or thiram sprays give effective control.

Damping off

This fungal disease, caused by *Pythium spp.,* occurs in nurseries in all parts of East Africa. The stems are attacked at soil level; affected seedlings fall over and die. Damping off only spreads in very humid conditions, e.g. if water is applied too frequently or if the seedling population is too dense. It can be checked by copper fungicides.

Granville wilt

This bacterial disease, caused by *Pseudomonas solanacearum,* occurs in East Africa only near Gulu in Uganda. The symptoms are wilting of the lower leaves first, followed by yellowing and death. Sometimes only half of a leaf wilts. The vascular tissue appears blackened when the stem is cut and produces a milky exudate if squeezed. The bacteria is soil-borne, lives on a large number of crops and weeds and can survive in the soil for many years. No method of control is known other than the use of resistant varieties.

Mosaic

This is a virus disease, transmitted by mechanical contact, which causes a number of symptoms, the most common of which is a yellow mottling of the leaves. The virus can survive after curing and can be transmitted via the hands of workers who smoke cigarettes or take snuff. Smoking and taking snuff must be prohibited in the nursery and workers should be made to wash their hands before entering the nursery. All crop residues should be burned.

Leaf curl

This viral disease is transmitted by white flies (*Bemisia spp.*). The symptoms are a thickening of the veins, a hardening of the leaves and a downwards curling of the leaf margins. The most important precaution is the destruction of all crop residues at the end of the season.

Rosette

This viral disease is transmitted by aphids (*Myzus persicae*). The symptoms are stunting and the production of many small leaves. Sometimes many axillary shoots are produced, in which case the disease may be called 'bushy top'. The only precautions are routine sprays of systemic insecticides in the nursery and the destruction of all crop residues at the end of the season.

Bibliography

1 **Akehurst, B. C.** (1966). *Flue-cured tobacco in the Iringa District of Tanzania.* E. Afr. agric. for. J., *31*:383.

2 **Akehurst, B. C.** (1968). *Tobacco.* Longman.

35
Wattle

Acacia mearnsii

Introduction

Wattle, sometimes called black wattle, is a tree crop of long standing importance in East Africa. In Kenya and Tanzania the first plants were established by missionaries in the first few years of the century. In Uganda, however, hardly any wattle has been planted because most of the country is climatically unsuited to it. Wattle was originally planted to provide firewood and building materials and to combat soil erosion but it is now grown primarily for the tannins in its bark. Tannins are a complex mixture of polyphenols which are used for tanning leather, and they are extracted at three factories situated at Eldoret and Thika in Kenya and at Njombe in Tanzania. Tanning extract from East Africa is exported mainly to the Indian subcontinent and to the Middle and the Far East either as a solid or as a dry powder. In either form it is guaranteed to contain at least 60% tannins.

Income from wattle bark can almost be equalled by that from the wood; this can be sold as firewood or can be made into charcoal, fenceposts or building poles. If sited correctly, wattle plantations form useful windbreaks or provide shade for livestock.

Kenya's wattle acreage is difficult to estimate owing to the large number of small plots in the smallholder farming areas of Central Province. Large scale plantings were between 65 000 and 70 000 acres (c. 26 000–28 000 ha) during the 1950s, but the acreage dropped considerably during the 1960s due to the poor state of the market. It has been estimated that about 5 000 acres (c. 2 000 ha) per year should be felled in Kenya in order to satisfy the demands of her buyers. This means that the total area should be maintained at about 50 000 acres (c. 20 000 ha) because wattle takes between eight and ten years to mature.

Plant characteristics

The wattle tree is a legume and it may grow as high as 50 or 60 ft (c. 15–18 m). Young plants have well developed tap roots but the great majority of roots of mature trees are concentrated in a dense surface mat. Trees growing in isolation branch profusely, but in a correctly spaced plantation very few major branches develop (see Fig. 145); this leads to straight unbranched trunks which are easy to strip. Trees bear a profusion of pale yellow flowers and subsequently pods, which split freely allowing their seeds to fall to the ground.

Ecology

Rainfall, and water requirements
Wattle needs a well distributed annual rainfall of 45 in (c. 1 150 mm) or more.

Altitude and temperature
Wattle must be grown higher than 6 000 ft

Fig. 145: A wattle plantation.

(c. 1 800 m) above sea level in Kenya if reasonable yields of good quality bark are to be obtained. The most important wattle growing area, the Uasin Gishu Plateau, is over 7 000 ft (c. 2 100 m) and large acreages have been established above 8 500 ft (c. 2 600 m). The highest large scale planting of wattle was at 9 000 ft (c. 2 700 m) but this has now been felled. The tan content of the bark is highest at high altitudes but the bark is thin and lacks weight. Quality tends to fall below acceptable limits under 6 000 ft. Njombe has a pronounced cool season because it is well south of the Equator, so wattle in that area can be grown below the 6 000 ft limit which has been set in Kenya.

Soil requirements

Wattle tolerates almost any soils provided they are free draining. The Uasin Gishu soils, for example, are notoriously infertile yet they produce trees which give satisfactory yields of bark. It is well-known, however, that better yields occur when wattle follows crops which have received generous applications of fertilisers.

Varieties

There are no named varieties of wattle.

Seed preparation

Wattle seed must be collected from the ground beneath trees which are at least three years old; seed from younger trees gives poor germination and weak plants. In practice fairly mature plantations are usually chosen, partly because larger quantities of seed can be gathered but more because the dense canopy suppresses almost all weed growth, thus making seed collection easier.

A combination of leaf mould, split pods, twigs and seed is gathered from the ground and the seeds are later separated from the remainder by winnowing and hand sorting. The seeds are then soaked in an equal volume of hot water. It is essential that the water should be just below boiling point when the seeds are immersed, so the usual practice is to bring the water to the boil, remove the fire from beneath the container and then pour in the seeds. It is also essential to leave the seeds immersed for a considerable period while the water cools down so they are usually left overnight. After soaking, the seeds are washed very thoroughly to remove the sticky mucus which would otherwise make handling and sowing

difficult; they are then dried in the sun. Unless wattle seeds are given this hot water treatment germination is so erratic that some of them may still be lying in the soil as long as ten years after sowing. Untreated seedlings are much less vigorous than those which have been grown from properly treated seed.

Field establishment

Wattle should be sown as early as possible in the rains. There are several different ways of sowing; each is suitable for different circumstances. Smallholders are recommended to undersow wattle in maize which is about 9 in (c. 23 cm) high. About ten seeds should be sown in each hole at a spacing of 10 ft × 7 ft (c. 3·0 m × 2·1 m). Maize can be grown between the wattle rows again in the second season but, as in the first, it should be kept far enough away from the young trees to allow them to develop healthily. Large scale wattle planters also undersow wattle to maize using an adapted maize planter but the most convenient method of establishment is to sow wattle and wheat together. This is done by partitioning off a small compartment in the middle of the seed box of a wheat drill; this compartment is so placed that it feeds the central coulter of the drill with wattle seed. When the wheat is harvested the cutter bar of the combine is raised a little; even if the tops of a few wattle seedlings are removed these can be thinned out at a later stage.

Rows are usually 9 or 10 ft (c. 2·7–3·0 m) apart but 6 ft (c. 1·8 m) rows may be preferred to smother a heavy infestation of couch grass.

Field maintenance

Thinning

Thinning is one of the most important operations in growing wattle. Continuous lines of seedlings emerge after mechanical sowing or after re-establishment. It is most important that these should be thinned neither too quickly, which would encourage the growth of weeds and grasses, nor too slowly, which would lead to excessive competition between the young plants. Undesirably high plant populations have been one of the main faults of wattle growers in the past; the inevitable results are poor yields of poor quality bark and a high proportion of un-strippable trees, i.e. trees from which the bark is not easily separated. Many considerations affect the timing of thinning programmes, e.g. whether the crop was undersown or was established on its own, whether there is a great deal of weed competition and whether the growing conditions have been favourable during the first few years. Rigid schedules are therefore undesirable. It is important, however, that thinning is done gradually and there must be several separate operations. These may include the following: a) a 'clumping' when the seedlings are only 1–2 in (c. 2·5–5·0 cm) tall (this is done by alternately removing and leaving a 'jembe's' width in the line of the seedlings), b) a hand pulling to leave only one seedling in each clump when the plants are between 6 in and 9 in (c. 15–23 cm) high, and c) a slashing with a 'panga' when they are waist high to leave them at their final spacing. The distance between plants should allow a plant population of about 600 trees per acre (c. 1 500 per hectare). If growing conditions are good the final thinning is done sometime towards the end of the second season or early during the third season. During all thinning operations weaker plants are removed and the stronger are allowed to remain.

Fertilisers
Being a legume, wattle should provide its own nitrogen. It is usually considered that it needs no direct applications of fertilisers provided that the crop with which it is sown receives a generous application of phosphate.

Weed control
In the first two or three years it is essential that weeds are well controlled; this is normally done by cultivation. Chemical control has been tried but firm recommendations are not yet available. The method of cultivation depends on the type of farm. Smallholders use hand implements but large scale wattle planters usually rely on extra-heavy disc harrows although they also use hand labour to control weeds within the rows. The number of weedings depends mostly upon the thoroughness of seedbed preparation and upon the speed of canopy formation. Large scale growers hope that only one inter-row cultivation will be necessary during the year of sowing or reestablishment. Weeding is seldom necessary after the third year because wattle forms a dense canopy which suppresses all weed growth. Even couch grass disappears although it reappears as soon as the trees are felled.

Fire breaks
Fire is the greatest threat to wattle plantations.

Even a light brush fire can kill large numbers of trees because they are unusually sensitive to heat. For this reason maximum precautions must be taken; fire breaks should be maintained not only around all blocks but within them as well.

Bark quality
Wattle bark consists of tannins, soluble carbohydrates (mostly sugars, gums and starch), fibre and moisture. During extraction the tannins and soluble carbohydrates are dissolved and are thus separated from the fibre to form the final product. The East African wattle industry demands that its extract should contain a minimum of 60% tannins and in order to ensure this only good quality bark is accepted at the factories. The commonly accepted criterion for judging bark quality is the tan:non-tan ratio, i.e. the relative proportions of tannins and soluble carbohydrates in the bark. The average ratio must be at least 3:1 per tree to produce an extract of not less than 60% tannins.

There are several factors which affect the tan:non-tan ratio, some of which are not fully understood. The ones which are important to the grower are mentioned below.

Age of the tree
The tan:non-tan ratio rises steadily until it reaches and maintains a peak from 8–10 years onwards. Trees must not be felled, therefore, until they are at least 8 years old; ideally they should be about 10 years old. A useful visual test for judging maturity is to inspect the colour of the bark; it must be grey, all trace of greenness having disappeared. Felling trees younger than eight years old leads to reduced yields of bark and wood as well as to reduced quality. In plantations much older than ten years there is usually an increasing number of un-strippable trees, sometimes as large a proportion as 50% at 15–16 years old, and an increasing number of plant deaths. This decline in plantations depends largely on environmental conditions and with favourable weather and soils, e.g. at Sotik in Kenya, good results have been obtained in plantations as old as 18 years.

Soil moisture
In wet conditions the soluble carbohydrates in the bark are reduced and the tan:non-tan ratio therefore increases. Quality is consequently higher during the rains, although there may be a delay of two or three weeks after the rains begin before it reaches a satisfactory level. There is also a variation in quality

from year to year; this depends upon the rainfall.

Part of the tree

The bark at the base of the trunk has a higher tan:non-tan ratio than the bark of the upper part. The generally agreed limits are that bark should be stripped up to the point where the trunk is only three inches (c. 7·5 cm) in diameter or up to the point where the bark changes colour from grey to green. Branches are normally too small but when, as occasionally happens, they are sufficiently large and have grey bark they may be stripped.

Diameter of the trunk and thickness of the bark

These two factors are closely correlated. Trees with thin trunks, such as those in un-thinned plantations, yield thin bark with a low tan:non-tan ratio. In trees with thick trunks the position is reversed. Poor quality bark from thin trees can thus be identified by its thinness during inspection at the factory.

Oxidation

After stripping, the tannins slowly oxidise giving a slightly lower tan:non-tan ratio and resulting in a rather darker and less desirable extract. Fresh bark which is sent to the factories immediately after stripping therefore gives the best quality. This is called green bark, although the name is misleading because the colour must be grey, as noted above. Two considerations prevent the exclusive processing of green bark. Many plantations are situated at a considerable distance from a factory and transport costs are so high that it is more economical to deliver dried bark, usually called stick bark. Stick bark is considerably lighter than green bark, the ratio being 1 ton of green to 0·6 tons of stick. The other consideration is that green bark can only be delivered in the stripping season, i.e. the time when the tan:non-tan ratio is sufficiently high and when the sap is running so that the bark can easily be removed from the wood; this is restricted to about four months each year. In order to run economically, however, factories must operate for about six months each year. If there is a sufficient acreage of wattle to keep the factory running for this period, large quantities of stick bark must be delivered. When stripping and drying, oxidation of tannins must be reduced to a minimum; this subject is considered in more detail in a later section.

Diseases.

Bark diseases tend to occur in old plantations, especially if the growing conditions are not ideal. The bark becomes corky or scaly and the tan:non-tan ratio is greatly reduced.

Felling and stripping

Felling is done with axes or power saws. Axes are still the most common implements and when they are used it is important that the first operation is the stripping of the lowest three or four feet (c. 0·9–1·2 m) of bark. If this is not done, much valuable bark is lost during chopping. As short a stump as possible should be left in order to ensure maximum yields of wood, although an exception to this rule occurs on some of the large estates on the Uasin Gishu Plateau when arable crops are planned to follow wattle. In these circumstances $2\frac{1}{2}$ ft (c. 0·75 m) stumps are left; these can later be pushed over at an angle by a bulldozer, then removed by two crawler tractors which pull a chain in the opposite direction.

After felling, all brushwood and the discarded part of the trunk under three inches in diameter are removed and piled in rows along the stump lines; these rows should be as high and as narrow as possible. The main trunk is then measured off into 3–4 ft (c. 0·9–1·2 m) sections and a ring cut is made at each point; the length depends on the requirements of the factory. After one longitudinal cut has been made, the bark can be removed in one piece from each section of the trunk (see Fig. 146). The bark is then cut into strips three inches (c. 7·5 cm) wide, after which it is ready for drying. The aim must be to present to the air as small a cut surface as possible, thus reducing oxidation to a minimum. For this reason the strips of bark should be no narrower than three inches, nor should the ends be ragged or cut at an angle. Corky bark, green coloured bark and thin bark must be discarded and no shoots or twigs should be included. If the strips are much wider than three inches each one tends to curl up into a roll and this leads to uneven drying. It is important to strip on the same day as felling, otherwise it is difficult or even impossible to remove the bark from the wood.

Drying

Practically all stick bark is sun dried in the field. Drying in the shade results in a better quality extract owing to less oxidation but it has proved to be unduly expensive. It may be a last resort, however, in the event of wet weather.

Fig. 146: Stripping wattle.

Mould formation on the bark causes discolouration of the extract and should be avoided by efficient drying. The strips of bark are placed across three long wattle trunks which have just been felled; these are placed parallel to each other so that the outside ones support the ends of the strips whilst the centre one supports the middle of the strips. It is essential that each strip is placed with its outer side upwards so that any rain is shed; if the inner side is placed uppermost rainwater is trapped in the bark and moulds inevitably develop. Leaching also occurs. Moulds also develop if any parts of the strips are allowed to rest on the ground. To ensure even drying the strips must be laid side by side rather than on top of each other.

Drying takes 1½–3 weeks, depending upon the weather and the bark is considered sufficiently dry when it is so brittle that it snaps when bent; at this stage it should contain about 12% moisture. When stick bark is removed from the field it is tied into bundles which weigh approximately 45 lb (c. 20 kg) each.

Yields

Average yields in Kenya are about 3 tons of stick bark per acre (c. 7·5 t/ha) in ten years, although with reasonable management this figure should be as high as 4·5 tons per acre (c. 11 t/ha). If 4·5 tons of stick bark are obtained per acre, about 35 tons of wood per acre (c. 90 t/ha) are usually produced simultaneously. Under normal conditions

8-10 tons of green bark per acre (c. 20–25 t/ha) are the highest yields that can be expected per acre, although in one unprecedented instance 18 tons per acre (c. 45 t/ha) were obtained.

Re-establishment

Wattle is re-established easily and cheaply by burning the brushwood which was piled along the stump lines; vast numbers of seed which were lying dormant in the soil are thus stimulated to germinate. The timing of burning is important; it must be done at the end of the dry season as the main rains start. By this time the brushwood should be dry enough to burn well, but more importantly the seedlings get a good start because the soil should be moist for the first few months of their life. The dense mats of seedlings are first reduced to narrow bands about 9 in (c. 23 cm) wide along the stump lines; on large scale plantations this is usually done with disc harrows. The bands are then ready for clumping and the other thinning operations follow in the same way as if wattle was being established for the first time. The stumps present no problems because they are in the new line of seedlings and therefore do not obstruct inter-row work.

Wattle can be re-established several times. In Kenya three or four crops have sometimes been grown in succession but in other parts of the world as many as seven have been obtained.

Pests and diseases

Wattle is attacked by very few pests and diseases. Spraying against leaf eating caterpillars is only occasionally necessary.

Bibliography

1 **Elmer, J. L.** and **Smith, R. G.** (1956). *Some pests of black wattle in Kenya, with a list of other insects inhabiting the plantations.* E. Afr. agric. J., *21*:230.

2 **Elmer, J. L.** *et al.* (1963). *The role of boron and rainfall on the incidence of a wattle dieback in East Africa.* E. Afr. agric. for J., *29*:31.

3 **Elmer, J. L.** *et al.* (1964). *A consideration of factors affecting quality in wattle bark.* E. Afr. agric. for J., *29*:350.

4 **Gosnell, J. M.** (1963). *The effects and control of couch in wattle plantations in West Kenya.* E. Afr. agric. for J., *29*:7.

5 **Nickol, D. A.** (1959). *The Bena Wattle Scheme, Njombe District, Tanganyika.* E. Afr. agric. J., *25*:53.

6 **Sykes, R. L.** and **Simon, T. D.** (1954). *Vegetable tannins in East Africa.* E. Afr. agric. J., *20*:59.

7 **Vail, J. W.** *et al.* (1957). *Dieback of wattle—a boron deficiency.* E. Afr. agric. J., *23*:100.

36
Wheat

Triticum spp.

Introduction

Wheat is a cereal crop which is grown in the East African highlands. It is ground into flour for making bread; almost all is consumed internally. Exports were limited until 1967 and 1968 when considerable surpluses were produced in Kenya. Self-sufficiency, however, is the present wheat policy in East Africa and there are no plans for large scale exports in the future.

Kenya

Wheat is the most important crop in the large scale, mixed farming areas of Kenya. The lower part of the Trans Nzoia District, which has a warm, humid climate, is the only part of these which is unsuitable for wheat. During the late 1960s over 400 000 acres (c. 160 000 ha) of wheat were sown annually in Kenya. The area with the largest acreage is the Uasin Gishu Plateau but its yields are usually low, partly owing to infertile soils. The area with the greatest potential is the wetter part of Masailand, from Mau Narok to Narok. There are an estimated 300 000 acres (c. 120 000 ha) of potential wheat land in this area and in 1968 over 20 000 acres (c. 8 000 ha) of wheat were grown there, partly by share-croppers and partly by Government. Kenya's wheat production rose steadily during the 1960s and in 1968 it reached approximately 220 000 tons. There is a little subsistence wheat production in northern Kenya around Marsabit.

Tanzania

Tanzania grows considerably less wheat than Kenya; in 1969 the acreage was approximately 145 000 (c. 60 000 ha). The main large scale wheat growing areas are in northern Tanzania at Karatu, Oldeani, the western slopes of Mt. Kilimanjaro, Monduli, Babati and Basuto (near Mt. Hanang). There is also some large scale wheat growing at Uwemba in the Southern Highlands. The main area where wheat is grown on smallholdings for subsistence is Njombe, where an estimated 65 000 acres (c. 26 000 ha) are sown annually. There is

also subsistence wheat production around Mbulu, Mbeya and Mbozi and in the Usambara Mountains, the Matengo Highlands and the Mwezi Highlands. In 1966 the large scale farms produced about 32 000 tons of wheat.

Uganda

Most of Uganda is climatically unsuited to wheat growing. There is a limited amount of large scale production on the north-west slopes of Mt. Elgon and a little subsistence production in the Ruwenzori Mountains and in the extreme south-west of Kigezi District.

Large scale production and smallholdings

The great majority of wheat in East Africa is sown, sprayed and harvested mechanically. There is scattered subsistence production, however, in the areas mentioned above where all operations are done by hand. Attempts to mechanise wheat growing on small plots have almost always been unsuccessful because drills, sprayers and combine harvesters do not work efficiently or economically in small inaccessible plots.

Plant characteristics

Wheat is an annual cereal crop. It grows to a height of 3–4 ft (c. 0·9–1·2 m), largely depending on the variety. There are several tillers on each plant; the number depends on such factors as variety, seed rate, soil fertility and the rainfall at the time of tillering. Many East African wheat varieties are bearded, i.e. each lemma has a long awn (see Figs. 147 and 148). The colour of the seed coat ranges from red to white. In East Africa the percentage of cross pollination is usually only 1·0–1·5%, although in some seasons it may be a little higher.

Life cycle

Wheat matures in four to seven months; the time depends on the variety and the altitude. At 7 000 ft (c. 2 100 m) most varieties mature in about $4\frac{1}{2}$ months whilst a few mature in $5\frac{1}{2}$ or 6 months. A rise or fall in altitude of 1 000 ft (c. 300 m) causes,

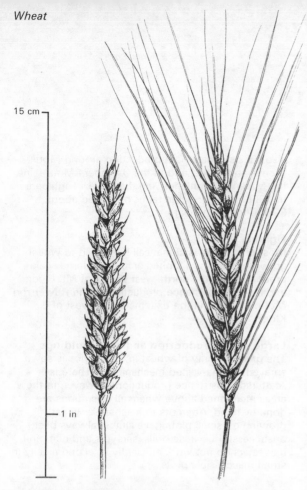

15 cm

1 in

Fig. 147: Beardless wheat.   Fig. 148: Bearded wheat.

respectively, a 15 day lengthening or shortening of the time taken to maturity.

Ecology

Rainfall, and water requirements

Wheat is a moderately drought resistant crop but for best yields it needs a well distributed supply of soil moisture. During the early stages of growth its demands are small although a drought at this stage retards growth permanently. When it reaches about 9 in (c. 23 cm) high it begins to tiller and the water demand consequently rises steeply until flowering. A drought at the time when tillering should occur may restrict the number of tillers; sometimes none develop. Ear initiation occurs towards the end of

tillering; a drought at this stage causes the initiation of few flowers and the heads, when they later emerge, are therefore short. After flowering the water demand almost levels off. A drought soon after flowering causes low grade grain, i.e. grains which are narrow and misshapen. After the grain has reached the cheese-like stage heavy rain is damaging; it slows grain drying, may cause sprouting (i.e. germination of grain in the head), discolouration or lodging and may create muddy conditions which prevent combine harvesters working.

With stored moisture in the soil at the time of sowing, good crops of wheat have been obtained with as little as 4–6 in (c. 100–150 mm) of rain during the growing period. In Kenya wheat is grown in areas with an average annual rainfall of 30 in (c. 750 mm) or more. On the western slopes of Mt. Kilimanjaro in Tanzania, however, it is grown with an annual average of only 23 in (c. 580 mm); in this case the fields are fallowed in alternate years to conserve moisture.

Altitude and temperature

Wheat is grown in Kenya from 6 000 ft to 9 500 ft (c. 1 800–2 900 m). In Tanzania it is grown from 4 000 ft to 8 500 ft (c. 1 200–2 600 m). The disadvantage of places at lower altitudes is either that they have too low a rainfall or that they are too humid, causing a high incidence of diseases. Under irrigation and with low humidity wheat has been grown successfully as low as 2 000 ft (c. 600 m) at Arusha-Chini near Moshi in Tanzania.

Soils

For best yields wheat needs well drained soils with a high nutrient content. It yields very poorly in waterlogged conditions.

Varieties

In most temperate countries both winter and spring wheats are grown. Winter wheats must be subjected to a severe cold period by sowing them before the onset of winter otherwise they continue to grow vegetatively; only spring wheats are therefore grown in East Africa.

Varietal resistance is the only practical way of countering stem rust attacks and the main aim of the Plant Breeding Station at Njoro in Kenya has been to issue rust resistant varieties. The rapid appearance of new races of stem rust is one of the main problems of wheat growing in East Africa; new races can often attack varieties which had previously proved to be resistant. It is always necessary, therefore, for the wheat breeders to have material available that is likely to be resistant even to new races. During the late 1940s many new rust races appeared and in the 1950s a policy of introducing and screening large numbers of varieties was adopted. The most useful introductions were from Central America and by the 1960s more than half of the East African varieties were of Central American origin. From 1966 onwards Kenya's farmers have been successfully discouraged from sowing some of the more susceptible varieties because the Wheat Board has contracted to buy them only at a considerably reduced price. At the same time planting loans have only been approved if suitable varieties were used.

Until the 1960s yield was only a minor consideration for the breeders. During the 1960s, however, with the release of satisfactorily rust resistant varieties, more emphasis was placed on yielding capacity. The third important consideration for breeders is quality (see section on quality below). In the past the quality of East Africa's wheats has not been particularly high and imports of good quality wheat have been necessary for mixing with the local varieties in the mills. The situation has recently improved, however, owing mostly to greater emphasis on quality in the breeding programme but also owing to reduced prices for low quality varieties as from 1964 and to the refusal of planting loans for unsuitable varieties.

Varietal recommendations for Kenya are published each year by the Plant Breeding Station; they indicate the susceptibility of varieties to diseases and to sprouting, time to maturity, recommended altitude, straw strength and quality. Several varieties should be sown on each farm as an insurance against the occurrence of different races of rust. In the past the recommended number has been five to seven varieties per farm. These numbers are likely to be reduced in the future as, to facilitate handling, only a few varieties, those with increased resistance, are released.

Several factors influence the choice of varieties. Varieties with a high stem rust index (i.e. with moderate susceptibility to this disease) should be kept to a minimum. Varieties which are sown early, and which therefore risk being harvested in wet weather, should not be susceptible to sprouting. It may be necessary to plan ahead to prevent many fields maturing at the same time; this can be achieved by sowing early maturing and late maturing varieties at appropriate times.

There is a limited market for *Triticum durum*, a wheat which is only suitable for making macaroni and related products. Two varieties were released for limited acreages in Kenya in 1969.

Rotations

Wheat has been the most attractive crop for the majority of large scale farmers in the highlands of Kenya and Tanzania. This has led many of them to adopt a policy of wheat mono-culture or of growing wheat for many years in succession without resting the land. At Oldeani in Tanzania some fields have been under wheat for thirty to forty years. With conventional methods of cultivation this policy usually results in a deterioration of the soil structure, leading to reduced water infiltration and increased soil erosion. Prolonged cropping with wheat also causes an increased incidence of soil borne diseases. To prevent this situation and the low yields which inevitably result, a three or four year cropping period should be followed by a three or four year rest under grass. Using minimal tillage (see section on seedbed preparation below) soil structure can be maintained almost indefinitely and longer cropping periods are possible provided that soil borne diseases are not too damaging.

Growing two crops of wheat in a year on the same farm is actively discouraged in Kenya. Farmers who follow this practice are refused planting loans. The danger of growing two crops in the same year is that it allows rust to multiply rapidly, passing from one crop to the next. Only on the northern slopes of Mt. Kenya and the western slopes of Mt. Kilimanjaro is it considered acceptable to grow two crops of wheat in the same year. In these areas there are two well defined wet seasons, neither of which is particularly reliable.

Field operations

Burning stubble and straw
This operation, although it follows immediately after harvesting, is often the first in preparing the land for the following crop. In temperate countries straw is valued as a bedding for livestock but in East Africa, where livestock are virtually never housed indoors, it is seldom removed from the fields. The choice lies between burning the straw and incorporating it into the soil. Neither course of action is necessarily correct; the decision whether

or not to burn may be affected by one or more of the following considerations:

1 The presence of straw may impede cultivations. This is especially true when disc ploughs are being used; these implements usually penetrate poorly when there is much straw on the ground.

2 The presence of soil borne diseases. Glume blotch and leaf blight may be carried over to the following season if infected straw is ploughed in. Burning prevents any such carry over.

3 Soil structure. Farmers who have light powdery soils with little organic matter sometimes prefer to incorporate the straw in order to improve soil structure.

4 The weather. Rain may make burning impossible. If the farmer has to wait for a few weeks until the straw dries out, weeds and volunteer wheat often grow up amongst the straw and prevent efficient burning.

5 The speed of straw decomposition. Decomposition is slow or absent if the soil is dry, has a low microbial population or, at high altitudes, is cool. The presence of undecomposed straw in the soil often restricts the growth of the following crop. To avoid this farmers may plough the straw in as early as possible, they may add nitrogen when sowing to compensate for that which is locked up or they may burn the straw.

Seedbed preparation

Harrowing is sometimes done shortly after harvesting in order to encourage germination of grains which have dropped from over-ripe heads or which have passed through the combine. The seedlings are later killed during ploughing.

Early ploughing is essential for several reasons. Its advatages are as follows:

1 Ploughing is the slowest of the operations in mechanical wheat cultivation. It must therefore be started early to allow all the land to be ready for sowing at the correct time.

2 If grassland is being broken or if straw is being incorporated early ploughing allows a reasonable time for decomposition.

3 Ploughing during the dry season helps to dry out Kikuyu grass and couch grass.

4 Early ploughing allows plenty of time to prepare a clean seedbed by several harrowings. Each harrowing kills the seedlings which have germinated since the previous operation and encourages some of the remaining weed seeds to germinate for the next harrowing.

5 Very early ploughing, immediately after harvesting the previous crop, is sometimes practised. It has two advantages: the soil is easier to work because there is usually a little moisture left, and soil moisture is conserved if the seedbed is subsequently kept clean because there is less moisture loss from a bare surface than from one with a cover of vegetation. The only disadvantage of ploughing soon after harvesting is that stubbles become unavailable for grazing.

Wheat seeds are small and need a fine seedbed for even and rapid germination. Opinions vary on how fine the seedbed should be; some farmers, usually those with fairly flat fields, prefer to do many operations, creating a clean, fine seedbed whilst others prefer to do as few operations as possible, leaving a rather coarse seedbed which resists soil erosion and which is relatively cheap to prepare.

A system of minimal tillage has proved to be beneficial at West Kilimanjaro in Tanzania. The sequence of operations is as follows: after harvesting, the straw is chopped up, either by a chopper spreader or by a disc harrow which is not fully offset. Until the next season's sowing the fields are kept clean by using heavy duty cultivators with two to five ducksfoot blades, i.e. blades which kill weeds by cutting them below the ground without disturbing the soil surface. Sowing is usually done by broadcasting and harrowing; special drills are available, however, for sowing directly into uncultivated soil. Costs of seedbed preparation are reduced, the soil structure is maintained, soil and water conservation are greatly improved, the crop grows more quickly in the early stages of growth, it is deeper rooting and yields are increased, on a field scale, by about 400 lb per acre (c. 450 kg/ha).

Time of sowing

Recommended sowing times differ from district to district. In general, the earlier wheat is sown in the long rains the better, provided that it ripens and can be harvested during a dry period. At Rongai in Kenya, for example, the usual time is March to May; this area is low lying and has a marked dry period, suitable for harvesting, in September and October. At Molo, however, the usual time is June and July; this high altitude area normally has dry weather only in January and February.

In Kenya farmers are refused planting loans unless they sow at the correct time, as specified by scheduled planting dates.

Seed

Good quality seed should be sown. Farmers can use the seed which they produced in the previous year or, if they want new varieties or are growing wheat for the first time, can buy their seed. In Kenya they can take advantage of the Field Approval Scheme; Field Approved Seed is inspected in the field and is tested for purity and germination; it is sold at a higher price than wheat for milling.

If soil borne pests or diseases are present a seed dressing should be used. The insecticide should be aldrin or dieldrin whilst the fungicide should be an organomercurial.

Sowing

Most of East Africa's wheat is mechanically sown, using drills (see Fig. 149). Most drills sow rows 7 in (c. 18 cm) apart; they should be set to place the seed 1–2 in (c. 2·5–5·0 cm) deep. Wheat is occasionally sown mechanically by broadcasting and harrowing. Smallholders in the areas mentioned in the introductory section of this chapter broadcast it by hand.

Seed rate

The seed rate determines the spacing within the rows. The usual seed rate is 100 lb per acre (c. 110 kg/ha), although many farmers are now putting on 110 or 120 lb (c. 120–130 kg/ha) whilst others in high altitude areas, where the high rainfall encourages tillering, use a seed rate of only 80 lb per acre (c. 90 kg/ha). Recent experiments, spread throughout the wheat growing areas of Kenya, have shown hardly any differences in yield from seed rates varying from 50 to 150 lb per acre (c. 55–170 kg/ha).

Fertilisers

A comprehensive series of fertiliser experiments in

Fig. 149: Drilling wheat.

Kenya, starting in 1967, have shown that fertiliser responses can only be predicted to any extent by studying the previous cropping history of the field concerned. They cannot be predicted by soil analyses except in extreme cases of deficiency or surplus, nor can recommendations be given for different districts, as was previously done.

Nitrogen usually gives small responses in the first or second year after the land has been broken from grass. In the first year this is only true if ploughing is done early and if there is enough moisture in the soil for straw decomposition. The reason for the small responses (i.e. additional yields of only about 200 lb of grain per acre from 20 lb of nitrogen) is that after the land has been broken from grass the soil is fairly rich in nitrogen. Later in the cropping period, from the third year onwards, nitrogen responses are considerably greater because most of the nitrogen from the grass has been utilised by the previous crop or has been leached. At this time additional yields from 20 lb of nitrogen per acre are usually about 600 lb of grain. The current recommendation is to apply no nitrogen in the first two years of the cropping period and thereafter to apply 20 lb per acre (c. 22 kg/ha). Heavy applications of nitrogen risk serious lodging. Nitrogen fertilisers are usually applied during sowing; top-dressing is only justified if waterlogging in the early stages of growth causes yellowing of the leaves.

It has long been known that wheat in East Africa needs additional phosphate for best yields. Phosphate fertilisers are especially necessary after breaking land from grass; at this time the soil has a low phosphate content and applications of 40 lb of P_2O_5 per acre usually increase grain yields by about 600 lb per acre. Owing to the residual effect of phosphate, responses decrease in successive years provided that phosphate is applied each year. In the third and fourth years of cropping the response to 40 lb of P_2O_5 per acre is usually only about 200 lb of grain. The current recommendation is to apply 80 lb of P_2O_5 per acre (c. 90 kg/ha) to the first crop after breaking the land from grass and to apply 50 lb (c. 55 kg/ha) in succeeding years. Phosphate fertilisers are applied through the seed drill.

Experimental applications of potassium and calcium have shown that these elements are very seldom limiting. Potassium has increased yields significantly in a few areas, e.g. at Kipkabus, but as often as not it reduces yields on other sites. It is possible that it may be needed more in the future if high yielding crops for several years in succession cause a considerable reduction of the potassium reserves in the soil.

Wheat needs a small supply of copper. Copper deficiency has been recognised for many years near Nakuru in Kenya, between Menengai and Njoro. Recently, however, copper deficiency has been noticed in Mau Narok and is suspected at Londiani; this is possibly due to the depletion of the element by several years of heavy wheat crops. Symptoms occur in patches within a field; the leaves appear burned; there are few tillers; the plants are more susceptible to drought and the grains are either small or absent. Heavy applications of phosphate increase the likelihood of these symptoms by tying up the copper in unavailable forms. The first recommendation for improving the supply of copper was to include copper sulphate with the fertiliser. This proved to be an inefficient way of applying copper, probably because it was readily tied up with the phosphate. The current recommendation is to apply $\frac{3}{4}$ lb of copper per acre (c. 0·85 kg/ha) as a seed dressing, i.e. $1\frac{1}{2}$ lb of copper oxychloride mixed with 100 lb of seed, and then to add a further $\frac{3}{4}$ lb of copper per acre as a foliar spray by mixing copper oxycholoride with the herbicide at the rate of $1-1\frac{1}{2}$ lb per acre (c. 1·1–1·7 kg/ha). This method has proved more effective than mixing copper sulphate with the fertiliser.

Rolling

Some farmers roll their fields after drilling. Rolling ensures that all seeds are in good contact with the soil. It encourages rapid and even germination, especially in rather dry conditions and on pulverised or puffy soils when germination may be very poor without rolling. The danger of rolling is that it may encourage soil erosion; it is not, therefore, a common practice except on relatively level fields.

Weed control

Smallholders weed their wheat by hand but this is impracticable on large acreages. Mechanical cultivations are impossible after the crop has been sown owing to the close spacing; large scale farmers must therefore rely primarily on herbicides for killing weeds in wheat. Two methods of prevention, however, are very important: the preparation of a clean seedbed and the use of weed-free seed.

The most commonly used herbicides are the phenoxyacetic acid derivatives, 2,4-D and MCPA. Their main advantages are that they are relatively cheap and they kill most of the common

Fig. 150: Combine harvesting wheat.

broad-leaved weeds. Some weeds, e.g. cleavers, spurrey, bindweed, campion and mallow, are resistant to 2,4-D and MCPA; they can be killed by using the more expensive phenoxyproprionic acid derivatives MCPP and 2,4-DP.* Residual herbicides, e.g. diuron, have been tried in wheat; they kill many of the weeds that are resistant to 2,4-D and MCPA but as yet there is insufficient experience of their use to justify firm recommendations. They have given poor results in northern Tanzania.

The hormone-type herbicides, e.g. 2,4-D, MCPA, MCPP and 2,4-DP, must be sprayed on to the foliage when the wheat is at the 4–6 leaf stage. Either tractor spraying or aerial spraying may be chosen. Tractor spraying is cheaper but needs careful supervision and does a certain amount of damage to the crop where the wheels have passed over it; aerial spraying is more expensive but it avoids direct contact with the crop and needs

*MCPP is also called CMPP or mecoprop-2, 4-DP is also called dichlorprop.

almost no supervision by the farmers. Timing of spraying is important. The operation must be done with a low wind speed, with moist soil for rapid absorption and translocation of the herbicide and preferably in sunny weather for the same reason. Spraying when the crop has less than four leaves risks damaging the crop; spraying when it has more than six allows the weeds to grow past the stage when they are easily killed by herbicide and also risks damage to the crop.

CCC

CCC is an abbreviation for chlorocholine chloride, a chemical which has recently appeared on the market in East Africa. When applied as a spray at the beginning of stem extension in temperate countries it can cause a considerable shortening and strengthening of the straw with a consequent reduction in lodging. It may be applied with the herbicide. A series of experiments in Kenya, covering all the important wheat growing areas,

243

showed that reductions in straw length were very small: only 2–3 in (c. 5·0–7·5 cm). Effects on lodging and yield were very variable; in one instance CCC gave a 35% yield increase but in other instances yields were reduced. At present, therefore, there is no general recommendation for CCC.

Harvesting

In the smallholder wheat areas of East Africa harvesting is done by cutting individual heads with hand knives. In all other areas wheat is cut and threshed by means of combine harvesters (see Fig. 150). Wheat must be dry before it is combined otherwise it is susceptible to rotting, sprouting and insect damage during storage. A moisture content of 14% or less is ideal, although it may be harvested at a higher moisture content if there are facilities for grain drying.

Lodging may hinder harvesting. It may be caused by strong winds and heavy rain in the later stages of growth, by using an excessive seed rate, an excessive application of nitrogen or a weak-strawed variety. Using combines with pick-up reels and pick-up guards on the cutter bars enables lodged crops to be harvested but the operation is much slower than if the crop is standing upright. Additional disadvantages of lodged crops are as follows: that they dry slowly because they are close to the ground and that they often ripen poorly, the grains failing to fill out properly. Early lodging usually leads to the production of many late tillers whose heads are still green at the time of combining; these heads result in an unclean sample of grain which may have to be passed through a seed cleaner.

Yields

The average yield of wheat in Kenya has generally been about 1 000 lb per acre (c. 1 100 kg/ha). In 1967, however, it reached a record of 1 420 lb per acre (c. 1 590 kg/ha) owing to favourable weather and a low level of rust infection. On good soils 2 000–3 000 lb per acre (c. 2 250–3 350 kg/ha) should be obtained with good husbandry. At high altitudes, in years of good rainfall and low rust infection, yields of 4 000 lb per acre (c. 4 500 kg/ha) are regularly obtained. The highest yield recorded on a field scale in East Africa is 5 600 lb per acre (c. 6 300 kg/ha). Yields are usually quoted in bags per acre: one bag contains 200 lb (c. 90 kg) of grain.

Quality

Baking quality
Baking quality, i.e. the ability of the dough to retain carbon dioxide bubbles and thus to produce a good, light loaf, depends on genetically governed grain characteristics. Gluten, which is a mixture of proteins, determines the baking quality. Strong wheats have a high content of good quality gluten and their dough is able to retain the carbon dioxide bubbles. Weak wheats, on the other hand, have insufficient or poor quality gluten.

Milling quality
Milling quality depends on whether the endosperm is hard, enabling it to pass through sieves readily, or soft, causing it to block sieves.

Pests

Dusty brown beetle. *Gonocephalum simplex*
The adults are about $\frac{1}{2}$ in (c. 1·3 cm) long. The main damage is done by the hard yellow larvae which eat the roots of wheat seedlings; the adults occasionally do a little damage by eating the leaves. Insecticidal seed dressings should prevent serious damage but if there is a heavy infestation BHC or aldrin baits may be necessary.

Shiny cereal weevil. *Nematocerus spp.*
The adults are shiny, bronze coloured weevils which eat irregular patches from the margins of the leaves. The main damage, however, is done by the larvae which eat the stem just below the soil surface. The adults can be killed by spraying with DDT but no control is known for the larvae.

Black wheat beetle. *Heteronychus consimilis*
This pest, which is also known as the Ol Kalou beetle, is particularly damaging in the area round Ol Joro Orok and Ol Kalou in Kenya. The adults eat through the stems of seedlings at soil level. Aldrin or BHC seed dressings give reasonable control of this pest.

Barley fly. *Hylemia arambourgi*
Details of the barley fly are given in Chapter 2. Wheat is attacked less severely than barley by this pest.

Aphids
Several species of aphids attack wheat; they

usually attack in the early stages of growth during a drought. They can be seen on the leaves and in the funnels. They are best controlled by spraying with dimethoate, menazon, diazinon or fenitrothion.

Diseases

Stem rust
This is the most important wheat disease in East Africa; it is caused by the fungus *Puccinia graminis tritici*. It is more of a problem in East Africa than in any other part of the world for three reasons. Firstly, there is no winter, as there is in the temperate wheat growing countries, to check the diseases. Secondly, because of the different rainfall regimes in different areas there is wheat at the susceptible stage of growth in some part of East Africa in almost every month of the year. These two facts enable stem rust to survive from one year to the next. The third reason why stem rust is such a problem in East Africa is that the unusual conditions mentioned above encourage the production of new physiological races and sub-races of the disease; these are often able to attack varieties which had previously proved to be rust resistant. This means that new resistant varieties are constantly being issued whilst ones which have recently become susceptible enough to constitute a risk are withdrawn.

Wheat becomes susceptible to stem rust at the 'boot' stage, i.e. before the ear emerges but when it is visible as a bulge inside the uppermost leaf sheath. Stem rust attacks wheat most readily in damp cloudy weather. The lesions are most commonly seen on the stems and the leaf sheaths although they sometimes occur on the leaf blades and the glumes. In the early stages of an attack the lesions consist of red or brown pustules; they tend to be elongated. Each lesion produces vast numbers of uredospores which remain viable for about two weeks and which can be spread for hundreds of miles by air. As wheat ripens the pustules produce black teleutospores; these spores are important in some temperate countries as they can survive the winter; in East Africa, however, with no winter and with a continual supply of plant hosts (either wheat, barley or grasses) stem rust can remain in the uredospore stage without relying on teleutospores for survival.

Lesions on the stems are particularly damaging because they restrict, or in the case of severe attacks, prevent the passage of nutrients to the heads; this results in shrivelled grain or, sometimes, in no grain being produced at all. During the 1950s there were some very bad rust years in which wheat yields were drastically reduced. In the 1960s, however, rust outbreaks were less severe, largely owing to more resistant varieties.

The only practical way in which the farmer can guard against stem rust is by growing resistant varieties. The disease is spread by air-borne uredospores so practices such as using rotations or seed dressings are of no benefit. Early planting is no insurance against attacks although late planted crops are often very heavily attacked. Chemical control, e.g. spraying with zineb, maneb or nickel salts, is effective but is too expensive to be economic.

Brown leaf rust
This disease is caused by the fungus *Puccinia recondita*. It is sometimes called brown rust or leaf rust. It seldom does much damage and is therefore less important than stem rust. It is more common at lower altitudes.

The lesions are brown at first and are most easily distinguished from those of stem rust by their size and shape: they are usually small and circular. They turn black as the crop matures. They occur on the leaf blades and the leaf sheaths and may appear at any stage of the crop's growth. Brown leaf rust is spread and controlled in the same ways as stem rust.

Yellow rust or Stripe rust
This disease is caused by the fungus *Puccinia striiformis*. The lesions consist of yellow pustules which are arranged in parallel lines along the leaves; they may appear at any stage of the crop's growth. In the later stages, and in severe cases, yellow spores may be found on the inner surface of the glumes; this stage is known as yellow ear rust.

Yellow rust spreads and is controlled in the same ways as stem rust and brown leaf rust. It can be very damaging in areas over 8 000 ft (c. 2 400 m) above sea level.

Glume blotch
This disease is caused by the fungus *Leptosphaeria* (*Septoria*) *nodorum*. It can cause considerable damage in wet years, especially where wheat has been grown for several years in succession.

The lesions are brown markings on the glumes and around the nodes; in the advanced stage of the disease black spots (pycnidia) can be seen on the lesions. The leaves become shrivelled, with light brown patches on them. Glume blotch is spread in three ways: from plant to plant by rain splash and from season to season by infected crop residues or infected seed. It can best be prevented by rotations, seed dressing and stubble burning. Some varieties show a certain amount of resistance.

Leaf blotch and leaf blight

These are caused, respectively, by the fungi *Leptosphaeria* (*Septoria*) tritici and *Pyrenophora* (*Helminthosporium*) trichostoma. Neither of these leaf diseases causes as much damage as does glume blotch. Both cause brown speckled lesions of the leaves and, later, death of the leaves. Both are spread and controlled in the same way as glume blotch.

Take all

This disease is caused by the fungus *Ophiobolus graminis*. It is soil borne and affects the roots. Affected roots are short and thick and offer little resistance when plants are pulled upwards. Stems are black and shiny just above soil level; this symptom can only be seen by peeling away the leaf sheaths. The ears are usually pale and empty. Rotations and the preparation of a firm seedbed are the only two precautions against take all.

Loose smut

This disease is caused by the fungus *Ustilago nuda*. The main symptom of loose smut is the appearance of a mass of black spores in place of the inflorescence. It is seldom seen nowadays in East Africa, largely owing to the release of resistant varieties.

Bibliography

1 **Anderson, G. D.** et al. (1966). *Effects of soil, cultivation history and weather on responses of wheat to fertilisers in Northern Tanzania*. Expl. Agric., 2:183.

2 **Clayton, E. S.** (1956). *Land use and grain yields in the Kenya highlands*. E. Afr. agric. J., 22:32.

3 **Graham, J. F.** et al. (1959). *Seed-dressing against the black wheat beetle in Kenya*. E. Afr. agric. J., 25:23.

4 **Hurd, E. A.** et al. (1969). *New emphasis in wheat breeding in Kenya*. E. Afr. agric. for. J., 35:213.

5 **Jones, R. J.** (1961). *The control of stem rust in wheat by aerial application of fungicides*. E. Afr. agric. for. J., 26:210.

6 **Pinkerton, A.** et al. (1965). *A note on copper deficiency in the Njoro area, Kenya*. E. Afr. agric. for. J., 30:257.

7 **Seitzer, J. F.** (1970). *Effects of CCC on bread wheat in Kenya*. Expl. Agric., 6:255.

8 **Weiss, E. A.** (1967). *Fertiliser trials on maize and wheat*. E. Afr. agric. for. J., 32:326.

Cereal crops

Italian millet. *Setaria italica*. (See Fig. 151)
This small-grained, drought resistant crop is also
called foxtail millet. It is occasionally grown in the
low altitude areas surrounding Mt. Kenya.

Oats. *Avena sativa*. (See Fig. 152)
Oats are an important fodder crop in the large scale,
mixed farming areas of Kenya. Only a very few
farmers grow them as a cereal. The grain is used for
making oatmeal.

Tef. *Eragrostis tef*
This is an important crop in Ethiopia, where an
estimated 8 400 000 acres (c. 3 400 000 ha) are
grown annually. A little is grown near Marsabit in
Kenya.

Essential oil crops

Essences, i.e. odiferous substances, are extracted
from essential oil crops, usually by distillation.

Geranium. *Pelargonium graveolens*
Essential oil geraniums have been grown to a
limited extent on large scale, mixed farms in Kenya.
At one time there were about 500 acres, mostly near
Naivasha, Njoro and Lumbwa. The oil is obtained
from the leaves and is used for cheap perfumery.

Rosemary. *Rosmarinus officinalis*
Rosemary has never been as popular as geranium
although it takes root from cuttings more readily.
It has been grown near Lumbwa in Kenya. The oils
are obtained in the same way and used for the same
products as geranium.

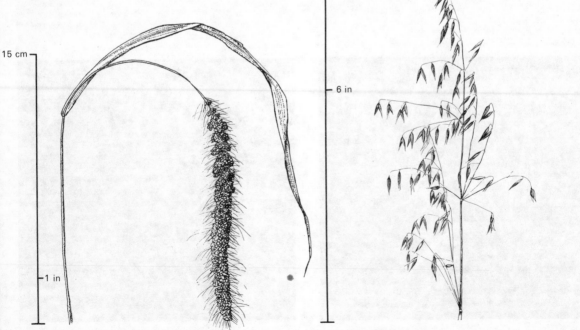

Fig. 151: Italian millet.

Fig. 152: Oats.

Fibre crops

Flax. *Linum usitatissimum*
Fibre varieties of this species have been grown in the Kenya highlands on large scale, mixed farms during both world wars at subsidised prices. All but one estate, near Kaptagat, abandoned production shortly after the second world war. The fibres are obtained from the bark of the long slender stems after retting, i.e. immersing the bark so that all but the fibres decompose. For oilseed varieties of this species, see Linseed below.

Kenaf. *Hibiscus cannabinus*
Kenaf, like flax, is a bast fibre, its fibres being obtained from the bark after retting. It has been grown commercially on one sisal estate at the Kenya coast and has been tried experimentally in many other areas. Ecological problems, the difficulty and expenses of providing large tanks of clean water, mechanisation problems and widespread damage by nematodes have prevented the expansion of this crop.

New Zealand flax. *Phormium tenax*. (See Fig. 153)
This is a leaf fibre which was once grown around Soy in Kenya and is still grown on one estate at Kaptagat. The fibre, like kenaf, is a jute substitute.

Miscellaneous

Bixa. *Bixa orellana*
A red dye, anatto, is obtained from the seed coats of this crop. It is used for colouring dairy products and cosmetics. It has been grown as a cash crop on the East African coast during the 1960s but its cultivation is no longer being encouraged owing to a recent decline in price.

Gourd. *Lagenaria siceraria*. (See Fig. 154)
Gourds are grown by smallholders in most parts of East Africa. Their fruits are almost always used as containers but those of some varieties may be eaten when immature, especially by Asians.

Passion fruit. *Passiflora edulis*. (See Fig. 155)
Passion fruit grow semi-wild in some of the warmer, wetter areas of East Africa such as the lakeshore areas of Uganda. Their fruits are often seen in local markets. In Kenya a passion fruit juice industry started in the 1960s when a factory was opened in Sotik; the fruits were mostly provided by smallholders in the Kisii highlands. More recently passion fruit has been encouraged for juice in the Taita Hills and near Thika.

Fig. 153: New Zealand flax.

Fig. 154: Gourd.

Fig. 155: Passion fruit.

Fig. 156: A soya bean plant.

Oilseed crops

Linseed. *Linun usitatissimum.*
Oilseed varieties of this species, which are shorter and flower more prolifically than the flax varieties, became popular in the wheat growing areas of Kenya in the late 1960s. About 1 000 acres (c. 400 ha) were grown in 1969. One of the main problems is caused by wet weather at the time of harvest' which prevents efficient combining because the stems remain green. Linseed is drought resistant and is likely to be most popular in the lower altitude wheat areas.

Rape. *Brassica napus*
Rape is an oilseed crop which, like linseed, can be harvested by combines. It is suitable for the higher altitude wheat areas in Kenya. It has never been grown on such a large scale as linseed.

Soya beans. *Glycine max.* (See Fig. 156)
Soya beans are grown for oil extraction by smallholders in Ankole and Bunyoro Districts in Uganda. Their seeds have a high content of good quality protein but they are virtually never eaten by African smallholders owing to the difficulty involved in cooking them. Soya beans have seldom given good results in trials in Kenya although it is likely that suitable varieties could be produced by breeding or further introduction.

Pulse crops

Bambarra groundnuts. *Voandzeia subterranea.* (See Fig. 157)
Bambarra groundnuts are popular in the area south of Lake Victoria in Tanzania. They are grown to a limited extent in Kakamega and Busia Districts in western Kenya. They give reasonable yields on infertile sandy soils.

Chick peas. *Cicer arietinum.*
Chick peas are also popular to the south of Lake Victoria and on the Mwea plains in Kirinyaga District in Kenya. They are one of the few crops that can be grown successfully on black cotton soils. They are planted near the end of the rains and utilise the residual moisture left in the soil.

Lima beans. *Phaseolus lunatus*
Lima beans are occasionally grown near the homestead in the lower altitude areas of East Africa.

Fig. 157: A bambarra groundnut plant.

Runner beans. *Phaseolus coccineus*
White seeded varieties of the runner bean are the common butter bean in Kenya. Until recently they were generally thought to be Lima beans. They are grown at high altitudes in Nakuru and Nyandarua District.

Peas. *Pisum sativum*
Peas are sometimes grown in the higher altitude areas of East Africa, especially in Central Province in Kenya.

Root crops

Yams. *Dioscorea spp.* (See Fig. 158)
Yams are grown sporadically in many of the wetter areas of East Africa, e.g. near the shores of Lake Victoria in Uganda, on the southern slopes of Mt. Kilimanjaro and in Central Province in Kenya. Some varieties produce aerial tubers on their vines.

Tannia. *Xanthosoma sagittifolia.* (See Fig. 159)
Tannias closely resemble cocoyams except that

their petioles join the lamina of the leaf at the margin. They are occasionally grown in areas where cocoyams are found in East Africa except in Central Province in Kenya.

Spice crops

Cloves. *Eugenia caryophyllus.* (See Fig. 160)
Cloves are an important crop in Zanzibar. Annual exports have been as high as 12 500 tons although they have declined during the last few years. The final product consists of the flower buds; these are picked from the trees whilst green and are then dried.

Coriander. *Coriandrum sativum*
Coriander is an important constituent of curry powder. It is grown on a field scale on the heavy soils of the Mwea plains in Kenya.

Vanilla. *Vanilla sp.*
Vanilla is grown on one estate near Kampala in Uganda. It is a member of the orchid family. It climbs up trees which provide shade as well as support. The harvested portion is the fully developed, green pods which are cured in the sun.

Fig. 158: Yams.

Fig. 159: Tannia.

Fig. 160: An undried clove inflorescence.

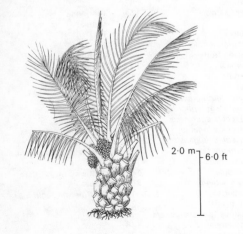

2·0 m ⌐ 6·0 ft

Fig. 161: An oil palm.

Fig. 162: Macadamia nuts.

Tree crops

Cocoa. *Theobroma cacao.*
Cocoa is grown on a few estates and by smallholders near Kampala in Uganda, in the Western Usambara Mountains and near Tukuyu in Tanzania. The pods grow directly from mature bark on the trunk and branches. The beans are removed from the pods and fermented to remove the mucilage. They are then dried.

Miraa. *Catha edulis*
The green shoots of miraa are popular as a stimulant in northern Kenya and in Somalia. Miraa is an important cash crop in Meru District in Kenya.

Kapok. *Ceiba pentandra*
The pods of kapok trees produce a fibre called 'floss' which can be used for upholstery. Kapok estates have been planted on the East African coast, both in Kenya and Tanzania. They have fallen into disuse or the trees have been replaced by other crops such as cashew nuts since a serious decline in the market.

Macadamia nuts. *Macadamia spp.* (See Fig. 162)
Macadamia nuts, sometimes called Queensland nuts, have recently been encouraged as a diversification crop in the coffee areas in Kenya. Several thousand acres have been planted. The nuts are popular for dessert purposes in the U.S.A.

Oil palm. *Elais guineensis.* (See Fig. 161)
There are groves of oil palms near Kigoma and on the shores of Lake Nyasa in Tanzania. Two types of oil are extracted: one from the pericarp and one from the kernel.

Index

The chapters on the major crops are arranged in alphabetical order